Process Technology

Safety, Health, and Environment

PROCESS TECHNOLOGY

SAFETY, HEALTH, AND ENVIRONMENT

THIRD EDITION

CHARLES E. THOMAS

Australia • Brazil • Japan • Korea • Mexico • Singapore • Spain • United Kingdom • United States

Process Technology: Safety, Health and Environment, Third Edition
Charles E. Thomas

Vice President, Editorial: Dave Garza
Director of Learning Solutions: Sandy Clark
Executive Editor: David Boelio
Managing Editor: Larry Main
Senior Product Manager: Sharon Chambliss
Editorial Assistant: Jillian Borden
Vice President, Career and Professional Marketing: Jennifer Baker
Marketing Director: Deborah Yarnell
Marketing Manager: Kathryn Hall
Production Director: Wendy Troeger
Production Manager: Mark Bernard
Content Project Management: PreMediaGlobal
Art Director: Joy Kocsis
Technology Project Manager: Christopher Catalina
Production Technology Analyst: Joe Pliss
Compositor: PreMediaGlobal

Library of Congress Control Number: 2011920385

ISBN-13: 978-1-1110-3635-5

ISBN-10: 1-1110-3635-7

Delmar
5 Maxwell Drive
Clifton Park, NY 12065-2919
USA

Cengage Learning is a leading provider of customized learning solutions with office locations around the globe, including Singapore, the United Kingdom, Australia, Mexico, Brazil, and Japan. Locate your local office at: **international.cengage.com/region**

Cengage Learning products are represented in Canada by Nelson Education, Ltd.

To learn more about Delmar, visit **www.cengage.com/delmar**

Purchase any of our products at your local college store or at our preferred online store **www.cengagebrain.com**

Printed in the United States of America
2 3 4 5 6 21 20 19 18 17

Preface xiii

Chapter 1 Introduction to Process Safety 1

Key Terms 2
Process Safety 3
The History of the Labor and Safety Movement in the United States 4
Personal Protective Equipment (PPE) 10
The Process Technician and the Chemical Processing Industry 11
Attitudes and Behaviors 13
New Technology and Potential for Catastrophic Events 14
General Plant Safety Rules 16
Job Hazard Analysis 17
Types of Fire Extinguishers 18
Types of Permits 19
Weapons of Mass Destruction 19
Tornados and Hurricanes 20
Summary 20
Review Questions 22

Chapter 2 Hazard Classification 23

Key Terms 24
Common Industrial Hazards 25
Physical, Chemical, Ergonomic, and Biological Hazards 25
Electrical Hazards 26
Industrial Noise Hazards 26

Radiation Hazards 26
Hazard Recognition 27
Accident Prevention 29
Accident Investigation 31
Summary 31
Review Questions 33

Chapter 3 Routes of Entry and Environmental Effects 35

Key Terms 36
Routes of Entry 37
Dose-Response Relationship 38
Environmental Effects 39
Exxon Valdez 39
Air Pollution 40
Air Pollution Control 41
Agencies 41
Air Permitting 41
Water Pollution Control 42
National Water Quality Standards 43
Water Permitting 43
Hazards: Scenarios 45
Summary 47
Review Questions 48

Chapter 4 Gases, Vapors, Particulates, and Toxic Metals 49

Key Terms 50
Physical Hazards Associated with Gases, Vapors, Particulates, and Toxic Metals 51
Health Hazards Associated with Gases, Vapors, Particulates, and Toxic Metals 53
Asbestos 54
Bhopal—Union Carbide 55
Particulates 60
Dust and Gases 61
Dust Explosions 64

Flammable Gases 64
Compressed Gas Cylinder 64
Metallic Substances 65
Metallic Compounds 67
Metals That are Fire Hazards 67
Summary 67
Review Questions 70

Chapter 5 Hazards of Liquids 71

Key Terms 72
Introduction to the Hazards of Liquids 72
Handling, Storing, and Transporting Hazardous Chemicals Safely 73
Physical Hazards Associated with Liquids 74
Health Hazards Associated with Liquids 75
Pressure and Pressurized Equipment 75
Process Systems 76
Flammable Liquid Storage 76
Spontaneous Combustion 78
Oxidizers 79
Hazards of Steam 79
Hazards of Water 79
Hazards of Light-Ends 82
Water Hazards 83
Acids and Caustics 84
Solvents 85
Paints and Adhesives 85
Hematopoietic System Toxins, Hepatotoxic Agents and Other Harmful Agents 86
Summary 86
Review Questions 89

Chapter 6 Hazardous Chemical Identification: Hazcom, Toxicology, and DOT 91

Key Terms 92
Introduction to Hazardous Chemical Identification 92
The Hazard Communication Program—The Workers' Right-to-Know Act 92

Material Safety Data Sheet (MSDS) 95
Toxicology 97
Safety Signs, Tags, and Warning Labels 98
Department of Transportation Labeling System 98
Hazardous Materials Identification System (HMIS) 103
National Fire Protection Association (NFPA) 103
Summary 105
Review Questions 106

Chapter 7 Fire and Explosion 107

Key Terms 108
Fire, Explosion, and Detonation 109
Chemical Explosions 109
Polymers and Fire 118
Flammable, Explosive, and Radioactive Hazards 119
Flammable and Explosive Materials 119
Fundamentals of Fire Prevention, Protection, and Control 120
The Chemistry of Fire 121
The Hazard of Air 122
Fire Stages 123
Flashpoint, Flammable Limits, and Ignition Temperature 123
Fire Classification System 124
Types of Fire Extinguishers 124
Fire Extinguisher Use 128
Fighting Fires 129
Summary 131
Review Questions 133

Chapter 8 Electrical, Noise, Heat, Radiation, Ergonomic, and Biological Hazards 135

Key Terms 136
Plant-Specific Hazards 136
Electricity 137
Bonding and Grounding 139
Heat and Radiation 139
Hearing Conservation and Industrial Noise 140

Ergonomic Hazards 143
Hazards of Confined Spaces 144
Hazards of Lifting 144
Biological Hazards 144
Blood-Borne Pathogens 145
Summary 145
Review Questions 147

Chapter 9 Safety Permit Systems 149

Key Terms 150
Types of Permits 150
Confined Space Entry 153
Control of Hazardous Energy (Lockout/Tagout) CFR 29 1910.147 158
Opening or Blinding Permits 161
Routine Maintenance Permits 162
Summary 164
Review Questions 165

Chapter 10 Personal Protective Equipment 167

Key Terms 168
Personal Protective Equipment 169
Hazards in the Workplace 175
Emergency Response—Four Levels of PPE 175
Written Respiratory Protection Programs 178
Summary 186
Review Questions 188

Chapter 11 Engineering Controls 189

Key Terms 190
Risk Evaluation 190
Design and Operation of Plants for Safety 192
Alarms and Indicators 193
Fire Alarms and Detection Systems 194
Toxic Gas Alarms and Detection Systems 195

Redundant Alarm and Shutdown Devices 195
Interlocks and Automatic Shutdown Devices 195
Process Containment and Upset Controls 196
Closed Systems/Closed-Loop Sampling.................................. 197
Floating Roof Tank and Ventilation Systems 198
Effluent Control and Waste Treatment...................................... 199
Noise Abatement 201
Flares...201
Pressure Relief Devices 202
Deluge Systems and Explosion Suppression Systems.............. 203
Summary 203
Review Questions 205

Chapter 12 Administrative Controls 207

Key Terms 208
Introduction to Administrative Controls...................................... 208
Community Awareness and Emergency Response 209
Job Safety Analysis.. 209
Hazards and Operability Study 210
Training and Mandated Training 210
Housekeeping 211
Safety Inspections and Audits..211
Monitoring Equipment.. 212
First Aid.. 214
Summary 215
Review Questions 218

Chapter 13 Regulatory Overview: OSHA, PSM, and EPA 219

Key Terms 220
Occupational Safety and Health Act 221
Process Safety Management (CFR 29 1910.119)...................... 222
Environmental Protection Agency 228
Summary 229
Review Questions 231

Chapter 14 HAZWOPER 233

Key Terms 234
Fall Protection 235
Hoisting Equipment.. 237
HAZWOPER 237
Unit Monitors and Field Survey Instruments—Identifying Hazardous Materials 244
Bunker Gear 245
Cutting, Welding, and Brazing 245
Summary 246
Review Questions 247

Chapter 15 Process System Hazards 249

Key Terms 250
Operating Hazards 252
Equipment- and System-Related Hazards 252
Steam Generation 265
Flare System 267
Weather-related Hazards 267
Chemicals- and Chemistry-related Hazards 269
Reactors 269
Distillation System 275
Human Factors 276
Summary 279
Review Questions 280

Chapter 16 Weapons of Mass Destruction, Hurricanes and Natural Disasters 281

Key Terms 282
Weapons of Mass Destruction 282
Terrorist Flies an Airplane into Chemical Plant, Refinery, or Oil Rig 283
Terrorist Enters CPI Workforce 285
Use of Conventional Bomb 285
Use of Nuclear Weapons 287
Suicide Bomber—Dirty Bomb 287

Use of Chemical Weapons 287
Use of Biological Weapons 291
Tornado Hits 292
Fast-Forming Hurricanes 294
Use of Military Weapons 298
Flooding 299
Emergency Preparedness 300
Summary 301
Review Questions 306

Glossary 307

Index 321

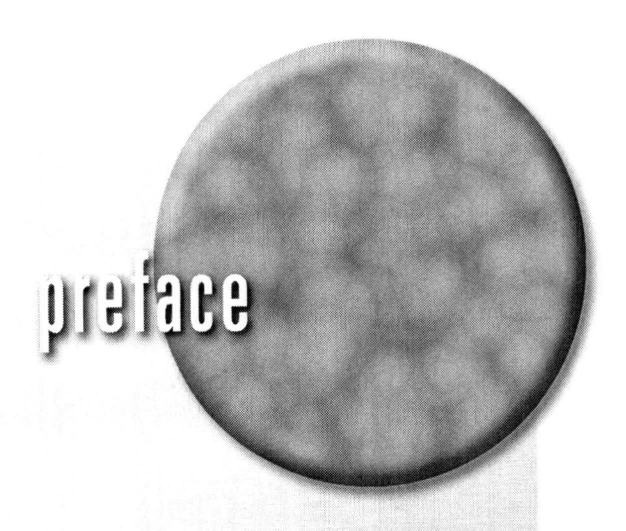

preface

Over the last three decades, technology has advanced faster than our ability to disseminate it into our existing workforce. The potential for catastrophic events is greater today than at any other time in our history. Areas of specialization, such as process operations and troubleshooting, require new approaches to safety program development and management. New threats from foreign terrorists have put the chemical processing industry on high alert. Most safety textbooks are designed to focus on the broad occupation of the safety professional and manager. This text is written for the process technician who operates and maintains chemical plants and refineries and works in pharmaceuticals, food processing, paper and pulp, or utilities. This text also has applications as an introductory course for safety management programs.

Safety, Health, and Environment, Third Edition covers most government-mandated training and foundational aspects of apprentice technician safety training. Several new sections have been added: review of the hazards associated with process systems, weapons of mass destruction, and preparation for natural disasters like hurricanes, tornados, or flooding. Essential topics discussed in this text include:

- Hazard Communication CFR 29 1910.1200
- HAZWOPER CFR 29 1910.120
- Respiratory Protection CFR 29 1910.134
- Permit Systems-Hot Work, Confined Space Entry
- Control of Hazardous Energy (Lockout-Tagout) CFR 1910.147
- Process Safety Management CFR 29 1910.119
- Personal Protective Equipment CFR 29 1910.133
- Fire Prevention, Protection, and Control CFR 29 1910.157
- Department of Transportation (DOT) 49 CFR 171-177
- Environmental Standards—RCRA, CERCLA, TSCA, and so on
- Occupational Noise Exposure CFR 29 1910.95
  - Weapons of mass destruction
  - Hurricanes, tornados, and flooding
  - Process equipment and system hazards
  - Job hazard analysis
  - Science and chemistry

Acknowledgments

The author and the publisher would especially like to thank the expert reviewers who read the manuscript and made recommendations to improve the final product:

Max Ansari, Houston Community College
and Robert Smith, Texas State Technical College-Marshall.

Introduction to Process Safety

Objectives

After studying this chapter, the student will be able to:

- Describe the chemical processing industry.
- Describe the significant events of the safety movement.
- Classify the safety roles and responsibilities of process technicians.
- Identify the basic principles of safety.
- Describe the general safety rules used in the industry.
- Explain the difference between the terms *process safety* and *occupational safety and health*.
- Explain the key elements of safety.
- Describe the basic elements of a hazard analysis.
- Explain the typical permits used by the chemical processing industry.
- List the various types of firefighting equipment.
- Describe the principles associated with production, transportation, and storage of chemicals.
- Describe the Occupational Safety and Health Act.
- Describe Process Safety Management.
- Identify the key elements of HAZCOM.
- Explain the principles of hazard classification and recognition.
- Explain the key aspects of personal protective equipment.
- Describe administrative and engineering controls.
- Explain the key issues associated with weapons of mass destruction, hurricanes, and natural disasters.

Key Terms

- **Administrative controls**—can be described as the programs and activities used to control industrial hazards.
- **Biological agents and weapons**—include hazardous bacteria like anthrax, cholera, pneumonic and bubonic plague, tularemia, and Q-fever. Include the use of viruses like Smallpox, Venezuelan equine encephalitis, and viral hemorrhagic fever. Include the use of biological toxins like Botulism, Ricin, Staphylococcal Enterotoxin B, and Tricholthecene Mycotoxins.
- **Chemical asphyxiants**—are designed to remove or displace oxygen and prevent the victim from utilizing oxygen after they enter the respiratory system.
- **Chemical processing industry (CPI)**—broad term used to describe chemical plants and refineries, power plants, food processing, paper and pulp, pharmaceuticals, and city utilities.
- **Hazard Communication (HAZCOM) Standard**—a program designed to increase plant worker awareness of chemical hazards and give instructions on appropriate safety measures for handling, storing, and working with these chemicals (1983).
- **Hurricanes**—are powerful, swirling storms with tentacles reaching out from a singular eye. Hurricanes have sustained wind speeds between 74 and 155 mph, spread heavy rainfall and flooding, and produce tremendous storm surges from 4 to 18 feet, spin-off tornados, downed trees and broken limbs, damaged homes, loss of electricity, utilities, and basic commodities.
- **National Council of Industrial Safety**—formed in 1913 to promote safety in the workplace.
- **Occupational safety and health**—deals with items like personal protective equipment, HAZCOM, permit systems, confined space entry, hot work, isolation of hazardous energy, and so on.
- **Occupational Safety and Health Administration (OSHA)**—independent inspectors are allowed to enter and inspect the workplace, cite violations, and set deadlines (1970).
- **Pittsburgh Survey of 1907**—provided the first statistical data on how many people had been killed or injured as a result of hazardous working conditions (1906).
- **Potable water**—drinkable water.
- **Process safety management (PSM)**—is designed to keep the process in the pipes and not in the environment (1990).
- **Process safety**—the application of engineering, science, and human factors to the design and operation of chemical processes and systems.
- **Process technology**—the study and application of the scientific principles associated with the operation and maintenance of the chemical processing industry.
- **Safety**—often described as an attitude that includes careful planning, following safety rules, safe work practices, and the use of personal protective equipment.
- **Terrorism**—an attempt to change a belief or point of view through the use of a violent act. Terrorist acts are designed to generate fear and focus world-wide media attention on the terrorists' cause.

Process Safety

This text focuses on the key elements that contribute to the subject of **process safety**, personnel safety, occupational health, and safety in general. Process safety is described as the application of engineering, science, and human factors to the design and operation of chemical processes and systems. The primary purpose of process safety is to prevent injuries, fatalities, fires, explosions, or unexpected releases of hazardous materials. Process safety focuses on the individual chemical processes and operational procedures associated with these systems. A process safety analysis is used to establish safe operating parameters, instrument interlocks, alarms, process design, and startup, shutdown, and emergency procedures. Process safety programs cannot completely eliminate risk; they can only control or reduce them.

Occupational safety and health deals with items like personal protective equipment, HAZCOM, permit systems, confined space entry, hot work, isolation of hazardous energy, and so on. Process technicians are required to work with existing systems and programs. Understanding the key elements of process safety and occupational safety allows technicians to contribute significantly to the reduction of risks and hazards associated with the **chemical processing industry (CPI)**. The CPI is a business segment composed of refinery, petrochemical, paper and pulp, power generation, pharmaceuticals, and food processing. The roles and responsibilities of a process technician include knowing the equipment and technology they are working with, being able to maintain and operate this equipment, complying with safety and environmental regulations, applying quality techniques to operational processes, collecting and analyzing samples, and troubleshooting system problems. The responsibilities of most technicians focus on the design and operation of the equipment and systems they operate.

The operational hazards identified in Figure 1-1 show a typical arrangement of equipment and systems. This figure is designed to acquaint you with the hazards, points of failure, and operational risks.

Safety is often described as an attitude that includes careful planning, following safety rules, safe work practices, and the use of personal protective equipment. Many technicians believe that safety is the single most important aspect of running a process. Safety-conscious people are invaluable assets that cannot be replaced. Good characteristics associated with the safe operation of a chemical plant or refinery include: having a familiarity with the system; being alert, serious, and exact; and visualizing potential problems or risks. The foundation of any safety program is the people who work in the plant.

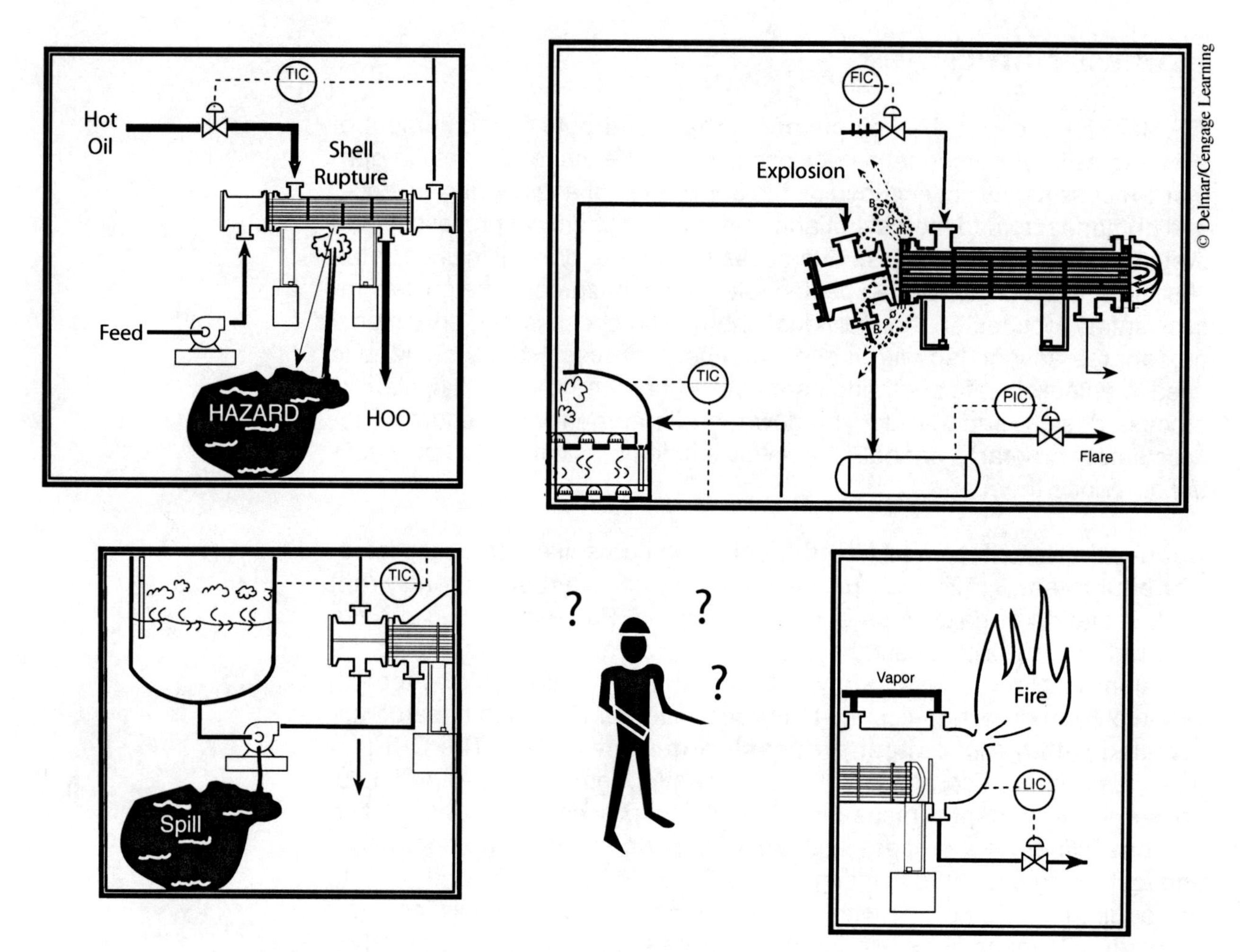

Figure 1-1 *Operational Hazards*

The History of the Labor and Safety Movement in the United States

In the United States, the industrial safety movement closely parallels the rise of the labor movement. Unfortunately, this period of our history is bloody and tragic as well as an era of hope and progress. The American economic system developed slowly, as workers battled for a voice in the workplace and a piece of the economic pie. There is an impressive contrast between modern workers and owners and those who held the same positions two hundred years ago. (See Figure 1-2.)

The safety movement in the United States has links to the English Industrial Revolution that occurred earlier than the U.S. version. By 1850, British politics, community problems, and the social fabric of daily life were

Figure 1-2 *History of the Safety Movement*

determined by the factory. Almost overnight, the Industrial Revolution transformed the English commercial society into an industrialized nation. The English factory became a powerful and uncongenial social tiger that set the pattern followed by U.S. industrial society.

During the early years of America's industrialization, organized labor was weak. Company management crushed any type of work stoppage by appealing to the court systems and portraying laborers as economic blackmailers. This proved to be a successful tool in crushing labor's request for accident prevention, guarding of rotating and hazardous equipment, and compensation for accident victims. Strikes were broken up as the courts sided with the owners. During this time, safety and compensation were linked by industry management to workers' requests for higher wages, better working conditions, and shorter working hours. During the early days of industrialization, management was not concerned about the deplorable conditions their people worked in, much less the safety and health of individual employees.

During the late 1800s, the textile industry was replaced as the leading manufacturer by coal mining and railroading. Like working in the textile industry, coal mining and railroading were extremely hazardous occupations. Coal mine cave-ins, explosions, fires, and toxic or flammable gases accounted for 566 deaths and 1,655 injuries over seven years at Schuylkill County coal mines in Pennsylvania. In 1877, railroad workers made $1.75 for a twelve-hour work day. Breakmen frequently lost hands or fingers.

The invention of the Westinghouse air brake in 1869 and the automatic coupler in 1880 made railroading safer.

Workers in the 1870s believed their jobs were hazardous. A brass finisher identified the risks of metallic dusts and dangerous machinery. A carriage painter was afraid of lead poisoning. Most workers were concerned about noise, heat, burns, dusts, and proper ventilation. Phossy jaw was a disease workers contracted from breathing the fumes of white or yellow phosphorus. This disease appeared to flourish among workers in match factories. Phossy jaw was characterized by a painful swelling of the jaw that produced severe scarring and disfigurement. In extreme cases, the jaw or portions of the face had to be removed. Black lung was a disease that coal miners contracted from long exposure to coal dust in the mines. Life expectancy among coal miners was short as workers were exposed to a variety of industrial hazards. Accurate safety records were never recorded during the early years of the Industrial Revolution.

The **Pittsburgh Survey of 1907** provided the first statistical data on how many people had been killed or injured as a result of hazardous working conditions in the United States. This effort was primarily funded by the Russell Sage Foundation for the workers in and around Pittsburgh, Pennsylvania. Sage investigators were shocked when they discovered that between 1906 and 1907, 526 people had lost their lives and 500 had been seriously injured or disabled. Over half of the families involved in this tragedy had been left without any means of support.

As the Pittsburgh Survey expanded outside the Pittsburgh area, the investigation revealed that there had been over 30,000 fatalities from industrial accidents across the United States during 1906. As news of this report became known, industry, government, and the public recognized a need for change. The Russell Sage investigators had unknowingly opened a new chapter for the safety movement. This new safety climate ushered in the development of organized safety programs and workers' compensation benefits.

In 1913, the **National Council of Industrial Safety** was formed to promote safety in the workplace. A few years later, its name was changed to the National Safety Council as representatives from colleges and universities, industry, labor unions, government, public utilities, and insurance companies voted to broaden the scope of the organization to include accident prevention in schools, home, industry, and on the roads and highways. By 1927, the National Safety Council published safety pamphlets, collected statistical data, printed annual facts on accidents in all areas, and sponsored the popular National Safety Congress. Another important organization from this period was the National Fire Protection Association, which produced a variety of information and materials on fire protection and prevention. During the 1920s, organized industrial safety programs became popular as companies competed for safety awards. The steel industry led the way in this area since Joliet Works of the Illinois State Steel Company had created the first in-plant safety department in 1892. This may be the primary reason that in 1926, Illinois Steel reported that they had gone three million worker-hours without a lost-time accident. (A lost time accident is a work-related injury that causes a worker to lose more than one day of work.)

During the early decades of the 1900s, the federal government became interested in public safety. In 1906, the Pure Food and Drug Act was passed; in 1908 federal employees were given workers' compensation; and in 1910, the United States Bureau of Mines was established. By 1913, the Department of Labor was organized, and by 1921, almost every state had some type of workers' compensation.

The period between 1914 and 1920 reflects World War I, which brought thousands of women and minorities into the industrial plants to replace those who were fighting in the war. In 1914, the Clayton Act was passed, which limited the use of injunctions during labor disputes. In this same year, the Ludlow Massacre (an event in which the wives and children of striking miners in Colorado were set on fire by National Guardsmen) forced the President to appoint a commission to review working conditions in the mines. In 1916, an eight-hour work day was created for railroad workers by the Adamson Act. During this same year, a federal child labor law was passed and struck down. In 1918, President Wilson created the National War Labor Board. A special division was also created in the Department of Labor for "Women in Industry." In 1920, the 19th Amendment to the Constitution was ratified, giving American women the right to vote.

The 1930s and 1940s saw the rise of unions as management and union officials squared off in adversarial positions. In 1935, the Social Security Act was approved. During the next year, minimum wage, overtime pay, safety standards, and child and convict labor provisions were established by the Walsh-Healey Act. During this same timeframe, the eight-hour day and forty-hour week were established for industry. Another important event during 1937 included the passage of the National Apprenticeship Act, which created the Bureau of Apprenticeship inside the Department of Labor. The Taft-Hartley Act in 1947 closed the door on the ever-expanding power of northern unions as union activities were restricted by states passing "right-to-work" laws. In 1948, the federal government hosted the first national conference on industrial safety, in Washington D.C. Before the close of this decade, child labor was directly prohibited by an amendment to the Fair Labor Standards Act.

The 1960s brought about a number of changes that directly affected the safety movement: the Equal Pay Act of 1963, the Civil Rights Act of 1964, and the Age Discrimination in Employment Act of 1968.

Occupational Safety and Health Act of 1970

The 1970s brought sweeping changes to the industrial safety movement with the passage of the Occupational Safety and Health Act of 1970, the Resource Conservation and Recovery Act of 1976, the Clean Water Act, the Safe Drinking Water Act, the Refuse Act, and the Toxic Substances Control Act. Under the **Occupational Health and Safety Administration (OSHA)**, inspectors were allowed to enter and inspect the workplace, cite violations, and set deadlines. The primary purpose of the **Occupational Safety and Health Act of 1970** is to (1) remove known hazards from the workplace that could lead to serious injury or death and (2) ensure safe and healthful working conditions for American workers. The Occupational Safety and Health Act applies to four broad categories: agriculture, construction, general industry, and maritime. Three agencies are primarily responsible for the administration of the Occupational Safety and Health Act: National Institute for Occupational Safety and Health (NIOSH), Occupational Safety and Health Administration (OSHA), and Occupational Safety and Health Review Commission (OSHRC).

1. **NIOSH—National Institute for Occupational Safety and Health.** This agency is responsible for safety and health research.
   - Develops toxic substance handling criteria
   - Researches safety and health issues
   - Recommends new safety and health standards
2. **OSHA—Occupational Safety and Health Administration**
   - Investigates catastrophes and fatalities
   - Establishes new safety and health standards and penalties
   - Inspects workplaces
3. **OSHRC—Occupational Safety and Health Review Commission**
   - Functions as an independent agency
   - Conducts hearings for situations on noncompliance that are contested
   - Can assess penalties, conduct investigation, and support modify, or overturn OSHA findings

Hazard Communication

In 1983, OSHA's **Hazard Communication (HAZCOM) Standard** was implemented. HAZCOM increased plant worker awareness of chemical hazards and gave instructions on appropriate safety measures for handling and working with these chemicals. Chemical manufacturers are required by the HAZCOM standard to perform the following functions:

- Analyze and assess the hazards associated with chemicals
- Develop written procedures for evaluating chemicals
- Document hazards and develop material safety data sheets (MSDSs) and warning labels
- Disseminate the information to affected individuals
- Label, tag, and attach warning documentation to chemicals leaving the workplace

Employers are responsible for the following:

- Preparing a written HAZCOM program
- Organizing a hazardous chemical inventory list
- Designing individual MSDSs

Process Safety Management

In the early 1990s, **process safety management (PSM)** provided the fangs that HAZCOM had needed. In response to a number of catastrophic events that had occurred in the process industry, OSHA and the Environmental Protection Agency (EPA) implemented the PSM standard, which was designed to keep the process in the pipes and not in the environment. PSM targets highly hazardous chemicals, and its objective is to prevent catastrophes from these chemicals. Key elements of the PSM standard include the following:

- Employee participation
- Process safety information requirements
- Process hazard analysis
- Development of operational procedures
- Employee training
- Emergency planning and response
- Pre-startup safety review
- Analysis of mechanical integrity
- Hot work permit
- Procedures for the management of change
- Incident investigation requirements
- Responsibilities of contractors
- Compliance audit requirements
- Trade secret regulations

In response to this standard, the chemical processing industry, government, business, and education communities joined forces to better educate the technical workforce. Chemical technology and technician college programs formally provided the platform for apprentice training programs for process technicians, laboratory technicians, and engineering technicians. The Bureau of Apprenticeship inside the Department of Labor still works with many of the leading manufacturers in the chemical process industry.

The 1980s and 1990s saw the introduction of the personal computer (PC), video cassette recorder (VCR), microwave oven, and the Internet into most U.S. homes. Rapid changes in technology coupled with enhanced environmental regulations characterized this period. Prior to 2000, a number of major company mergers began to take place as many smaller companies were forced out of the market. New technology and a smaller, more technically educated workforce began to replace the baby boomers. Never at any period of time has modern technology controlled such vast arrays of equipment and systems. Process safety is an essential aspect of operation in the chemical processing industry.

Administrative Controls

Administrative controls can be described as the programs and activities used to control industrial hazards. This includes the policies, procedures, plans and agreements, principles, rules, agreements, and systems used in administrative control. Policies are guiding principles. A procedure is a sequential list of steps included to carry out an action. Plans and agreements are constructive methods used to carry out an action between different groups. **Mutual Aid Agreements** are written agreements between industry and outside emergency response organizations in the event of a catastrophic release or situation. Principles are used to establish a set of rules or guidelines. These rules are statements of how something is to be completed. Systems are associated with principles and rules and are described as organized sets of related principles.

Written programs are influenced by government and regulatory guidelines, company-specific guidelines, and unit-specific requirements. Examples of written programs include:

- Hazard Communication Program (HAZCOM) CFR 29 1910.1200
- Community Awareness and Emergency Response (CAER)
- Process Hazards Analysis (PHA)
- Hazards and Operability Study (HAZOP)
- Incident command systems (ICS)
- Plant permit systems
- Operator training
- Housekeeping
- Audits and inspections
- Mutual Aid Agreements
- Accident investigations
- Industrial hygiene monitoring
- Fugitive emissions monitoring

Personal Protective Equipment (PPE)

Most of the hazards found in the chemical processing industry have been assessed and have PPE that will provide some degree of protection. Personal protective equipment provides an effective means for protecting technicians from hazardous situations. Typical outer wear worn by technicians includes:

- *Safety hats* provide protection to the head.
- *Safety glasses* provide protection to the eye from front impact.
- *Fire-retardant clothing* is designed to protect a technician for a limited amount of time from a flash fire or heat source. The clothing is specially treated and will resist bursting into flames upon contact but will not protect a technician during sustained periods of exposure.

- *Safety shoes* provide protection from falling or rolling objects. A typical safety shoe policy requires all technicians to wear safety shoes that have all leather uppers and steel toes.
- *Hearing protection* protects technicians from noise that can permanently damage the inner ear. Earplugs and earmuffs are commonly used to keep exposures below 82 decibels for 12 hours or 85 for 8 hours.
- *Gloves* are designed to provide protection from chemicals or abrasive or cutting surfaces.
- *Face shields* are designed to prevent chemical splashes and contaminants from getting into the eyes.
- *Chemical monogoggles* are designed to prevent chemical splashes and contaminants from getting into the eyes.
- *Slicker suits* provide protection from rain and inclement weather.
- *Radios* are used to provide emergency and operational information to other technicians and supervisors.
- *Respirators* provide protection from chemicals and airborne contaminants. They come in two basic designs: air-purifying and air-supplying.
- *Chemical suits* provide protection from specific chemicals.
- *Totally encapsulating chemical protective suits (TECP)*

The Process Technician and the Chemical Processing Industry

It is easy to describe the role of a doctor and the medical profession, a lawyer and the criminal justice system, or a school teacher and the education community because we have grown up watching them on television and hearing our friends and parents describe them. We know which occupations make the most money and the type of educational credentials required. Because the role of the process technician or operator is such a specialized occupation, geographically and occupationally, it is a little more difficult to explain what a process technician is.

Process operators are typically found in areas that are heavily industrialized. The Silicon Valley for process operations appears to be the Gulf Coast, with a focal point near Houston. There are numerous other areas in the United States where large numbers of process technicians can be found: Alaska, Alabama, New Jersey, Texas, Louisiana, Oklahoma, Southern California, Colorado, Delaware, South Carolina, Ohio, Kentucky, West Virginia, Michigan, Illinois, Wyoming, Washington State, Montana, Alabama, Utah, and North Dakota.

The occupation of process operators is best defined by the chemical or process they are working with. Process technicians can be found in chemical plants and refineries, the pulp and paper industry, food processing plants power plants, city utilities, and many other areas. The roles and responsibilities of a process technician change depending on which area they are working in. Common relationships across the occupation include the following:

- Equipment and technology—valves, piping, pumps, compressors, tanks, steam turbines, motors, heat exchangers, cooling towers, boilers, furnaces, reactors, distillation columns, flare systems, wastewater treatment systems, lubrication systems, refrigeration systems, plastics plant systems, auxiliary systems, hydraulic systems, specialized equipment, symbols and diagrams, and so on.
- Instrument and electrical—instrumentation, control loops, computers, electrical systems, interlocks, and permissives
- Maintenance—use hand tools, perform minor maintenance, coordinate equipment maintenance activities, fill oil reservoirs, and perform housekeeping
- Operation—start and stop processes, control industrial processes with computer automation, catch samples, make rounds, fill out checklists, fill out logbooks, and troubleshoot problems
- Math and science—basic math, statistics, energy, heat, heat transfer, pressure, temperature, fluid flow, level, applied chemistry, applied physics, statistical process control, control charts, and so on
- Language—reading, writing, and communicating
- Computer technology—word processing, spreadsheets, databases, and the Internet
- Safety, health, and environmental—general safety rules, process safety management (PSM), Hazard Communication (HAZCOM), Hazardous Waste Operations and Emergency Response (HAZWOPER), respiratory protection, personal protective equipment (PPE), fire protection, Department of Transportation (DOT), environmental regulations, permits, and so on
- Human relations—interpersonal skills, working in self-directed work teams, attitudes and behaviors, listening skills, communication, motivation, goal setting, and time management
- College skills—well-rounded, broad-based education (future technicians will need to have a variety of educational experiences in order to address the rapidly changing workplace)

According to the American Chemical Society, there are over 500,000 process technicians in the United States and over 250,000 laboratory technicians. Process operators fall into an educational category referred to as

process technology. Process technology is defined as the study and application of the scientific principles associated with the operation and maintenance of the chemical processing industry.

Records indicate that process technicians are safer at their worksites than in their own homes. The chemical processing industry (CPI) has established a number of safety systems to protect their employees, facilities, and community. These safety systems will be discussed during the course of this text.

Attitudes and Behaviors

Most prospective employers have a screening device in their preemployment interviews that checks for specific safety attitudes and behaviors. Strong candidates for employment who demonstrate poor attitudes and behaviors toward safety will find it difficult to pass initial interviews. The prospective employer will typically ask the candidate, "If you are given a direct order by your immediate supervisor to perform a task that you feel is unsafe, how will you respond?" The candidate's answer to this question may determine whether he or she will be hired by this company.

Industrial employers are also concerned about environmental issues and a prospective employee's attitude toward the environment. A typical question could be, "You have been assigned to the environmental control section of our plant that treats wastewater before releasing it into the river. Three days after your initial assignment a heavy storm moves over the area, flooding neighboring communities and overwhelming the plant's waste treatment system. The plant's sewer system is flooded, the biological system saturated, primary and secondary clarifiers are at full capacity, and the lagoons are almost full. Your supervisor calls you in for a private meeting to discuss options:

- Open the flood gates and release a small amount of the untreated effluent before the dikes break.
- Hold up all material and treat according to environmental regulations.

Which option do you choose?

Environmental issues are typically associated with fines and prison sentences. Informed technicians know how to respond to the preceding scenario. The flood gates should not be opened under penalty of law. Interviewers could ask questions about littering, disposing of motor oil, new environmental regulations, and housekeeping. An applicant's answers to these questions can give an industrial employer insight into his or her environmental values.

The attitudes and behaviors for safe and environmentally sound work habits include the following:

- Working safely while performing job tasks
- Complying with environmental rules and regulations
- Responding to environmental spills and releases
- Reporting unsafe acts and conditions
- Maintaining good housekeeping
- Obeying safety rules
- Studying safety and environmental standards
- Following standard operating procedures
- Performing minor maintenance on leaking equipment

During routine operation a process technician will encounter safety, health, and environmental issues. (See Figure 1-3.) Leaking pumps, pipes, and drums create environmental and safety problems. Spills could be in reportable or nonreportable quantities and must be addressed immediately by the technician. Poor housekeeping has been associated with increased safety incidents. Putting off reporting of unsafe acts and conditions directly affects a plant's safety record. It is important for process technicians to obey all safety rules and follow their companies' operational procedures.

New Technology and Potential for Catastrophic Events

Over the past 20 years, many advances have taken place that have revolutionized the chemical processing industry. The number one advance has been in the area of automation. The CPI can now produce more with fewer people because of new technology. Several companies that employed over 12,000 employees during the 1980s are now producing two to three times the product with only 2,000 employees. Two of these companies were sold because of environmental issues, modernized, resold, right-sized, and automated.

As equipment and technology improved, older workers made way for the new generation. Job descriptions for process technicians were torn up and rewritten. It quickly became clear to the chemical processing industry that a highly educated and adaptable workforce was needed. Industrial manufacturers turned to their local community colleges and universities. Initial programs developed in 1992 and 1993 were crude, but by 1998–1999, a standardized state curriculum was adopted in Texas and Louisiana. Several other large state efforts are underway and should provide major steps toward standardization. The development of this curriculum had its roots in an industrial partnership initiated by educational alliances, the American Chemical Society, and the industry. It should be emphasized that this was a national effort and not a single state's effort.

Figure 1-3
Process Technicians

Along with this highly technical workforce come a number of risks. In order to describe these risks, we will need to look at this problem in two parts: (1) the make-up of the workforce and (2) the technology boom. Because the new work force is a mix of the older "baby boomer" generation and a sprinkling of recent graduates of process technology programs, an unbalanced arrangement exists. Over the next five to ten years huge numbers of this workforce will retire, leaving a void that will be filled by our local community colleges and universities. During this transitional hand-off, advances in technology will continue. New processes will be developed and technology will quickly fill the void with the latest and greatest. Global competition is so intense that many companies will be forced to apply cutting edge technology in order to stay in business. Most of this technology will be integrated (patched) with existing plant systems and applied for the first time by this mixed workforce. Because of increased capacity and productivity, it will be possible for a technician to simply push a button and cause a major incident.

The American Chemical Society projects that the current responsibilities of process technicians, chemists, and engineers will begin to merge. Another concept that is beginning to gain in popularity is the merging of process, electrical, instrument, and mechanical technicians. The process technician of the future will be a degreed individual interfacing with advanced electronic systems and equipment. Process technology programs currently exist in Texas, Louisiana, Illinois, Michigan, Utah, California, Delaware, South Carolina, North Dakota, Ohio, Alabama, Kentucky, Colorado, Alaska, New Jersey, Oklahoma, Delaware, West Virginia, Wyoming, Washington State, and Montana.

General Plant Safety Rules

General safety rules are designed to protect human life, the environment, and physical equipment or facilities. Before entering a refinery or chemical plant, a simple overview of the general plant safety rules is conducted. (See Figure 1-4.) These rules include:

- Do not go to a fire, explosion scene, accident, or vapor release unless you have specific duties or responsibilities.
- Obey all traffic rules.
- Do not park in designated fire lanes.
- Report injuries immediately.
- Stay clear of suspended loads.
- Smoking and matches are not permitted in most sections of a plant.
- Drink from designated water fountains and **potable water** outlets.
- Use the right tool for the right job.
- Report to the designated equipment owner before entering an operating area. Stay in your assigned area.
- Illegal drugs and alcohol are not permitted in the plant.
- Firearms and cameras are not allowed in the plant.
- Take steps to remove hazardous conditions.
- Review and follow all safety rules and procedures including: personal protective equipment, hazard communication,

Figure 1-4
Plant Safety Rules

respiratory protection, permit system, hazardous waste operations and emergency response, housekeeping, and fire prevention.

- Know and understand the following alarms and rules associated with them: vapor release, fire or explosion, evacuation, and all clear.

Job Hazard Analysis

A job hazard analysis is often referred to as a process used to assess risk. The basic steps include the identification of unacceptable risks and the process used to eliminate or control these risks. A hazard is defined as a condition or practice that could contribute to an undesirable or unplanned event or as the potential for harm. A hazard analysis reviews and analyzes an operating process from start to finish. The primary focus is on the identification of hazards associated with each job task. Other areas include the work environment, required tools, and relationship between the technician and the task.

In order to conduct a job hazard analysis the following detective work is required:

- List what can go wrong.
- Identify the consequences.
- Explain how it could arise or occur.
- Select the contributing factors.
- Determine the frequency with which the hazard occurs.

Fire Classification System

The fire classification system is designed to simplify the selection of fire-fighting techniques and equipment.

- Class A fires involve the burning of combustible materials such as wood, paper, plastic, cloth fibers, and rubber.
- Class B fires involve combustible and flammable gases, liquids, and grease.
- Class C fires are categorized as electrical fires.
- Class D fires cover combustible metals.

Types of Fire Extinguishers

Fire extinguishers come in a variety of shapes and designs. The selection and use of a fire extinguisher depend on the type of fire being fought. Process technicians need to be familiar with the classification of fire system in order to select the proper extinguisher in an emergency situation. Portable, hand-held fire extinguishers are carefully placed around the process unit. These devices are a fundamental part of the fire prevention, protection, and control standards. Because most fires require split-second reactions, the portable fire extinguisher is an excellent first line defense. Portable fire extinguishers can be used to stop a fire before it gets out of control, provide an escape route through a large fire, and contain a fire until help or the fire team arrives.

The carbon dioxide extinguisher is composed of a cylinder filled with compressed carbon dioxide. CO_2 extinguishers are effective on Class B and C fires because they displace oxygen. *Dry chemical fire extinguishers* are composed of a cylinder, dip tube, pressure gauge, hose and nozzle, BC or ABC dry chemical agent, carrying handle, operating lever, locking pin, and a compressed nitrogen or carbon dioxide cartridge. *Foam fire extinguishers* are used to control flammable liquid fires. The foam forms an effective barrier between the flammable liquid and the oxygen needed for combustion. *Water fire extinguishers* are composed of a cylinder, dip tube, pressure gauge, carrying handle, locking pin, operating lever, and overfill tube. Portable, hand-held water-filled fire extinguishers are designed for use on Class A fires only.

Types of Permits

Three permit systems are common in the industry: the control of hazardous energy (lockout/tagout), confined space entry, and hot work. These three permitting systems are common within the CPI because they are government mandated programs. Examples of a permitting system could include the following:

- Hot work permit—any maintenance procedure that produces a spark or excessive heat, or requires welding or burning.
- Energy isolation procedure, lockout/tagout—isolates potentially hazardous forms of energy: electricity, pressurized gases and liquids, gravity, and spring tension. (The standard is also designed to shut down a piece of equipment at the local start or stop switch, turn the main breaker off, attach a lockout adapter and process padlock, try to start the equipment, and tagout and record in lockout logbook.)
- Confined space entry, permit to enter—designed to protect employees from oxygen-deficient atmospheres, hazardous conditions, power-driven equipment, and toxic and flammable materials.
- Opening or blinding permit—removing blinds, installing blinds, or opening vessels, lines, and equipment.
- Unplugging permit—barricades area, clears lines for unplugging, informs personnel, issues opening blinding permit, and issues unplugging permit.
- Routine maintenance permit—general maintenance and mechanical work that does not involve hot work or opening up a vessel.

Weapons of Mass Destruction

Benjamin Franklin once stated, "failure to prepare is preparing to fail." The use of weapons of mass destruction can have a catastrophic affect on the chemical processing industry and the communities that surround them. Weapons of mass destruction can be classified as biological agents, chemical agents, and bombings. In light of the events that occurred on September 11 the world has changed and the chemical processing industry has revised its approach to safety preparedness. Other incidents like the Bhopal, India disaster, the Phillips explosion, Exxon Valdez oil release, and the 2010, BP gulf oil release have illustrated how vulnerable the industry is to unexpected events.

Potential risks include:

- Terrorist flies an airplane into a chemical plant or refinery
- Terrorist enters CPI workforce

- Use of conventional bomb
- Use of nuclear weapon
- Suicide bomber
- Use of chemical weapons
- Use of biological weapons
- Tornado hits
- Fast forming hurricanes
- High winds
- Use of military weapons
- Flooding

Tornados and Hurricanes

Tornados are classified as extremely violent, turbulent, rotating columns of air that maintain contact with the cumulonimbus cloud and the surface of the earth. The CPI utilizes the Fujita Scale to classify Tornados. Tornados produce wind speeds between 40 and 110 mph and tend to gather or collect debris, rocks, and dust (flying projectiles). These destructive forces of nature are around 250 feet across and travel for several miles on the ground before dissipating.

Hurricanes are powerful, swirling storms with tentacles reaching out from a singular eye. These storms are characterized with having pulsing wind gusts from 74 to over 155 mph, heavy rainfall and flooding, tremendous storm surges from 4 to 18 feet, spin-off tornados, downed trees and broken limbs, damaged homes, loss of electricity, utilities, and basic commodities.

Summary

Process safety is described as the application of engineering, science, and human factors to the design and operation of chemical processes and systems. The primary purpose of process safety is to prevent injuries, fatalities, fires, explosions, or unexpected releases of hazardous materials. Process safety focuses on the individual chemical processes and operational procedures associated with these systems. A process safety analysis is used to establish safe operating parameters, instrument interlocks, alarms, process design, and startup, shutdown, and emergency procedures. Occupational safety and health deals with items such as personal protective equipment, HAZCOM, permit systems, confined space entry, hot work, isolation of hazardous energy, and so on. Understanding the key elements of process safety and occupational safety allows technicians to contribute significantly to the reduction of risks and hazards associated with the chemical processing industry.

The industrial safety movement closely parallels the rise of the labor movement. Industrial workers battled for a voice in the workplace and a piece of the economic pie. There is an impressive contrast between past industrial worker safety and the modern workforce.

The Pittsburgh Survey of 1907 provided the first statistical data on how many people were being killed or injured as a result of hazardous working conditions in the United States. Russell Sage investigators were shocked when they discovered that between 1906 and 1907, 526 people had lost their lives and 500 had been seriously injured or disabled. Over half of the families involved in this tragedy had been left without any means of support. As the Pittsburgh Survey expanded beyond the Pittsburgh area, the investigation revealed that there had been over 30,000 fatalities from industrial accidents across the United States during 1906. As news of this report became known, industry, government, and the public recognized a need for change. The Russell Sage investigators had unknowingly opened a new chapter for the safety movement.

Process technicians can be found in chemical plants and refineries, the pulp and paper industry, food processing, power plants, city utilities, and many other areas. According to the American Chemical Society, there are over 500,000 process technicians in the United States. Process operators fall into an educational category referred to as process technology. Process technology is defined as the study and application of the scientific principles associated with the operation and maintenance of the chemical processing industry. The roles and responsibilities of a process technician include knowing the equipment and technology they are working with, being able to maintain and operate this equipment, complying with safety and environmental regulations, applying quality techniques to operational processes, collecting and analyzing samples, and troubleshooting system problems. The process operator of the future will be required to have formal educational credentials and advanced training prior to taking an entry-level position in the chemical processing industry.

Review Questions

1. Describe the importance of process safety.
2. Explain the key elements of occupational health and safety.
3. What is the definition of the term *process technology*?
4. What are the significant differences between industrial worker safety in the past versus the present?
5. List the roles and responsibilities of process technicians.
6. List five general safety rules used in the operation and maintenance of a typical chemical plant or refinery.
7. List the most important events in the safety movement from 1960 to the present.
8. Why did industrial workers in the 1870s believe their jobs were hazardous?
9. What is phossy jaw and how was it contracted?
10. How will understanding the key elements of process safety and occupational safety allow chemical technicians to contribute significantly to the reduction of risks and hazards associated with the chemical processing industry?
11. Why is the Pittsburgh Survey of 1907 a significant part of the safety movement?
12. Describe the relationship OSHA, HAZCOM, and PSM have to the safety movement.
13. Describe the fire classification system.
14. Describe the different types of fire extinguishers.
15. Explain how hurricanes are classified.
16. Describe the scale used to rate a tornado.
17. List the different types of permits found in the CPI.
18. Explain the hazards associated with the use of weapons of mass destruction.
19. List the steps associated with conducting a job hazard analysis.
20. List the key elements of HAZCOM.

Hazard Classification

OBJECTIVES

After studying this chapter, the student will be able to:

- Describe the hazard classification system.
- Apply the methods of hazard recognition and classification.
- Describe common industrial hazards found in the plant.
- Compare acute and chronic hazards.
- Describe the physical hazards associated with chemicals.
- Explain the activities that are associated with ergonomic hazards.
- Explain how biological hazards affect chemical technicians.
- Explain the principles of accident prevention.
- Analyze the principles of accident investigation.
- Describe the various types of ionizing radiation.

Key Terms

- **Accident prevention, basic principles of**—includes a safe working environment, safe work practices, and effective management.
- **Accidents**—an unplanned disruption of normal activity resulting in an injury or equipment damage.
- **Acute (immediate) poisons**—may be ingested, inhaled, injected, or absorbed; examples of these chemicals include chlorine, acids, and caustics.
- **Acute hazards**—materials from which symptoms develop rapidly after exposure.
- **Biological hazards**—described as any living organism capable of causing disease in humans; this includes insects, bacteria, fungi, and molds.
- **Chemical hazards**—terms associated are carcinogens, mutagens, teratogens (affects unborn fetus), reproductive toxins; those that cause asphyxiation, anesthetic, neurotoxic; those that cause allergic response, irritants, sensitizers, corrosives, toxic, highly toxic; and those that target organ effects.
- **Chronic (delayed) hazards**—asbestos fibers, coal dust, and toxic metals such as lead or manganese.
- **Ergonomic hazards**—activities that require chemical technicians to work in unusual or awkward positions for extended periods of time.
- **Ionizing radiation**—cannot be detected by any of the five human senses, classified as: alpha particles, beta particles, gamma rays, x-rays, and neutron particles.
- **Keys to accident prevention**—(1) Determine the cause. (2) Prevent its recurrence.
- **National Fire Protection Association**—has a standardized system used in chemical hazard identification.
- **Physical hazards associated with chemicals**—categorized as combustible liquid, compressed gas, explosive gas, flammable gas, flammable liquid, organic peroxide, oxidizer pyrophoric gas, unstable, and water reactive.
- **Physical hazards**—classified as electrical, noise, radiation, or temperature.
- **Radioactive substances**—metallic uranium, x-ray, and strontium 90. These hazardous materials break down the cells of exposed tissue.
- **Toxic hazards**—fuels, metal fumes, solvents, products, and byproducts.
- **Toxic substance**—a chemical or mixture that may present an unreasonable risk of injury to health or the environment.
- **Unsafe act**—any act that increases a person's chance of having an accident.
- **Unsafe condition**—a condition in the working environment that increases a technician's chance of having an accident.

Common Industrial Hazards

Chemical technology technicians are exposed to a variety of common industrial hazards, which can be flammable, toxic, and radioactive. When technicians may come in contact with these hazards, they must follow several steps to ensure their safety. The first step is the proper instruction in the identification, handling, and use of hazardous substances. The second critical step includes the use of safety procedures in working with any of these substances. The **National Fire Protection Association** has a standardized system used in chemical hazard identification. **Toxic hazards** include fuels, metal fumes, solvents, products, and byproducts. Flammable substances catch fire or explode easily. **Radioactive substances** include metallic uranium, x-ray, and strontium 90. These hazardous materials break down the cells of exposed tissue.

Acute and chronic hazards have an immediate or delayed effect on workers. **Acute (immediate) poisons** may be ingested, inhaled, injected, or absorbed. Chapter 3 will explore the routes of entry a chemical travels as it enters a physical body. Acute effects occur when high concentrations of chemical compounds come in contact with a chemical technician during a single exposure. Examples of these chemicals are chlorine, acids, and caustics. **Chronic (delayed) hazards** include asbestos fibers, coal dust, and toxic metals such as lead or manganese. Chronic hazards are cumulative and frequently occur over an extended period of time.

Physical, Chemical, Ergonomic, and Biological Hazards

Physical hazards are classified as electrical, noise, radiation, or temperature. **Physical hazards associated with chemicals** are categorized as a combustible liquid, a compressed gas, explosive, a flammable gas, a flammable liquid, organic peroxide, oxidizer, pyrophoric, unstable, and water reactive.

Chemical hazards are often referred to as health hazards by process technicians and can be categorized as carcinogens, mutagens, teratogens, reproductive toxins; those that cause asphyxiation, anesthetic, neurotoxic; those that cause allergic response, irritants, sensitizers, corrosives, toxic, highly toxic; and those that target organ effects.

Ergonomic hazards are activities that require chemical technicians to work in unusual or awkward positions for extended periods of time. This may include repetitive motions, monotony, work pressure, inability to match the standard of performance, console operations, or equipment and systems operation.

Biological hazards are described as any living organism capable of causing disease in humans. This includes insects, bacteria, fungi, and molds.

Electrical Hazards

Process technicians work with a variety of electrical energized equipment and systems. Electrical motors are used to drive pumps, compressors, generators, mixers, conveyors, fans, blowers, large valves, and a wide variety of other systems. Motor control centers (MCCs) are frequently used to turn equipment on and off. These substations have a variety of voltages coming in and out of local transformers. The chemical processing industry has initiated special procedures for the control and isolation of hazardous energy. Isolating and safely tagging and locking out process equipment is part of a process technician's job.

Industrial Noise Hazards

Industrial noise is described as valueless or unwanted sound that is measured in decibels. The trigger point for hearing protection is 85 decibels. Noise over 140 decibels can cause permanent hearing loss. When hearing loss occurs, it can never be recovered. Industrial equipment produces noise at a variety of decibel levels. Hearing protection is provided and required in areas where exposure will exceed recommended levels.

Radiation Hazards

Events that happened in places such as Three Mile Island and Chernobyl have made the world aware of the hazards associated with radiation. On April 26, 1986, a fire and explosion at the Chernobyl nuclear power plant sent out an invisible radioactive cloud that eventually encircled the globe. This major environmental event took place 80 miles north of Kiev in the Soviet Ukraine.

Of the 102 known elements listed on the periodic table, 50 are known to be radioactive. Recent developments in nuclear testing equipment, nuclear power, x-ray, and other nuclear technologies have the ability to seriously affect biological organisms.

Ionizing radiation cannot be detected by any of the five human senses. It is classified as alpha particles, beta particles, gamma rays, x-rays, and neutron particles. Alpha particles are the least penetrating and provide little threat to human tissue; however, these high-energy, high-velocity particles can damage specific internal organs if exposed. Beta particles are emitted from radioactive materials and penetrate much deeper than alpha particles. These particles can be ingested, inhaled, or absorbed through the skin. Energy from these hazardous particles is completely released in soft human tissue.

Unlike alpha and beta particles, gamma rays are not made up of atomic particles. Gamma rays are aggressively penetrating energy waves that can be stopped only by thick layers of concrete or lead and are extremely hazardous even from great distances.

X-rays are used in medical diagnostics and treatments, medical photographs, and to inspect vessels, packages, equipment, and piping. X-rays penetrate deeply and can change the molecular structure of tissue; therefore, extreme caution must be exercised through monitoring and shielding.

Neutron particles penetrate deeply and can be stopped only by heavy shielding. Neutron particles have no practical application; however, the military and certain research groups are experimenting with future uses.

Radiation is indirectly measured by the ionization footprint produced as it passes through a medium. Units for measuring radiation include rad (energy absorbed in a substance), rem (amount of biological injury), rep (radiation in human tissue), or roentgen (X or gamma radiation unit measurement). These four units refer to the energy, the charge, or biological effect. Figure 2-1 lists the four areas of hazardous agents.

Hazard Recognition

In a typical safety program, **hazard recognition** is considered to be a management function. In this text, hazard recognition is considered to be a process technician's primary function. **Accidents** are defined typically as an unplanned disruption of normal activity resulting in an injury or equipment damage. Controlling the number of accidents that occur in a plant is a function of worker performance, equipment and tools, and working environment. The **key to accident prevention** is linked to (1) determining the cause and (2) preventing its recurrence. By understanding the causes of accidents, process technicians can more easily defend against their occurrence. Accidents can be broken down into four parts:

1. Contributing causes—poor instructions
2. Immediate causes—did not use Personal Protective Equipment PPE
3. The accident—struck by truck for example
4. The consequences of the accident—fatality

Classification systems for accidents include struck by; struck against; caught in, on, or between; fall from above; fall at ground level; strain or overexertion; electrical contact; and burn.

1 PHYSICAL HAZARDS
- Electrical hazards
- Noise hazards
- Radiation hazards
- Temperature hazards
  - combustible liquid/gas
  - explosive
  - flammable gas/liquid
  - organic peroxide
  - oxidizer
  - pyrophoric
  - unstable/water reactive

2 CHEMICAL HAZARDS
- Health hazards:
  - carcinogens, mutagens
  - teratogens, reproductive toxins
  - asphyxiation, anesthetic
  - neurotoxic, allergic response
  - irritants, sensitizers
  - corrosives, toxic
  - highly toxic
  - those that target organ effects

3 ERGONOMIC HAZARDS
- Requires process technicians to work in unusual or awkward positions for extended periods of time:
  - repetitive motion
  - monotony
  - work pressure
  - console operations

4 BIOLOGICAL HAZARDS
- Any living organism capable of causing disease in humans:
  - insects
  - bacteria
  - fungi
  - molds

Figure 2-1 *Classification of Hazardous Agents*

Unsafe acts and **unsafe conditions** are responsible for many of the accidents that occur inside the plant. An unsafe act is defined as any act that increases a person's chance of having an accident. Unsafe acts can be eliminated by a no-tolerance policy. Because this is not the most effective way to do business, other alternatives need to be researched. Unfortunately, the negative reinforcement theory makes it difficult to convince someone to stop performing unsafe acts. Reinforcement theory is a process used to shape human behavior by controlling the rewards or punishments designed to draw out or reinforce a desired behavior. Many of these people think the shortcuts they take are worth it or that "it can't happen to me." Unsafe conditions are easier to handle than unsafe acts. An unsafe condition is defined as a condition in the working environment that increases a technician's chance of having an accident. (See Figure 2-2.) Unsafe conditions can be corrected simply by encouraging process technicians to

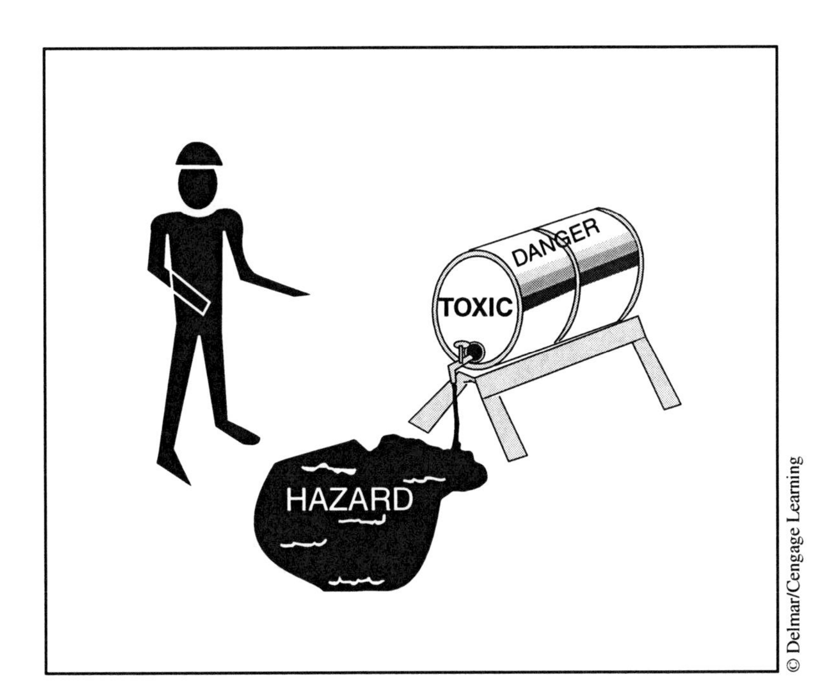

Figure 2-2
Hazard Recognition

identify and correct an unsafe condition. Incentive programs and training have proven to be an effective tool in preventing unsafe conditions.

Accident Prevention

The basic principles of **accident prevention** include a safe working environment, safe work practices, and effective management. If management does not support the safety program, if process technicians do not follow the rules of a safety program, or if the environment is riddled with unidentified hazards, the system will not work.

Plant managers are the central focus of safety program management. Managers design the company's safety policy, enhance safety awareness, and facilitate safe and healthful work conditions.

In the typical chain of command, a safety coordinator is the designated specialist who provides current safety supervision and expertise on the program. The safety coordinator takes much of the responsibility for day-to-day operation of a large safety program off the shoulders of the plant manager. The safety specialist receives training in specialized areas and carefully analyzes and compares existing and new governmental rules and guidelines to current plant operations.

The first-line supervisor has the greatest impact on a plant's safety program. These individuals typically have the most knowledge about a specific process and are looked upon as the primary contact for safety awareness.

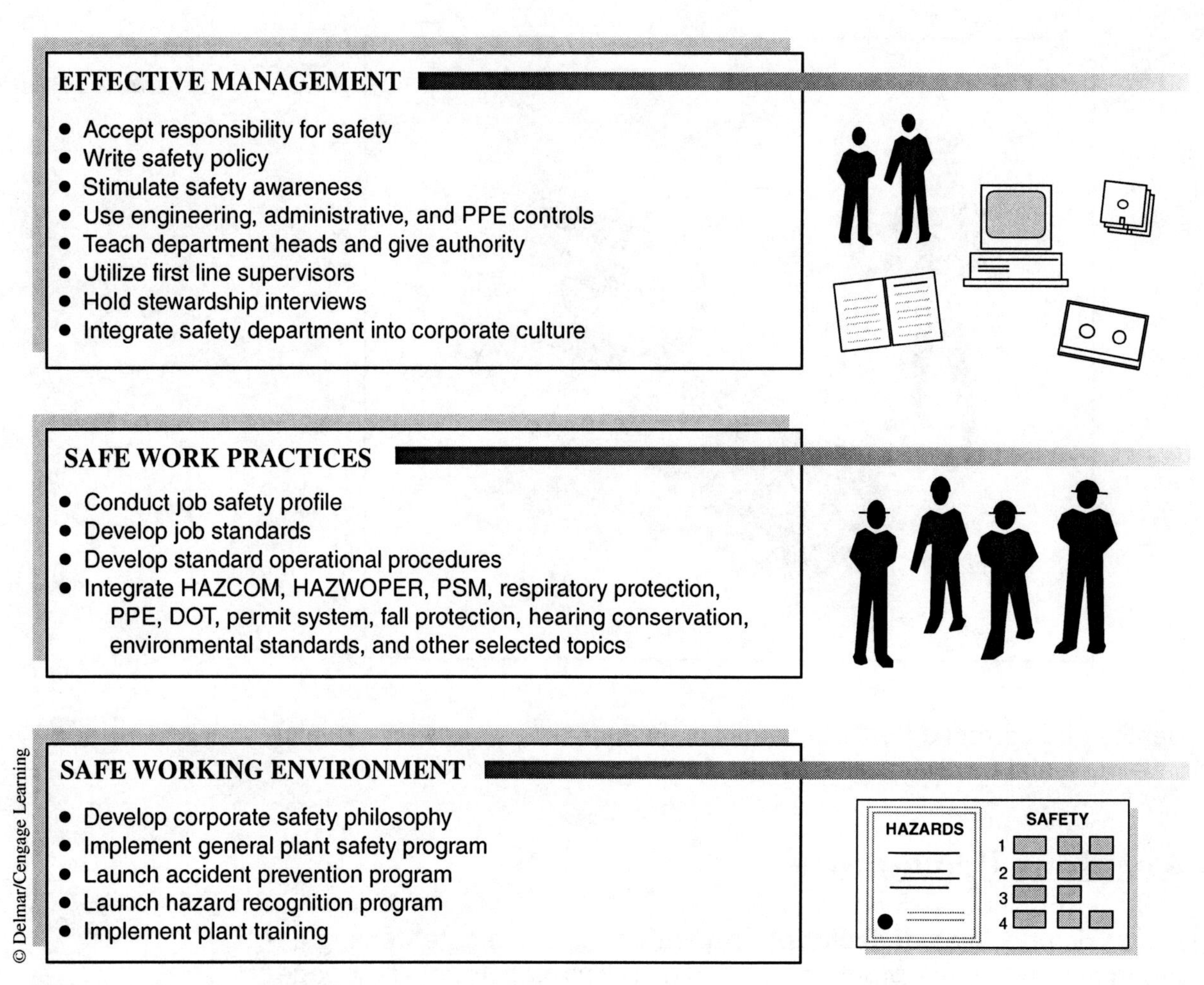

Figure 2-3 *Accident Prevention*

Team safety meetings are typically conducted by the first-line supervisor. First-line supervision provides the bridge between operations personnel and the safety supervisor or, in some cases, the plant manager. Figure 2-3 shows how accident prevention is divided into three parts.

Process technicians ensure that the rules of the safety program are maintained by all of the people entering the facility. Technicians are the backbone of any safety program because they are responsible for the following:

- Good housekeeping
- Hazard recognition
- Safe work practices
- Safe work environment

Accident Investigation

Accident investigation is designed to identify the point of failure and prevent recurrence. The point of failure typically falls into one of two groups: hazardous conditions or poor work practices.

First-line supervisors perform the following tasks:

- Conduct accident and near-accident investigations
- Visit accident site as soon after the accident as possible
- Take photographs
- Collect data and information from witnesses and those involved in accident
- Fill out accident report:
- Analyze what has happened
- Identify objectives

Summary

Chemical technology technicians are exposed to a variety of common industrial hazards that are classified as physical, chemical, ergonomic, and biological. Where technicians may come in contact with these hazards, several steps must be followed in order to ensure their safety. The first step includes the proper instruction in the identification, handling, and use of hazardous substances. The second step, a critical one, includes the use of safety procedures in working with any of these substances.

Physical hazards are classified as electrical, noise, radiation, or temperature. Physical hazards associated with chemicals are categorized as: combustible liquid, compressed gas, explosive, flammable gas, flammable liquid, organic peroxide, oxidizer, pyrophoric, unstable, and water reactive. Chemical hazards are often referred to as health hazards by process technicians and can be categorized as: carcinogens, mutagens, teratogens, reproductive toxins, those that cause asphyxiation, anesthetic, neurotoxic, those that cause allergic response, irritants, sensitizers, corrosives, toxic, highly toxic, and those that target organ effects.

Ergonomic hazards are activities that require chemical technicians to work in unusual or awkward positions for extended periods of time. This may include repetitive motions, monotony, work pressure, inability to match the standard of performance, console operations, or equipment and systems operation.

Biological hazards are described as any living organism capable of causing disease in humans. These include insects, bacteria, fungi, and molds.

Acute and chronic hazards have an immediate or delayed effect on workers. Acute (immediate) poisons may be ingested, inhaled, injected, or absorbed. Examples of these chemicals include: chlorine, acids, and caustics. Chronic (delayed) hazards include: asbestos fibers, coal dust, and toxic metals such as lead or manganese. The National Fire Protection Association has a standardized system used in chemical hazard identification.

Hazard recognition is considered to be a process technician's primary function. Accidents are typically described as an unplanned disruption of normal activity resulting in injury or equipment damage. The key to accident prevention is linked to (1) determining the cause and (2) preventing its recurrence. By understanding the causes of accidents, process technicians can more easily defend against their occurrence. Unsafe acts and unsafe conditions are responsible for many of the accidents that occur inside the plant. An unsafe act is defined as any act that increases a person's chance of having an accident. An unsafe condition is defined as a condition in the working environment that increases a technician's chance of having an accident. Unsafe conditions can be corrected simply by encouraging process technicians to identify and correct an unsafe condition.

The basic principles of accident prevention include a safe working environment, safe work practices, and effective management. Plant managers are the central focus of safety program management. Managers design the company's safety policy, enhance safety awareness, and facilitate safe and healthful work conditions. The first-line supervisor has the greatest impact on a plant's safety program. These individuals typically have the most knowledge about a specific process and are looked upon as the primary contact for safety awareness. The first-line supervisor typically conducts team safety meetings. First-line supervision provides the bridge between operations personnel and the safety supervisor or in some cases the plant manager. Process technicians ensure that the rules of the safety program are maintained by all of the people entering the facility.

Accident investigation is designed to identify the point of failure and prevent recurrence. The point of failure typically falls into one of two groups: hazardous conditions or poor work practices.

Review Questions

1. What are the physical hazards associated with chemicals?
2. Describe the hazard classification system.
3. Compare acute and chronic hazards.
4. Describe biological hazards found in chemical plants and refineries.
5. Describe ergonomic hazards found in chemical plants and refineries.
6. Describe physical hazards found in chemical plants and refineries.
7. Describe chemical hazards found in chemical plants and refineries.
8. The key to accident prevention is linked to what?
9. Accident investigation is designed to do what?
10. How is the National Fire Protection Association's standardized system used?
11. Describe an unsafe condition.
12. Describe the basic principles of accident prevention.
13. List the radioactive substances found in the chemical processing industry and describe the effects of exposure on the human body.
14. List the chemicals typically associated with acute hazards and (immediate) poisons and describe common entry into the human body.
15. Describe the term "ionizing radiation," and explain detection and classification methods.
16. Explain the risks associated with "electrical hazards."
17. Describe the decibel range associated with "industrial noise hazards."
18. How many elements on the periodic table are known to be radioactive?
19. List the categories used to describe toxic hazards.
20. List the elements of effective safety management.

Routes of Entry and Environmental Effects

Objectives

After studying this chapter, the student will be able to:

- Identify the attitudes and behaviors for safe and environmentally sound work habits.
- Contrast safety, health, and environmental issues with the performance of job tasks.
- Explain how hazardous chemicals can enter the human body.
- Discuss the dose-response relationship.
- Describe key elements of environmental awareness training.
- Identify the primary reasons for air pollution control.
- Explain the importance of water pollution control.
- Describe solid waste control.
- Explain how the toxic substances control standard protects chemical technicians.
- Describe the Resource Conservation and Recovery Act (RCRA).
- Explain the purpose of the Toxic Substances Control Act.
- Contrast the clean up of chemical waste and the Comprehensive Environmental Response, Compensation and Liability Act (CERCLA) or "Superfund Law."
- Describe the role and responsibilities of the Environmental Protection Agency (EPA).

Key Terms

- **Air permits**—permits must be obtained for any project that has the potential of producing air pollutants.
- **Air pollution**—the presence of contaminants or pollutant substances in the air that interferes with human health or welfare or produces other harmful environmental effects.
- **Biological system**—designed to remove hydrocarbons from wastewater. A biological system includes an aeration basin, clarifiers, lagoons, sewer systems, pumps, and so on.
- **Clean Air Act (CAA) of 1970**—designed to (1) enhance the quality of the nation's air; (2) accelerate a national research and development program to prevent air pollution; (3) provide technical and financial assistance to state and local government; (4) develop a regional air-pollution-control program. The CAA approved the establishment of National Ambient Air Quality Standards (NAAQS) to protect the environment and public health. In 1990, new standards were established that were designed to reduce acid rain, air toxins, ground-level ozone, and stratospheric ozone depletion.
- **Clean Water Act**—passed in 1898, initially focused on toxic pollutants; in 1972, adopted the Best Available Technology (BAT) strategy for all cleanups; in 1987, provisions for funded sewage treatment plants were provided as well as citizen suit provisions. Regulates the release of pollutants into lakes, streams, and oceans.
- **Community right-to-know**—increases community awareness of the chemicals manufactured or used by local chemical plants and businesses, involves community in emergency response plans, improves communication and understanding, improves local emergency response planning, and identifies potential hazards.
- **Earth Day, 1970**—held to educate the U.S. public about environmental concerns.
- **Federal Highway Administration**—regulates the transportation of hazardous materials and truck traffic.
- **Federal Railroad Administration**—regulates railroad traffic, including the transportation of hazardous materials.
- **Nuclear Regulatory Commission (NRC)**—established in 1974 to regulate the nuclear devices used in the chemical processing industry. This includes x-ray and measuring devices used to inspect vessels and equipment.
- **Pipelines**—regulated by the U.S. Department of Transportation, these are lines of pipe that convey liquids, gases, or finely divided solids.
- **Resource Conservation and Recovery Act (RCRA)**—enacted as public law in 1976. The purpose of the RCRA is to protect human health and the environment. A secondary goal is to conserve our natural resources. RCRA completes this goal by regulating all aspects of hazardous waste management: generation, storage, treatment, and disposal (a concept referred to as "cradle to grave").
- **Solid waste**—non-liquid, non-soluble material ranging from municipal garbage to industrial waste that contains complex and sometimes hazardous substances.

- **Toxic Substances Control Act (TSCA)**—a federal law enacted in 1976 intended to protect human health and the environment. TSCA was also designed to regulate commerce by (1) requiring testing and (2) necessary restrictions on certain chemical substances. TSCA imposes requirements on all manufacturers, exporters, importers, processors, distributors, and disposers of chemical substances in the United States.
- **Water pollution**—the human-made or human-induced alteration of physical, biological, chemical, or radiological integrity of water.

Routes of Entry

Figure 3-1 illustrates how hazardous chemicals can enter the human body.

Inhalation

Airborne chemicals enter the body through the mouth or nose and may irritate the nose, throat, bronchi, and deep lung tissue. Inhalation of a hazardous agent is the most common route of entry in the workplace. Some gases or vapors will not irritate the respiratory tract but will be absorbed into the blood system through the lungs. This process can affect the blood, brain, bone, liver, fatty tissue, kidney, and colon.

Hazardous chemicals can alter healthy cells, and they may become cancerous. Airborne chemical particulates may be retained in the respiratory tract and cause allergic reactions, lung scarring, cancer, or fibrosis.

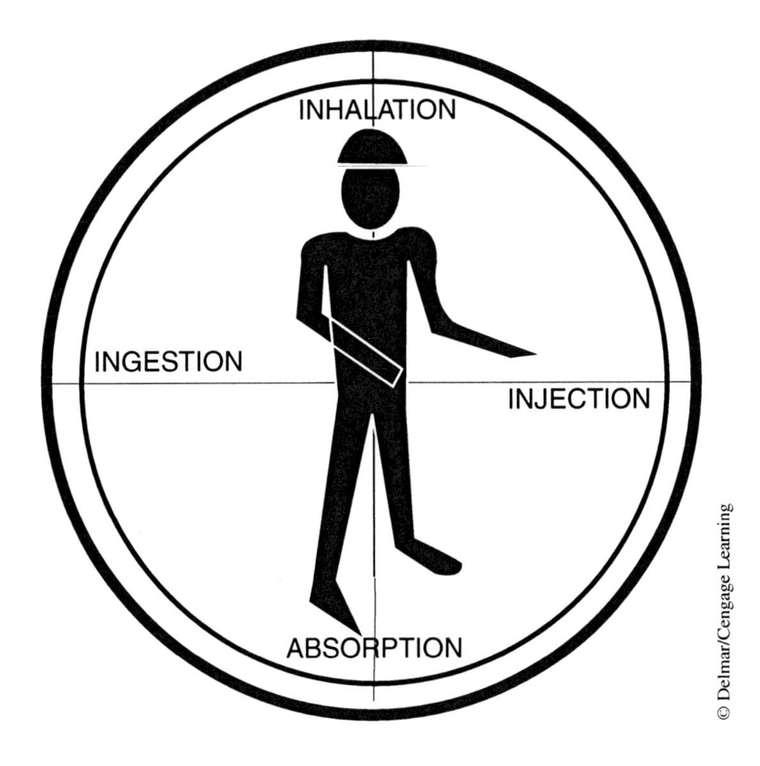

Figure 3-1
Routes of Entry

Absorption (skin contact)

Only a small number of chemicals are known to have serious impacts upon skin contact. These chemicals should be identified quickly if they are in your unit. When a hazardous chemical comes into contact with skin, there are four possible results:

1. Nothing will occur.
2. The chemical will react with skin and cause primary irritation.
3. The chemical will penetrate the skin and cause sensitization.
4. The chemical will penetrate the skin and enter the blood.

Ingestion

This is a process in which a hazardous chemical is transmitted to food, liquid, or cigarettes and then swallowed. Airborne chemicals may also be taken in through the respiratory system and then swallowed in the saliva or mucus. Typically, large quantities of hazardous chemicals are not ingested; however, small quantities of arsenic, lead, or mercury can be highly toxic over a period of time.

Injection

This is a process in which a toxic or hazardous material is injected into the body by a needle or sharp object, or through cuts in the skin.

Dose-Response Relationship

Chapter 6 will discuss in detail the topic of toxicology and the dose-response relationship. *Dose* is defined as the amount of chemical entering or being administered to a subject. *Response* is defined as the toxic effect the dose has upon the subject. This may be expressed under the following categories:

- Ingestion—the amount per unit of body weight
- Injection—the amount per unit of body weight
- Absorption—the amount per body surface area
- Inhalation—the amount per unit volume breathed

The dose-response relationship can be mathematically expressed as *Concentration (C)* × *Duration of exposure (T)* = *Constant (K)*. In this equation, the minimum dose required to produce a measurable effect is called the *threshold concept*. The term *lethal dose (LD)* is used to describe the amount or dose of a given substance that will likely cause death. In areas where gases and vapors are used, the term *lethal concentration (LC)* is used to identify how much of the airborne material would need to be inhaled in order to cause death.

Environmental Effects

Shortly after the industrial revolution, the effects of environmental problems began to become apparent to the public. In heavily industrialized areas, the walls of buildings were covered with the dark soot formed by burning coal. The clean waters of local rivers lost many of their pristine qualities as local industry openly dumped contaminants. The Federal **Clean Water Act** was passed in 1898. The **Resource Conservation and Recovery Act (RCRA)** was passed in 1976 to protect human health and the environment by ensuring the correct disposal of hazardous wastes, recycling practices, and **solid waste** disposal. **Air pollution** is produced by factories, homes, and motor vehicles that produce gases containing sulfur dioxide and nitrogen dioxide. Sulfur dioxide mixes with rain to form sulfuric acid. Nitrogen dioxide combines with hydrocarbons and sunlight to form smog.

When the *Exxon Valdez* spilled 11 million gallons of North Slope crude oil into the waters off Prince William Sound, Alaska, the public became aware of the potential environmental disasters that existed in international waters. Bhopal, India and the Union Carbide toxic release demonstrated how vulnerable local communities were to chemical releases from large chemical complexes. The Phillips and ARCO incidents in 1989 and 1990, near heavily populated communities, set in motion a series of events designed to keep these modern processes in the pipes and secure local environmental effects.

Exxon Valdez

On March 24, 1989 the *Exxon Valdez* (as shown in Figure 3-2) ran aground on a reef in Prince William Sound, Alaska. Approximately 11 million gallons of North Slope crude oil gushed out of the ruptured hull, contaminating the coastlines of the Alaska Peninsula, lower Cook Inlet, the Kenai Peninsula, the Kodiak Archipelago, and Prince William Sound. Oil from the Valdez was found over 600 miles southwest of the spill. The spill had catastrophic effects on five state parks, three national parks, one state game sanctuary, a national forest, four national wildlife refuges, and four state-critical habitat areas.

The Valdez incident was the largest tanker oil release in U.S. history. Clean-up and containment efforts started immediately, but five years after the spill, oil could still be found in sensitive areas. During the cleanup efforts thousands of people responded. A fleet of private fishing vessels went to the aid of critical fishing hatcheries by setting up an outer perimeter of skimmers to remove the oil. Exxon initiated a massive cleanup effort and funded millions of dollars in recovery efforts. Workers cleaned the beachfront with high pressure steam and bioremediation techniques. Biological

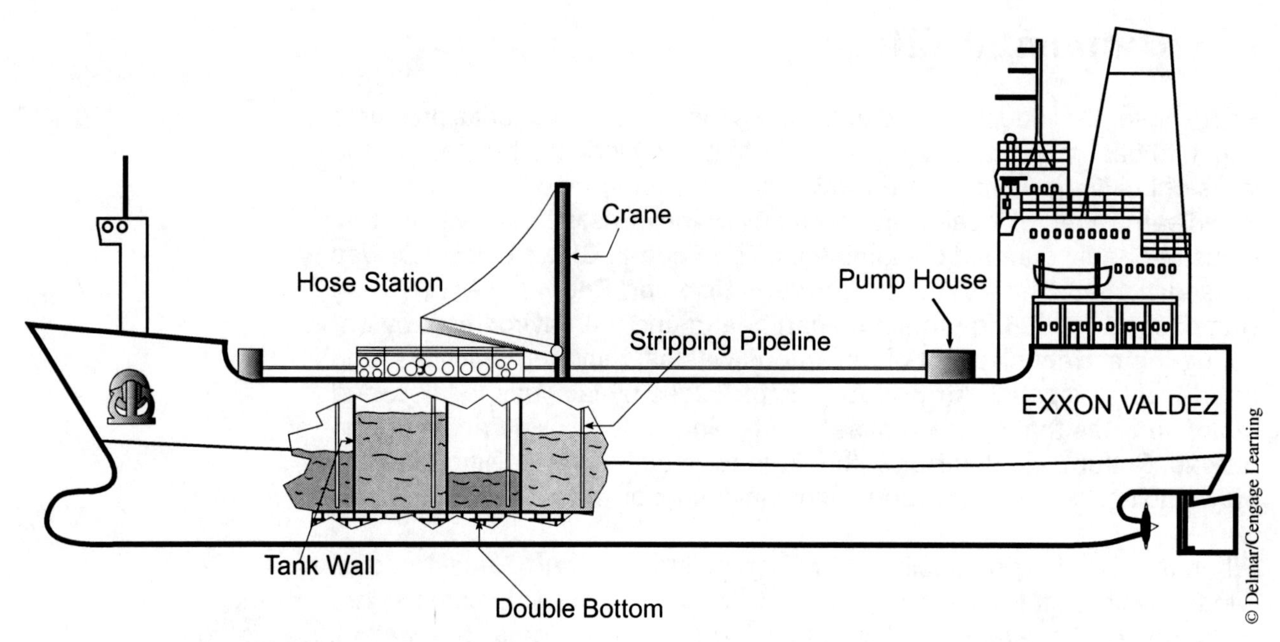

Figure 3-2 *Exxon Valdez*

systems are designed to remove hydrocarbons from salt water or wastewater. A biological system for controlled wastewater includes an aeration basin, microbes, clarifiers, lagoons, sewer systems, pumps, and so on. Oil spills from tankers, oil and natural gas rigs or pipelines utilize, boons, small boats, flares, biological microbes, and dispersants.

Air Pollution

Modern factories, office buildings, and homes produce smoke that contains sulfur dioxide and nitrogen dioxide. Sulfur dioxide and nitrogen dioxide are known to have harmful effects on the human respiratory system. As sulfur dioxide travels higher into our atmosphere, it turns into sulfur trioxide that mixes with rain to form sulfuric acid. As nitrogen dioxide rises, it combines with hydrocarbons and sunlight to form smog. Smog is an irritant that damages the eyes, the respiratory tract, and plants. Motorized vehicles produce unburned hydrocarbons, carbon monoxide, and nitrogen dioxide. This is bad for our environment because hydrocarbons injure plant life and human life and help form smog. Carbon monoxide produces headaches and dizziness in humans. Low concentrations of carbon monoxide can kill. Incineration units and wide-scale burning concentrates high levels of mercury and small particles in the atmosphere. This reduces visibility, injures the respiratory tract and nervous system, and affects the climate.

Modern technology produces a variety of useful products. This same technology produces byproducts that can harm the environment. Because of the potential hazards that exist with new technology, environmental laws and regulations have been passed to protect our future. The purpose of the **Clean Air Act** is to enhance the quality of the nation's air, accelerate a national research and development program to prevent air pollution, provide technical and financial assistance to state and local government, and develop a regional air pollution control program.

Air Pollution Control

In 1955 the original clean air act (Air Pollution Control Act of 1955) was passed. Since 1955, a number of modifications have been made:

- 1960 amendment directed Surgeon General to study vehicle pollution
- 1963 amendment directed research into fuel desulfurization and development of air quality criteria
- 1965 amendment to study new sources of pollution
- 1967 Quality Air Act (passed)
- 1970 Clean Air Amendment
- 1977 amendment to clean air act for emission standards
- 1990 reauthorization of Federal Clean Air Act covers the following:
    - Air toxins
    - Acid deposition
    - Job training for workers laid-off due to Clean Air Act
    - Air quality standards
    - Permits
    - Stratospheric ozone and global climate protection
    - Provisions for enforcement
    - Acid rain and air monitoring research
    - Provision to improve air quality and visibility near national parks
    - Provisions relating to mobile sources

Agencies

The air pollution control board maintains numerous regional offices throughout the United States. Each location receives public complaints, coordinates investigations, documents violations, and recommends enforcement actions.

Air Permitting

Permits must be obtained for any project that has the possibility of producing air pollutants. The Air Control Board (ACB) will place limits on emissions and will need about three to eight months to complete the permit

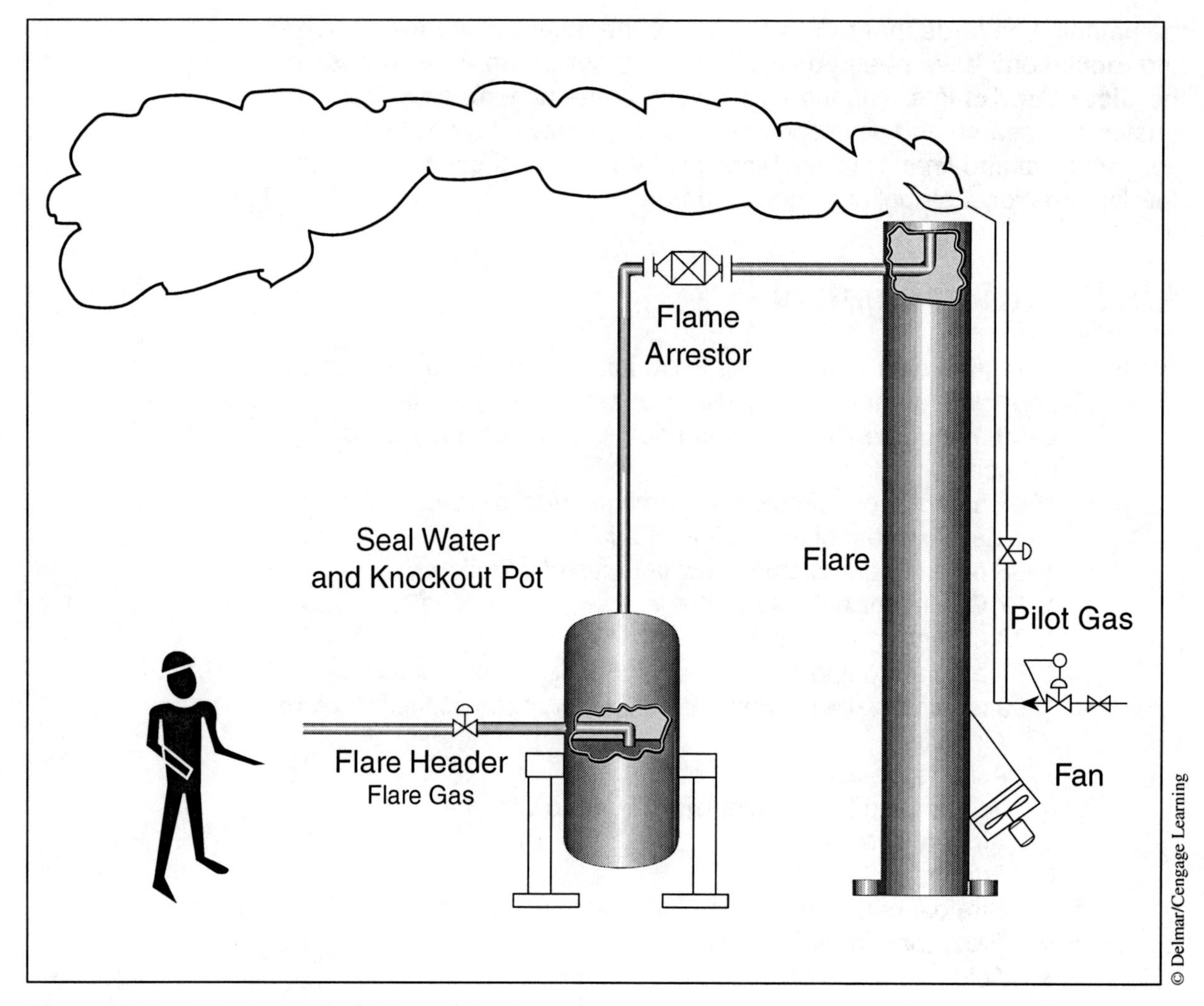

Figure 3-3 *Environmental Awareness*

process. After the ACB issues the permit, a yearly inspection will be scheduled. Penalties for civil and criminal abuses of the Clean Air Act range from $25,000 a day to $250,000 and 2 to 15 years in jail. An example of an ACB regulation is that smoking flares in excess of five minutes should be reported. Failure to report results in severe penalties. Environmental awareness is an important part of a process technician's job. (See Figure 3-3.)

Water Pollution Control

The Federal Clean Water Act was passed in 1898. Fifty years later, Congress provided funds for the construction of municipal wastewater treatment facilities. The Water Control Act of 1965 took a "water quality" approach and initiated close examination of receiving waters. States were required to establish standards for water quality. The Clean Water Act in

1972 adopted the Best Available Technology (BAT) strategy for all cleanups. Under the 1987 amendments, states are required to identify waters that are not expected to meet quality standards.

The Clean Water Act regulates wastewater. Wastewater standards are applied to the following:

- Process wastewater—process contact water, contaminated water from vessels and equipment, tanks, slab cleanup, and so on
- Rainwater—sewer system releases
- Cooling water—cooling tower blow-down, boiler blow-down

The Federal Clean Water Act is designed to protect water quality in the United States. The Environmental Protection Agency, State Water Commission, Army Corps of Engineers, State Parks and Wildlife, and U.S. Fisheries and Wildlife help enforce the U.S. Clean Water Act.

National Water Quality Standards

The National Water Quality Standards state the following:

- All United States waters shall be fishable and swimmable.
- No discharge of toxic pollutants in toxic quantities will be allowed.
- Technology must be developed to eliminate pollutant discharge.

Water Permitting

The Clean Water Act requires a chemical processing industry to have a water permit. In some states, a two-permit system exists. This requires the State and Federal government to approve the water permit.

Solid Waste Control

Solid waste is a byproduct of modern technology. It is typically described as non-liquid, non-soluble material ranging from municipal garbage to industrial waste that contains complex and sometimes hazardous substances.

The Resource Conservation and Recovery Act (RCRA) was enacted as public law in 1976. The purpose of RCRA is to protect human health and the environment. RCRA focuses on the correct disposal of hazardous wastes, recycling practices, and solid waste disposal. A secondary goal of RCRA is to conserve our natural resources. RCRA completes this task by regulating all aspects of hazardous waste management: generation, storage, treatment, and disposal. This concept is referred to as "cradle-to-grave."

The RCRA includes a civil penalty of $25,000 a day, a criminal penalty for willful endangerment comprising a $250,000 fine and 15 years in jail, and a $1 million fine for a company. Any person involved in breaking the law will be held liable by the enforcement agencies. This will include any process technician who knowingly breaks the law.

Solid waste is categorized as follows:

Class One:	Hazardous: ignitable, reactive, corrosive, toxic (For non-hazardous materials RCRA regulations do not apply.)
Class Two:	Examples: garbage, cured epoxy resin, biopond filter solid
Class Three:	Example: uncontaminated or inert material, such as wood

The cleanup of waste sites is regulated by the Comprehensive Environmental Response Compensation and Liability Act (CERCLA). This act is frequently referred to as the "Superfund Law." Under CERCLA, many large companies have been required to provide money for cleanup activities for materials dumped by them or their parent companies years ago. This effort supports the "cradle-to-grave" concept and forces chemical manufacturers to develop long-range procedures for the handling, storage, and disposal of chemicals. Within each state, The State Water Commission is the primary regulating agency charged with enforcement of solid waste disposal generated within US boundaries.

Permitting

More than 90 days of storage for a hazardous chemical requires a permit. An ideal facility would include the following:

- Coverage to prevent rainwater contamination
- No contact with soil
- Containment for all equipment
- Raised equipment to permit inspection for leaks

Toxic Substances Control

The **Toxic Substances Control Act (TSCA)** is a federal law enacted in 1976. TSCA was intended to protect human health and the environment. TSCA was also designed to regulate commerce by (1) requiring testing and (2) establishing necessary restrictions on certain chemical substances. TSCA imposes requirements on all manufacturers, exporters, importers, processors, distributors, and disposers of chemical substances in the United States.

Controls

The TSCA inventory (75,000) was established to trace and record all products manufactured, imported, sold, processed, or used for commercial purposes. Exemptions from TSCA regulations include research and development chemicals and byproducts without commercial purpose. TSCA also controls premanufacture review of new chemical substances, risk assessment by testing and information gathering, record keeping, reporting on health and environmental effects associated with chemical substances, and restrictions on known hazardous chemicals.

The Toxic Substances Control Act has severe penalties for those who break the law. Yearly penalties for current violations are estimated at over $40 million. The Environmental Protection Agency is the primary agency charged with enforcing toxic substance control.

Community Right-to-Know

The **community right-to-know** principle increases community awareness of the chemicals manufactured or used by local chemical plants and businesses, involves communities in emergency response plans, improves communication and understanding, improves local emergency response planning, and identifies potential hazards. In 1986, the Emergency Planning and Community Right-To-Know Act (EPCRA) was enacted by Congress. The purpose of the Right-To-Know Act is fourfold: to provide the community with information about chemicals and hazardous materials used near them, to establish a national system for reporting accidental releases, to provide a structured system for responding to emergency situations, and to give communities access to a central database of material safety data sheets (MSDSs) that describes how to respond to each emergency.

CERCLA holds generators and disposers of hazardous waste liable for past practices, and established the "Superfund" of $1.6 billion to pay for cleanup operations of abandoned hazardous waste sites. It also informs the public of these sites and the known hazards. Community right-to-know and Community Awareness and Emergency Response (CAER) programs work with CERCLA to protect the community. Under the community right-to-know principle, CERCLA, EPCRA, SARA, and HAZCOM work with agencies such as the Department of Health, the State Water Commission, and the U.S. Environmental Protection Agency.

Quality Standards

Industry believes that reducing and recycling wastes at their source is the first priority of responsible waste management. Industry has put in place environmental management systems to make use, handle, and dispose of its products safely. Industry is committed to making major expenditures in environmental technology to reduce emissions and protect the environment.

Hazards: Scenarios

Scenario #1

At the completion of a typical run in the plastics plant, the additive powder feeders are isolated, locked-out, and cleaned by maintenance technicians. The confined space procedure is followed closely, and a gas test is performed. The readings appear to be normal. A small ¼-inch nitrogen line is connected to the east end of the feeder. The rapid cleaning of this equipment is important prior to the startup of the next run. The maintenance technicians complain about headaches and near black-outs while working inside the ribbon blender feeder, near the east end. A quick review of the MSDS classifies nitrogen as a nonhazardous gas; however, safety concerns exist. The atmosphere is composed of 78% nitrogen, 21% oxygen, and 1% trace materials. Near the east end of the feeder the nitrogen concentration was higher than the standard.

1. What is the primary hazard associated with nitrogen?
2. What is the route of entry?
3. What is the response or effect?
4. Is the exposure acute or chronic?

Scenario #2

A process technician is making rounds through the unit when a small amount of nitric acid leaks out of the pipe-rack and down her arm. The technician is wearing her standard personal protection equipment (PPE): hard-hat, safety glasses, fire resistant clothing, safety boots, and radio.

1. What is the primary hazard associated with nitric acid?
2. What is the route of entry?
3. What is the response or effect?
4. Is the exposure acute or chronic?

Nuclear Regulatory Commission (NRC)

Nuclear Regulatory Commission (NRC) was established in 1974 to regulate the nuclear devices used in the chemical processing industry. This includes x-ray and measuring devices used to inspect vessels and equipment. The NRC's mission is to protect the environment, promote security and common defense, ensure public health, and regulate the country's civilian use of nuclear materials and byproducts. At the present time the United States has 104 power-producing reactors, and 36 non-power producing reactors. These facilities are potential Al-Qaeda and terrorist targets. The NRC oversees:

- **reactor security**
- **reactor safety**
- **reactor license renewal**
- **radioactive material safety**
- **spent fuel management, recycling, disposal and storage**

Pipelines

Pipelines in the United States are regulated by the U.S. Department of Transportation, these are lines of pipe that convey liquids, gases, or finely divided solids. Other examples of materials transferred through pipelines include beer, sewage, water, slurry, fuel oils, natural gas, hydrogen, biofuels, and many other gases. Oil pipelines are made from plastic or steel with inner diameters ranging from 4 to 48 inches. Pipes are typically buried under the ground, 3 foot to 6 foot deep. The origin of pipelines being used to transport materials can be tracked to oil fields in Oil Creek, Pennsylvania in the 1860s, when the Oil Transport Association constructed a 2 inch, 6-mile wrought iron pipeline. The Russians claim that Vladimir Shukhov and the Branobel company pioneered the transport of materials through a pipe.

Summary

Air pollution is defined as the contamination of the air, especially by industrial waste gases, fuel exhausts, or smoke. **Water pollution** is defined as the contamination of the water, especially by industrial wastes. Solid waste is technically defined as a discarded solid, liquid, or gas in a container. Solid waste is a byproduct of modern technology. This definition includes materials that have been recycled or abandoned through disposal, burning or incineration, accumulation, storage, or treatment. In an industrial environment, responding to an emergency follows a specific set of standards. Emergency response drills are carefully planned and include preparations for worst-case scenarios. Examples are: vapor releases, chemical spills, explosions, fires, equipment failure, hurricane, high winds, loss of power, bomb threats, and so on. The community right-to-know principle performs the following functions:

- Increases community awareness of the chemicals manufactured or used by local chemical plants and businesses
- Involves community in emergency response plans
- Improves communication and understanding
- Improves local emergency response planning
- Identifies potential hazards

The Resource Conservation and Recovery Act (RCRA) was enacted as public law in 1976. The purpose of RCRA is to protect human health and the environment. A secondary goal is to conserve our natural resources. RCRA completes this goal by regulating all aspects of hazardous waste management: generation, storage, treatment, and disposal. This concept is referred to as "cradle-to-grave." The Toxic Substances Control Act (TSCA) is a federal law enacted in 1976, intended to protect human health and the environment. The TSCA was also designed to regulate commerce by (1) requiring testing and (2) establishing necessary restrictions on certain chemical substances. The TSCA imposes requirements on all manufacturers, exporters, importers, processors, distributors, and disposers of chemical substances in the United States.

Review Questions

1. What is the technical definition of solid waste?
2. What is the community right-to-know principle designed to do?
3. Describe the purpose of the Resource Conservation and Recovery Act (RCRA).
4. Vapor releases, chemical spills, explosions, fires, equipment failure, hurricane, high winds, loss of power, and bomb threats fall under which main program?
5. "Cradle-to-grave" is a term associated with which government act? Describe how this applies to local chemical plants and refineries.
6. After how many minutes should smoking flares be reported? What are the penalties for not reporting them?
7. What must be obtained for any project that has the possibility of producing air pollutants?
8. What was the significance of the Exxon Valdez incident?
9. List the harmful products produced by air pollution. Include how they are produced.
10. In your opinion, list the importance of CERCLA. Defend your answer.
11. Describe the typical routes of a chemical entry to the body.
12. Describe the effect that the Exxon Valdez spill had on the local community and economy.
13. When was the Resource Conservation and Recovery Act (RCRA) passed?
14. Describe the purpose of the Toxic Substances Control Act.
15. What was the TSCA inventory (75,000) established to do?
16. Describe how smog is formed.
17. In what year was the Federal Clean Water Act passed and what was it designed to do?
18. Describe the dose-response relationship and the term "toxicology."
19. Explain the purpose of the Federal Railroad Administration.
20. Describe the purpose of the Nuclear Regulatory Commission (NRC).

Gases, Vapors, Particulates, and Toxic Metals

OBJECTIVES

After studying this chapter, the student will be able to:

- Describe the physical and health hazards associated with gases and vapors.
- Describe the physical and health hazards associated with particulates.
- Describe the physical and health hazards associated with toxic metals.
- Identify common poisonous metals and describe the hazards associated with each.
- List the metals that are fire hazards.
- Describe the hazards associated with toxic metals.
- Explain the hazards associated with toxic metal compounds.
- Describe the hazards associated with compressed gas cylinders.
- List the flammable gases that will mix easily with air.
- Describe the hazards associated with dust explosions.
- List special precautions used when working with particulates, dust, and gases.
- Describe the Bhopal, India vapor release.
- Describe the background and history of asbestos.
- Explain how to work safely with asbestos.

Key Terms

- **ACM**—asbestos-containing material.
- **Ambient**—surrounding or encircling.
- **Anesthetic gases**—have a numbing effect and causing the loss of sensation and unconsciousness. Affected workers will become dizzy, lose coordination, and fall asleep because the central nervous system has been chemically depressed. Death may occur from respiratory paralysis. All organic gases are considered anesthetics.
- **Asbestos**—a fire-retardant material used in brake linings, hair dryer components, insulation, and shingles. It was used by the ancient Greeks for lamp wicks and by the Egyptians as mummy wrap.
- **Ceiling level "C"**—the maximum allowable human exposure limit for an airborne substance. This amount cannot be exceeded, even momentarily.
- **Central nervous system**—the brain and spinal cord.
- **Chemical asphyxiants**—gases such as carbon monoxide and hydrogen cyanide that prevent cells from using oxygen or prevent the blood from supplying oxygen.
- **Cutaneous hazards**—a chemical that affects the dermal layer of the body.
- **Dermatitis**—skin irritation or inflammation.
- **Flammable gas**—any gaseous material that forms an ignitable mixture when combined with 13% air at 68°F and 14.7 psi.
- **Health hazard**—a chemical that has been statistically proven by one or more scientific studies to have acute or chronic health risks for humans.
- **Incompatible**—a term applied to chemicals that react violently when they come into physical contact.
- **Lower explosive limit (LEL)**—the lowest concentration at which a vapor or gas will produce a rich enough vapor concentration in air to ignite in the presence of an ignition source.
- **Milligrams per liter (mg/l)**—milligrams of material per liter of air.
- **Nonflammable compressed gas**—a material that does not conform to the definitions of a flammable or poisonous gas and exerts an absolute pressure of 41 psia at 68°F.
- **Olfactory**—a term used or associated with the sense of smell.
- **Parts per million (ppm)**—parts of material per million parts of air.
- **Permissible exposure limit (PEL)**—regulatory limits set by OSHA on the amount or concentration of a substance in the air, based on an eight-hour, time-weighted average exposure (TWA).
- **Physical hazard**—a term applied to a chemical that statistically falls into one of the following categories: combustible liquid, compressed gas, explosive or flammable, organic peroxide, oxidizer, pyrophoric, unstable, or water reactive.
- **Simple asphyxiants**—gases, such as nitrogen, helium, hydrogen, carbon dioxide, and methane, that will displace the oxygen content in the air.

- **Systemic poisons**—are formed when toxic gases enter the bloodstream through the lungs and migrate toward specific body organs and tissues.
- **Upper explosive limit (UEL)**—the highest concentration at which a vapor or gas will produce a rich enough vapor concentration in air to ignite in the presence of an ignition source.

Physical Hazards Associated with Gases, Vapors, Particulates, and Toxic Metals

A **physical hazard** associated with gases, vapors, particulates, and toxic metals is described as a chemical that falls into one of the following categories: compressed gas, explosive, **flammable gas,** oxidizer, pyrophoric, or unstable. Gaseous airborne contaminants can affect unprotected technicians as irritants, asphyxiants, anesthetics, and **systemic poisons**.

Gaseous irritants may injure eye, throat, nose, and lung tissue through corrosive action. Pulmonary edema is a condition resulting from the inflammation of tiny air sacs inside the lungs. In effect, fluid fills the lungs, drowning the patient. A partial list of chemical irritants includes: ammonia, hydrogen chloride, acetaldehyde, formaldehyde, hydrogen fluoride, sulfur dioxide, chlorine, bromine, fluorine, ozone, nitrogen dioxide, phosgene, arsenic trichloride, and phosphorous trichloride.

Some gases are known to interfere with the supply of oxygen to the body. This process is referred to as asphyxiation. **Simple asphyxiants** include inert gases such as nitrogen, helium, hydrogen, carbon dioxide, and methane that will displace the oxygen content in the air. **Chemical asphyxiants** such as carbon monoxide and hydrogen cyanide prevent cells from using oxygen or prevent the blood from supplying oxygen. **Anesthetic gases** have a numbing effect and will cause the loss of sensation and unconsciousness. Affected workers will become dizzy, lose coordination, and fall asleep because the **central nervous system** has been chemically depressed. Death may occur from respiratory paralysis. All organic liquids in the gaseous state are considered anesthetics: methane, ethane, ethylene, acetylene, methyl alcohol, ethyl alcohol, methyl and ethyl ether, carbon tetrachloride, toluene, benzene, and xylene. Hydrogen Sulfide H_2S is a gas that smells like rotten eggs in lower concentrations. In higher concentrations it has the ability to overpower the **olfactory system** or sense of smell and cause death.

Systemic poisons are formed when toxic gases enter the bloodstream through the lungs and migrate toward specific body organs and tissues. A partial list of these gaseous systemic poisons and what they affect includes the following:

- Arsine—blood cells and liver
- Benzene—bone marrow

Gas/Vapor	Lower-Upper Flammable Limit (vol.%)		Auto-Ignition Temperature (°C)	Boiling Point (°C)
Acetone	2.6	13.0	465	
Acetylene	2.5	100	305	−83.6
Ammonia	15.0	28	-	−33.5
Benzene	1.3	7.9	560	80.2
1,3-Butadiene	2.0	12	420	
n-Butane	1.8	8.4	405	
Carbon Monoxide	12.5	74	-	−191.5
Ethane	3.0	12.4	515	−88.3
Ethyl Alcohol	3.3	19	365	78.3
Ethylene	2.7	36	490	−103.8
Toluene	1.2	7.1	480	110.6
Gasoline	1.3	7.1	440	
n-Heptane	1.05	6.7	215	
n-Hexane	1.2	7.4	225	
Hydrogen	4.0	76	400	
Methane	5.0	15	540	−161.4
Methyl Alcohol	6.7	36	385	64.7
Propylene	2.4	11	460	
n-Pentane	1.4	7.8	260	
Propane	2.1	9.5	450	−44.5

Figure 4-1 *Physical Properties of Chemicals*

- Carbon tetrachloride—liver and kidneys
- Ethylene dichloride—liver and kidneys
- Hydrogen selenide—liver and spleen
- Hydrogen sulfide—respiratory system
- Mercury—nervous system, kidneys, glands
- Methyl alcohol—nervous system and optic nerve
- Methyl chloride—kidneys, heart, and nervous system
- Phosphorous—bone

Figure 4-1 is a table that lists the physical properties of chemicals. Physical hazards of gases include the following:

- Compressed gas—has a gauge pressure of 40 psig at 70°F (21.1°C)

- Explosive—a chemical characterized by the sudden release of pressure, gas, and heat when it is exposed to pressure, high temperature, or sudden shock
- Flammable gas—forms a flammable mixture with air at ambient temperature
- Oxidizer—a chemical that promotes combustion in other materials through the rapid release of oxygen, usually resulting in a fire
- Pyrophoric—a chemical that ignites spontaneously with air at temperatures below 130°F (54.4°C)
- Unstable—a chemical that will react (condense, decompose, polymerize, or become self-reactive) when it is exposed to temperature, pressure, or shock

Health Hazards Associated with Gases, Vapors, Particulates, and Toxic Metals

There are several **health hazards** associated with gases, vapors, particulates, and toxic metals. Health hazards are listed as: carcinogens, mutagens, teratogens, reproductive toxins, asphyxiation, anesthetic, neurotoxic, allergic response, irritant, sensitizer, corrosive, toxic, highly toxic, and those that target organs. The list that follows outlines these hazards:

- Carcinogens—are known cancer-causing substances. Carcinogenicity labels indicate that a substance may cause cancer upon contact or have a chronic effect over a large number of years. For example, coal tar is known to cause cancer on an exposed body part. Vinylchloride is known to have a chronic or long-term effect on the liver when it is inhaled over a long period of time. Most items listed on the MSDS will indicate whether a substance is a suspected carcinogen. Most carcinogen labels indicate that the substance has been proven to have an impact on laboratory animals.
- Mutagen—a chemical that is suspected to have the properties required to change or alter the genetic structure of a living cell.
- Teratogen—a substance that is suspected to have an adverse effect on the development of a human fetus.
- Reproductive toxin—a chemical that inhibits the ability of a person to have children. Chemicals are routinely tested for this property.
- Asphyxiation—simple and chemical. Simple asphyxiation occurs when oxygen is removed or displaced by a chemical. For example, a sudden release of nitrogen gas into an enclosed space will dissipate life-supporting oxygen, causing loss of consciousness and death. Chemical asphyxiation occurs when a chemical blocks or impedes the ability of a person to use oxygen. For example, carbon monoxide attaches to red blood cells and prevents oxygen from being utilized even if it is available in sufficient quantities.

- Anesthetic—dulls the senses. Example: alcohol.
- Neurotoxic—slows brain down. Example: lead and mercury.
- Allergic response—a person may not have an allergic reaction to a chemical immediately. In some cases, it takes multiple exposures. Example: poison ivy.
- Irritants—a chemical that causes temporary discomfort when it comes into contact with human tissue. Example: **Dermatitis**.
- Sensitizers—a chemical that affects the nerves. Example: Phenol will be absorbed through the skin and will sensitize the affected area.
- Corrosives—a chemical that causes severe damage to human tissue. Example: sulfuric acid.
- Toxic—a chemical that has been determined to have an adverse health impact. This term is linked to toxicology, dose, and response, and at least one experiment that has indicated the chemical has an adverse effect on laboratory animals.
- Highly toxic—a term applied to a chemical that requires only a small amount to be lethal.
- Target organ toxin—a term applied to a chemical that contacts the body at a remote location and is transferred to another area of the body where it has an adverse effect on a specific organ.

Asbestos

Asbestos Awareness and Gasket Removal—1910.1200/1001 and 1926.59/1101 covers the industry standards associated with asbestos. Industry uses the term, Asbestos-containing material (**ACM**) to identify materials containing asbestos fibers. **Asbestos** was used by the ancient Greeks and Egyptians. The Greeks used asbestos for wicks in their lamps while the Egyptians used it to wrap the bodies of their dead. By 1900, large-scale mining and production of asbestos vaulted to commercial levels. Modern yearly consumption rates had reached 800,000 to 900,000 tons. The mineral was used in brake linings, roof shingles, insulation, fire retardant materials, and in many other products.

Asbestos fibers are so small that they are invisible to the human eye. When compared to a human hair, an asbestos fiber is so small that it would be like comparing a pencil to a giant redwood tree. A single asbestos fiber dropped from a height of 8 feet would take 40 to 70 hours to reach the ground. When an abrasive, irritating asbestos fiber enters the human lung, the body's natural defense system activates by encapsulating it. This process creates scar tissue that will remain inside the lung for life. Over time, the scar tissue causes the lungs to lose their ability to move oxygen in and out of the lungs and supply oxygen to the bloodstream. This condition is referred to as asbestosis. Because asbestos is classified as a cancer-causing agent, asbestosis is considered serious.

Bhopal—Union Carbide

During the early morning hours of December 3, 1984, the chemical processing industry was rocked as a large vapor cloud composed of methyl isocyanate (MIC) escaped from the methyl isocyanate (MIC) plant and drifted over the sleeping town of Bhopal, India. The toxic vapor cloud became a nightmare from which nearby residents did not wake. Many died as they clutched their throats in terror, struggling for breath. Methyl isocyanate is a major component used in the production of Sevin and Temik, pesticides used to control insect infestations and crop production in India. By the time the sun came up, 1,400 people had died. Seven years later, the official body count had risen to 3,800 and the number of people who had disabling injuries had reached 11,000. Some estimates indicate that over 200,000 people were hurt during this incident. For the first time in modern chemical manufacturing history, an event had taken place near a populated area that demonstrated how vulnerable a community is to the hazards of the chemical processing industry.

Figure 4-2 is a simple flow diagram of the Union Carbide India Limited process that caused the disaster. The process was designed to produce a range of pesticides and herbicides. These products were essential to the economy of the Indian government. In order to produce these products, a number of steps were required. The first step was the reaction of carbon monoxide with chlorine to produce intermediate phosgene. The second step of the process mixed monomethylamine with phosgene to yield methyl isocyanate. Process technicians refer to this product as MIC. The third step of the process caused alpha napthol to react with MIC to produce carbaryl. Carbaryl is the end product that was used in various concentrations to produce a variety of herbicides and pesticides. Each of the chemicals used in this process is hazardous and requires specific safety handling procedures.

The Union Carbide India Limited process used three 40-foot tanks, with a storage capacity of 15,000 gallons each. The primary tanks were numbered 610 and 611, with 619 reserved as an emergency spare. Tank 619 was supposed to remain empty so that in the event of an emergency, the contents of 610 or 611 could be transferred.

Plant management decided to shut down the operation on November 14th to perform routine maintenance and to reduce chemical inventories. Plant startup was scheduled for November 26, 1984.

Problems within the Union Carbride India Limited (UCIL) process facility became apparent when the process operators started going through their plant's startup procedures. Vital safety systems and their backups were either faulty or had been shut down during the two-week maintenance

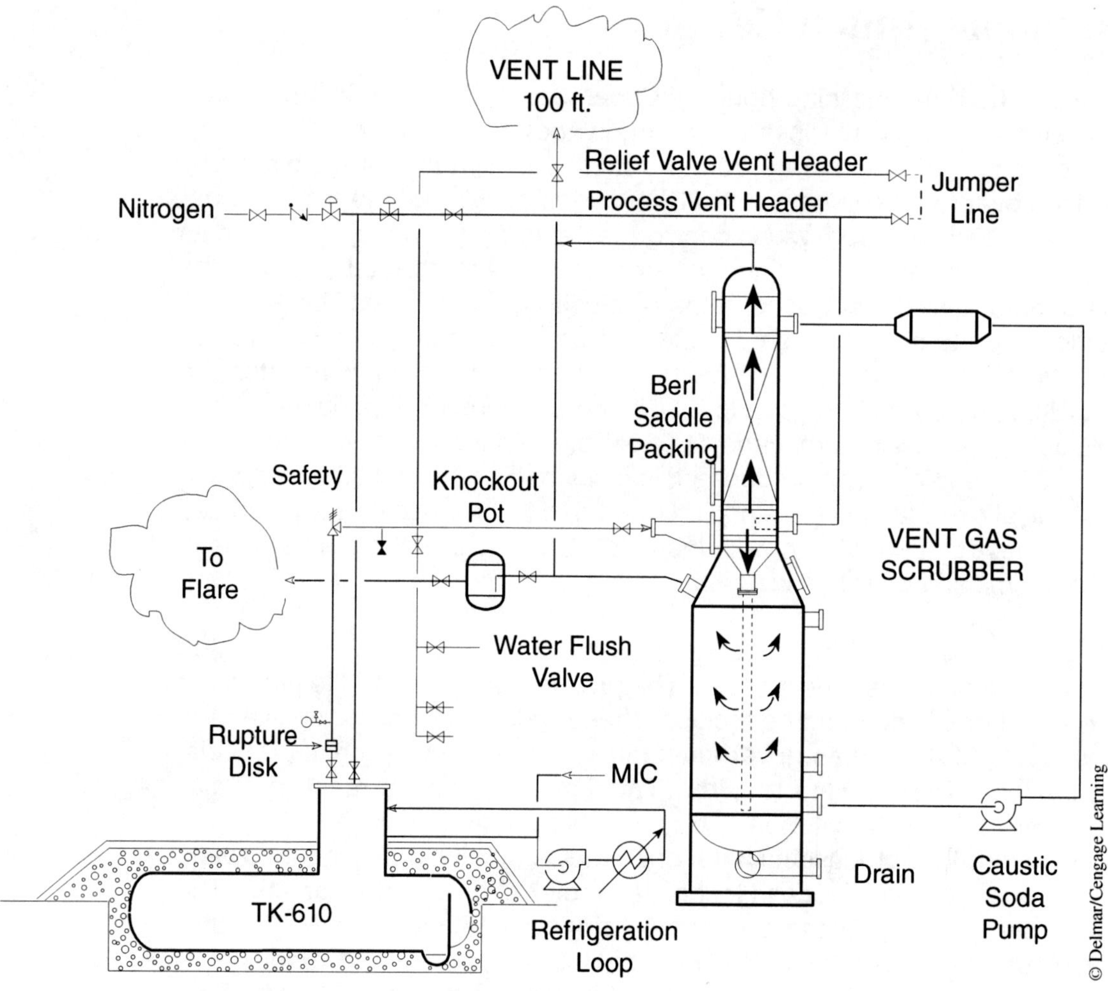

Figure 4-2 *Flow Diagram of the Process That Caused the Disaster*

operation. Four of these critical systems included the vent gas scrubber, flare, MIC tanks instrumentation, and refrigeration unit. The flare is a device designed to safely burn hazardous emissions before they leave the plant; the vent gas scrubber was designed to neutralize hazardous gases by causing them to react with caustic soda. According to plant operators, the temperature, pressure, and level alarms had not been working for several years. Plant management had also dismantled the MIC refrigeration system that was designed to keep the contents of tanks 610, 611, and 619 cool. Instead of repairing this equipment and ensuring that it was operational, management instructed the startup team to pressure up tank 610 with nitrogen. Tank 610 had approximately 42 tons of MIC stored in it and tank 611 had about 40 tons of MIC. When the operators attempted to pressure up tank 610 with nitrogen, they were unable to achieve transfer pressure. A leak was suspected on tank 610's nitrogen outflow, or blowdown, valve. This is probably accurate, because tanks 610, 611, and 619 remained under a 2-psig nitrogen blanket during the shutdown. This precaution was necessary because MIC reacts violently with water. Instead of fixing the valve, management decided to switch to tank 611, a fatal mistake.

At 9:30 PM, process technicians were water washing four downstream MIC relief valve vent header (RVVH) gas lines. Tanks 610, 611, and 619 were protected from water contamination by a series of safety valves and rupture disks, an isolation valve, and a slip blind. During this incident, the slip blind was not installed. While washing out the lines, the technicians noticed that the water going into the header was not coming out of the open bleeder valves at the same flow rate. It was determined that four bleeder valves were partially clogged, and two were completely clogged. The decision was made to keep water washing. Water began to accumulate in the upper sections of the RVVH, some going as high as 20 feet above the MIC tanks. It should be apparent at this point that safety took a back seat to production in the UCIL plant. From an experienced technician's point of view, this is an unusual way to start up a hazardous process.

Another key safety violation was management's decision to install a jumper line between the RVVH and the process vent header. This jumper header is illustrated in Figure 4-3. The jumper line appears to provide an access point for the water into tank 610.

As the water and iron entered the tank, a number of chemical reactions took place. According to chemists, MIC and water react violently after 23 hours of exposure at 60°F. The time frame and incidents do not support that this reaction was the first one to occur. According to Union Carbide Corporation's piping design specifications, the header should have been made out of stainless steel. Instead, it was made out of a cheaper material, carbon steel. Stainless steel, not carbon steel, is recommended because the iron in carbon steel (rust or corrosion) acts as a catalyst and will cause MIC to react with itself. When the wash water entered the tank, the exothermic reaction started.

Process technicians reported that the ground became hot around tank 610 as pressures rose and the runaway exothermic reaction reached temperatures of 725°F. Normal gas flow rates into the scrubber averaged 180 kg/hour at 95°F. During the runaway reaction, gas rates exceeded 36,000 kg/hour. Under these runaway conditions a fully operational flare, scrubber, and refrigeration system would have been overwhelmed by the release.

Several other key factors that were uncovered by the investigative teams were that the plant had been losing money because of a weak market for Sevin. Plant reductions had decreased operational teams from eleven process technicians and one supervisor to five process technicians and one supervisor. Work loads were redistributed, and people were assigned to work areas with which they were unfamiliar. The maintenance staff was reduced from six to two technicians. Plant workers were encouraged to take early

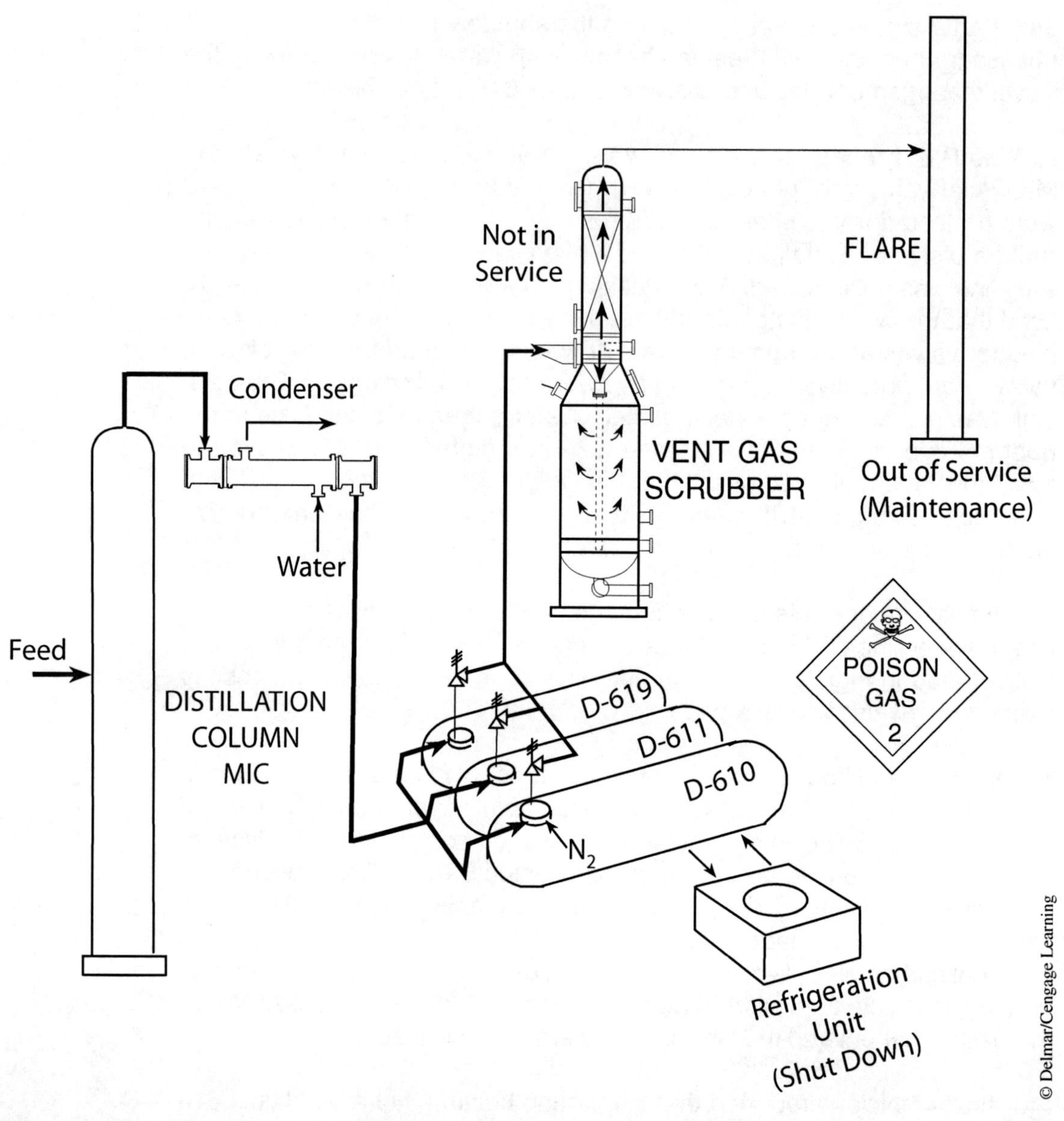

Figure 4-3 *Bhopal Incident Major Equipment*

retirements, 150 employees were placed in a temporary worker pool, and 300 temporary workers were laid off. Plant moral at the Union Carbide India Limited plant was at an all-time low. The sequential steps that followed the incident were:

- Union Carbide sent technical experts to assist in removing the methyl isocyanate (MIC) and investigate the incident.
- The Indian Central Bureau of Investigation sealed the plant and limited Union Carbide Corporation access to the incident site, documents, employees, and so on.
- All Union Carbide Corporation aid packages to victims were rejected by the government.

Contrary to an independent investigation, Union Carbide Corporation believed that the accident was caused by a disgruntled employee that had connected a water hose to tank 610 in an effort to ruin the product. This employee would have needed a crow's foot adapter to make the connection and would have needed to remove the device before the investigative teams arrived. Several operators remembered seeing a water hose in this area. Figure 4-3 shows the basic layout of all major equipment at Union Carbide.

Union Carbide India Limited was an operation run entirely by the government of India. The government had a national interest in promoting local control of plant operations and self-sufficiency. In 1984, UCIL was celebrating 50 years of operation. UCIL owned and operated 14 plants, split into 5 divisions with over 9,000 employees. Union Carbide Corporation (UCC) owned 50.9% of the company's stock, government-run insurance companies owned 24% of the shares, and the rest was owned by 23,000 shareholders. The Bhopal plant was a multicultural joint venture in India between UCC and UCIL. Most of the Indian workforce had over ten years of experience with methyl isocyanate operation. Union Carbide Corporation provided capital, equipment, and technical expertise to run the plant. The Indian government felt it was important for the Western influence to be removed and allow their Eastern counterparts to run the plant. The last American citizen had left the plant two years earlier.

Mistakes

The mistakes that were made are summarized as follows:

- The vent gas scrubber and flare system was undersized and not operational because of a management decision.
- The refrigeration system was dismantled.
- Tank 610 alarms and instrumentation were faulty.
- A jumper line was installed.
- Stainless steel was not used in areas where it was specified.
- Plant layoffs and lack of training caused poor morale and insufficient staff.
- A safe zone around the plant had not been established. The government (and company) allowed a densely populated shanty town to grow up around the plant by deeding land to its citizens. Most of these people were in the direct path of the release and died first.
- Communication between Union Carbide India Limited and the Indian government broke down. The initial meeting with the press was short. Many questions were asked with few answers given.
- The Indian government seized control of the plant and arrested the leader of the Union Carbide Corporation investigation team. The team was denied access to the plant and employees.

- Many of the employees moved after the plant shut down and left no forwarding address.
- Official plant documents were altered after the vapor release.
- The amount of $5 million sent to the victims of the release was withheld by the government and sent back to Union Carbide.
- A $2 million project funded by Union Carbide to aid the victims of the release was bulldozed by the Indian government.
- Samples of the MIC drum were difficult to obtain and delayed the analysis phase of the project by the Union Carbide Corporation.
- The Indian government filed a suit against the Union Carbide Corporation and was awarded $470 million. The government wanted $3 billion but settled with Union Carbide. A few months later, the government changed hands and the court's first ruling was thrown out.

The initial focus of the Bhopal incident was on the gas release and the people who died. The second phase of the incident evolved into a legal mess between Union Carbide Corporation and the Indian government. The third phase of the incident was the disaster at Bhopal detective story: What happened? Who was responsible? The fourth phase of the incident focused on the multicultural differences between the United States and India. The final phase is the evaluation of the disaster. What did we learn? What did we do about what we learned?

Outcomes

The Bhopal tragedy focused and elevated public concern about storing, transporting, and manufacturing hazardous chemicals near populated areas. The incident alarmed the chemical processing industry and initiated the enactment of laws requiring companies to notify the community and government about hazardous chemicals they use or produce. Bhopal helped refine the Environmental Protection Agency's Federal Superfund Reauthorization and Emergency Planning Councils, a program designed to work with communities and prepare them to respond to environmental disasters.

Particulates

Airborne particulates are defined as small liquid or solid particles dispersed through the air. These particulates may pose a breathing hazard to process technicians working in the area. Dusts that contain aluminum oxide or limestone are classified as nuisance or inert aerosols. Various concentrations of airborne particles will have differing effects upon unprotected process technicians. Respiratory protection programs are designed to protect workers from exposure to various types of airborne contaminants.

Some particulates have been found to cause growths in the lungs or pulmonary fibrosis. Fumes or dusts containing tin or iron cause minimal damage; however, silica and asbestos fibers cause extensive damage. The coal mining industry discovered that prolonged exposure of workers to coal dust caused the debilitating disease "black lung."

Mists or sprays containing corrosive substances such as acids or alkaline compounds would eat away at the soft tissue of the eyes and lungs. Systemic poisons are found in airborne particulates containing lead, manganese, pesticides, or radioactive materials. When these materials are absorbed by the lungs, they enter the bloodstream and produce a number of toxic reactions. Some fumes and dusts have been known to produce chills or intense fever in unprotected workers. Airborne agents expected of causing these reactions include zinc, copper, or sugar cane residue.

Dust and Gases

Chemical technicians may be exposed to a wide variety of dusts and gases during normal work routines in a plant. Respiratory protection programs are designed to provide two types of protection: air purifying and air supplying. The type of dust or gases found at the job site will determine the type of respirator selected. Any type of dust can be potentially dangerous if sufficient quantities are inhaled into the lungs.

Nine types of dusts pose a hazard for process technicians. A complete list of dust hazards includes: (1) inert dusts, (2) chemical irritants, (3) allergy-producing substances, (4) fever-producing substances, (5) coal dust, (6) fibrosis-producing materials, (7) cancer-causing agents, (8) systemic poisons, and (9) dust explosions. Figure 4-4 illustrates these nine hazards.

Dusts that are not poisonous and do not react or interfere with the body chemistry are referred to as inert dusts. Aluminum oxide, clay, gypsum, and limestone are common forms of inert dusts. Chemical irritants damage the soft tissue of the respiratory system by reacting with the fluids. Examples of chemical irritants include acids, alkalis, chromates, and peroxides. Dusts produced by plastic resins, pollen, and vegetable fibers may induce an allergic reaction. These dusts are called allergy producers and in large quantities may inhibit a person's respiratory system. Fever-producing dusts produce chills and fever that may last from one to two days. Unprotected chemical technicians exposed to copper and zinc dust may experience these symptoms.

- In the early 1900s, coal mining was a popular and hazardous way to make a living. Although many hazards existed in the coal mining industry, long-term exposure to coal dust appears to have been the most debilitative. Miners that survived the hazards of working underground often died horrible deaths associated with the suffocating effects of black lung. Fibrosis-producing dusts are typically associated with asbestos and silica. As these small fibers enter the lungs, scar tissue is formed that

Figure 4-4
The Nine Hazards of Dust

1. **Coal Dust Hazard**
   Example: Coal mining
   Effects: Black lung, a chronic respiratory condition that reduces the lungs' ability to absorb oxygen

2. **Chemical Irritant Dust Hazards**
   Example: Dusts from peroxides, chromates, acids, and alkalis
   Effects: Ulceration, inflammation, and irritation of throat and lungs

3. **Fibrosis Dust Hazard**
   Example: Asbestos, silica
   Effects: Scar tissue formation in lungs that reduces ability to absorb oxygen

4. **Systemic Poison Dust Hazard**
   Example: Manganese, pesticides, radioactive substances, lead, and cadmium
   Effects: Causes permanent organ damage; accumulates in liver, kidneys, or brain

5. **Allergy-Producing Dust Hazard**
   Example: Pollen, fur, or vegetable fibers; plastic resins
   Effects: Targets respiratory system and soft tissues in nose, eyes, and throat, increasing mucous flow and causing difficulty in breathing

6. **Cancer-Producing Dust Hazard**
   Example: Asbestos, chromates, radioactive dusts
   Effect: Formation of tumors in lungs and organs

7. **Inert Dust Hazard**
   Example: Aluminum oxide, clay, gypsum
   Effect: Minor irritation and discomfort to nose, eyes, and throat

8. **Fever-Producing Dust Hazard**
   Example: Sugar cane, copper, zinc
   Effect: Chills, severe fever, lasts 24 hours

9. **Dust Explosion Hazard**
   Example: Coal or grain dust
   Effect: Under right concentrations, it will ignite and explode

leads to a condition called pulmonary fibrosis. Cancer-causing dusts include asbestos, chromates, cigarette smoke, and radioactive materials. Symptoms of exposure to cancer-causing dusts typically occur 10 to 15 years after exposure. Systemic poisons include dusts that contain cadmium, lead, manganese, pesticides, and radioactive materials. Exposure to systemic poisons affects targeted organs such as the brain or liver. **Ceiling level "C"** is the maximum allowable human exposure limit for an airborne substance. Section 16 on the MSDS will provide additional information for different airborne substances. This amount cannot be exceeded, even momentarily.

The body is unable to remove the poisons from the system, and damage is often permanent and irreversible. The final hazard associated with industrial dusts is the risk of dust explosions. Many catastrophic incidents have occurred when dust concentrations have detonated from a single spark. Coal dust, grain dust, and fine sawdusts are typically associated with dust explosion hazards.

Working in the chemical processing industry places a technician in an environment surrounded by pipes and vessels filled with hazardous gases. The five classifications for hazardous gases include: chemical irritants, simple asphyxiants, chemical asphyxiants, anesthetics, and systemic poisons. Constructive training in this area is an essential element in the training process of a process technician. Figure 4-5 identifies the characteristics associated with gaseous respiratory hazards.

1. **Chemical Asphyxiant Hazard**
   Example: Hydrogen cyanide, carbon monoxide
   Effects: Interrupts body's ability to absorb or transport oxygen

2. **Simple Asphyxiant Hazard**
   Example: Nitrogen, carbon dioxide, helium, halon, Freon
   Effects: Displaces oxygen, causing oxygen starvation

3. **Chemical Irritant Hazard**
   Example: Ammonia, chlorine, sulfur dioxide
   Effects: Inflammation, increased mucous secretion

4. **Systemic Poisons Hazard**
   Example: Benzene, mercury vapor, arsine, carbon tetrachloride
   Effects: Irreversible damage to brain, liver, and kidneys; accumulates in target organs

5. **Anesthetic Gases Hazard**
   Example: Acetylene, butane, propane, toluene, xylene
   Effects: Dizziness, intoxication, unconsciousness, and respiratory paralysis

Figure 4-5
Examples of Gaseous Respiratory Hazards

The gas phase provides the perfect medium for absorption of hazardous materials into the bloodstream. Chemical irritants are classified as corrosive gases such as ammonia, chlorine, fluorine, and sulfur dioxide. When these gases are inhaled, they immediately attack the soft tissue of the respiratory system and in severe cases can interrupt blood circulation to the lungs.

Inert gases such as Freon, nitrogen, helium, and carbon dioxide are used in a variety of industrial applications. These gases are classified as simple asphyxiants and have the ability to displace oxygen in the event of a release. An understanding of basic chemistry is essential when working with these gases. Heavier elements listed on the periodic table will sink during a release in an enclosed space, whereas lighter elements will rise. For example, helium is lighter than air, whereas carbon dioxide is not. Unprotected technicians exposed to these simple asphyxiants will experience dizziness, disorientation, and fatigue.

Dust Explosions

As mentioned earlier, dust explosions occur when particle concentrations are detonated from a single spark. Coal dust, grain dust, and fine saw dusts are typically associated with catastrophic dust explosion hazards. The fire triangle has three sides: fuel, oxygen, and an ignition source. In order for a dust explosion to occur, the "fuel" must be divided into small enough pieces to burn. Almost anything will burn if it is broken into small enough parts that expose more surface area to oxygen. In operating areas where a process technician is exposed to dust hazards, explosion-proof devices, spark-proof tools, ventilation systems, respirators, PPE, and proper safety training are used.

Flammable Gases

Some substances mix readily with the air and quickly form mixtures that will explode or burn. These substances are referred to as flammable gases. This family of explosive and flammable gases includes acetylene, hydrogen, methane, ethane, butane, and propane. An important part of designing a chemical process is the use of a hazard analysis that is designed to keep dangerous gases like hydrogen in carefully controlled systems. A variety of safety devices is used to handle the release of flammable gases. A short list of this equipment would include pressure relief devices, piping, and flare systems.

Compressed Gas Cylinder

An important part of a process technician's job is the handling, storing, and moving of compressed gases. Moving and securing compressed gas cylinders require special operating procedures. Safety procedures require cylinders to be secured in a cylinder rack. Pressure relief devices are used to control excessive pressures. These devices include safety relief valves and fusible plugs.

When cylinders are heated up the compressed gases expand rapidly, increasing pressure. Direct flames will weaken the exposed side of the cylinder. Gas cylinders can rupture with catastrophic effects. Fire monitors are used to cool off cylinders exposed to open flames. The fine droplets of water keep the cylinders cool and reduce the mixture of flammable gases with air and heat during a release.

Cylinders have protective caps, overexposed valves, and regulators. Special hoses are used to connect the cylinder to the gas manifold. These hoses should be inspected frequently to ensure they are working properly. When a hose blows loose it will whip rapidly and discharge the contents of the cylinder with great force. Proper personal protective equipment must be worn when working near gas cylinders and hose manifolds. Compressed cylinders include hydrogen, nitrogen, chlorine, oxygen, helium, hydrogen sulfide, carbon dioxide, ammonia, sulfur dioxide, fluorine, Freon, Halon, and a large variety of hydrocarbon gases. This is a very small and incomplete list of the compressed gases found in the chemical processing industry.

Metallic Substances

Toxic metallic substances include systemic poisons such as lead, manganese, and mercury. As mentioned earlier, these poisons collect in specific organs and cause significant damage. This type of poisoning frequently occurs from inhaling dusts and vapors composed of toxic metals. Figure 4-6 shows a list of toxic metals and their harmful effects. A short list of **toxic metals** would include beryllium, copper, lead, manganese, mercury, and zinc.

Beryllium is an extremely light and hazardous metal commonly found in the chemical processing industry. It is used in the manufacture of electronic parts, specialty lightweight alloys, and x-ray tubes. Initial exposures can irritate the skin, eyes, and lungs, provoking spasmodic coughing, tightness in the chest, pain, and difficulty breathing. Beryllium has a catastrophic effect on the liver, gall bladder, and upper respiratory system. A single acute exposure may require four to six months of recovery. Technicians exposed to chronic levels of beryllium will experience constant pain in the joints, bones, and lungs and may develop cancer in the affected areas.

There are many industrial applications for lead that place process technicians in close contact with hazardous conditions. Lead is found in some industrial paints, ceramics, glass, batteries, solder, fishing materials,

Figure 4-6
Harmful Effects of Toxic Metals

1. **Beryllium**
   Acute: Pneumonia-like response; eye and skin irritation
   Chronic: Liver damage, respiratory system damage
2. **Copper**
   Acute: Fever, chills
   Chronic: None
3. **Lead**
   Acute: Anemia, paralysis, liver damage, stomach pains, brain damage
   Chronic: Liver, kidney, and brain damage
4. **Manganese**
   Acute: Brain damage
   Chronic: Brain damage
5. **Mercury**
   Acute: Headache, fever, chest pain, breathing problems, and mouth and gum inflammation
   Chronic: Brain, kidney, and liver damage
6. **Zinc**
   Acute: Fever, chills
   Chronic: None

and type metal. Acute lead poisoning may induce the following symptoms: anemia, paralysis, loss of higher-level motor skills, and stomach problems. Long-term chronic lead exposures accumulate in the brain, kidneys, and liver and cause permanent damage. Symptoms of chronic lead exposure include depression, irritability, loss of coordination, and uncontrollable tremors or shaking.

Manganese is a rare metal known to cause brain damage. This element causes scar tissue to form on the brain; this scar tissue may later develop into insanity. Symptoms include weakness, blank facial expressions, strange speech patterns, spasmodic laughter, and instability.

Mercury is a dangerous protoplasmic poison that destroys biological tissue. It is a liquid at room temperature and is deadly when inhaled, ingested, or absorbed through the skin. The acute effects of mercury poisoning include headaches, breathing problems, fever, and upper respiratory pain. The chronic long-term effects of mercury poisoning are directed at the liver, kidneys, and brain. Symptoms include uncontrollable tremors, coordination problems, and loss of temper.

Metallic Compounds

A compound is described as a substance formed by the chemical combination of two or more substances in definite proportions by weight. Hazardous metallic compounds combine one or more of the known toxic metal(s) in the chemical compound. These substances include: lead oxide, mercuric nitrate, cadmium salts, chromates, and manganese oxide. Hazardous metallic compounds can be as dangerous as the purer form of the toxic metal and can be more dangerous depending on how the chemical is introduced into the human system.

Metals That are Fire Hazards

Process technicians are carefully trained in the use and operation of fire extinguishers and firefighting techniques. Class D fires involve fires caused by metals. This type of fire presents a technician with a different list of firefighting techniques. Class D fires cover combustible metals such as aluminum, magnesium, zinc, potassium, sodium, titanium, and zirconium. When these metals are powdered or broken into small enough parts to expose enough surface area, they will burn. Magnesium and zinc burn at extremely high temperatures (1,000°F/ 540°C), and are difficult to put out once started. Class D fire extinguishers are used in areas where this type of hazard is present. It is important to mention that other types of fire extinguishers may make the fire worse. Sodium and potassium are toxic, react with water to form hydrogen gas, and may violently explode. **Incompatible** is a term applied to chemicals that react violently when they come into physical contact.

Storing and handling sodium and potassium require special procedures. Potassium is submerged in oil to reduce air contact and sodium is stored in a dry, moisture-proof container. Technicians working with metals that are fire hazards should have special training and access to Class D firefighting equipment.

Summary

A physical hazard associated with gases, vapors, particulates, and toxic metals is described as a chemical that falls into one of the following categories: compressed gas, explosive, flammable gas, oxidizer, pyrophoric, unstable, or water reactive. Gaseous airborne contaminants can affect unprotected technicians as irritants, asphyxiants, anesthetics, and systemic poisons.

Some gases are known to interfere with the supply of oxygen to the body. This process is referred to as asphyxiation. Simple asphyxiants include gases, such as nitrogen, helium, hydrogen, carbon dioxide, and methane, which will displace the oxygen content in air. Chemical asphyxiants such as carbon monoxide and hydrogen cyanide prevent cells from using oxygen or prevent the blood from supplying oxygen. Anesthetic gases have a numbing effect and will cause the loss of sensation and unconsciousness. Organic gases are considered anesthetics: methane, ethane, ethylene, acetylene, methyl alcohol, ethyl alcohol, methyl and ethyl ether, carbon tetrachloride, toluene, benzene, and xylene. Systemic poisons are formed when toxic gases enter the bloodstream through the lungs and migrate toward specific body organs and tissues.

There are a number of health hazards associated with gases, vapors, particulates, and toxic metals. Health hazards are listed as: carcinogens, mutagens, teratogens, reproductive toxins, asphyxiation, anesthetic, neurotoxic, allergic response, irritant, sensitizer, corrosive, toxic, highly toxic, and those that target organ effects.

Asbestos fibers are so small they are invisible to the human eye. When compared to the human hair, an asbestos fiber is so small that it would be like comparing a pencil to a giant redwood tree. A single asbestos fiber dropped from a height of 8 feet would take 40 to 70 hours to reach the ground. When an abrasive, irritating asbestos fiber enters the human lung, the body's natural defense system is activated and encapsulates it. This process creates scar tissue that will remain inside the lung for life. Over time, the scar tissue causes the lungs to lose their ability to move oxygen in and out and supply oxygen to the bloodstream. This condition is referred to as asbestosis.

On December 3, 1984, the chemical processing industry was rocked as a large vapor cloud composed of methyl isocyanate (MIC) escaped from the Union Carbide India Limited (UCIL) plant and drifted over the sleeping town of Bhopal, India. The toxic vapor cloud killed 1,400 nearby residents. Seven years later the official body count had risen to 3,800, and the number of people who had disabling injuries had reached 11,000. Some estimates indicate that over 200,000 people were hurt during this incident. For the first time in modern chemical manufacturing history an event had taken place near a populated area that demonstrated how vulnerable a community is to the hazards that exist inside the chemical processing industry.

Airborne particulates are defined as small liquid or solid particles dispersed through the air. These particulates may pose a breathing hazard to process technicians working in the area. Dusts that contain aluminum oxide or limestone are classified as nuisance or inert aerosols. Various concentrations of airborne particles will have differing effects upon unprotected process technicians. Respiratory protection programs are designed to protect workers from exposure to various types of airborne contaminants.

Some particulates have been found to cause growths in the lungs or pulmonary fibrosis. Fumes or dusts containing tin or iron cause minimal damage;

silica and asbestos fibers cause extensive damage. The coal mining industry discovered that prolonged exposure of workers to coal dust caused the debilitating lung disease "black lung."

Mists or sprays containing corrosive substances such as acids or alkaline compounds would eat away at the soft tissue of the eyes and lungs. Systemic poisons are found in airborne particulates containing lead, manganese, pesticides, or radioactive materials. When these materials are absorbed by the lungs they enter the bloodstream and produce a number of toxic reactions. Some fumes and dusts have been known to produce chills or intense fever in unprotected workers. Airborne agents expected of causing these reactions include zinc, copper, and sugar cane residue.

Nine types of dusts pose a hazard for process technicians. A complete list of dust hazards includes: (1) inert dusts, (2) chemical irritants, (3) allergy-producing substances, (4) fever-producing substances, (5) coal dust, (6) fibrosis-producing materials, (7) cancer-causing agents, (8) systemic poisons, and (9) dust explosions.

Some substances mix readily with the air and quickly form mixtures that will explode or burn. These substances are referred to as flammable gases. This family of explosive and flammable gases includes: acetylene, hydrogen, methane, ethane, butane, and propane.

When cylinders are heated up, the compressed gases expand rapidly, increasing pressures. Direct flames will weaken the exposed side of the cylinder. Gas cylinders can rupture with catastrophic effects. Fire monitors are used to cool off cylinders exposed to open flames. Cylinders have protective caps over exposed valves and regulators. Special hoses are used to connect the cylinder to the gas manifold. When a hose blows loose it will whip rapidly and discharge the contents of the cylinder with great force.

Toxic metallic substances include systemic poisons such as lead, manganese, mercury, zinc, and copper. These poisons collect in specific organs and cause significant damage. This type of poisoning frequently occurs from inhaling dusts and vapors composed of toxic metals.

Hazardous metallic compounds combine one or more of the known toxic metal(s) in the chemical compound. These substances include: lead oxide, mercuric nitrate, cadmium salts, chromates, and manganese oxide.

Class D fires are those that involve combustible metals such as aluminum, magnesium, zinc, potassium, sodium, titanium, and zirconium. When these metals are powdered or broken into small enough parts to expose enough surface area, they will burn.

Review Questions

1. List the simple asphyxiants identified in this chapter and describe how to safely work with these gases.
2. Carbon monoxide and hydrogen cyanide are described as chemical asphyxiants. Describe how these chemicals work.
3. Describe the effect an anesthetic gas will have on an exposed technician.
4. List the steps required to extinguish a flammable metal fire.
5. Describe the hazards associated with asbestos.
6. Explain why flammable gases are a serious concern in the chemical processing industry.
7. List the hazards associated with working around compressed gas cylinders.
8. Explain how a dust explosion occurs.
9. A compound is described as a substance formed by the chemical combination of two or more substances in definite proportions by weight. What are the hazards associated with a compound that has lead oxide, mercuric nitrate, cadmium salts, chromates, or manganese oxide as a major component?
10. Describe the effects of beryllium, copper, lead, manganese, mercury, and zinc on an unprotected technician.
11. Describe the pressure relief devices found on a compressed gas cylinder.
12. List the nine different types of dusts that pose a hazard for process technicians.
13. Describe the Bhopal, India vapor release in one or two paragraphs.
14. List the most critical mistakes made in the events that led up to and throughout the Bhopal, India release.
15. List the generic health hazards associated with working with gases, vapors, particulates, and hazardous metals.
16. List the generic physical hazards associated with working with gases, vapors, particulates, and hazardous metals.
17. A chemical that reacts violently when it comes into physical contact with another chemical is referred to as ______.
18. When the concentration of an airborne chemical reaches the level at which it kills 50% of the test animals, it is referred to as the ______.
19. The lowest concentration at which a vapor or gas will produce a rich enough vapor concentration in air to ignite in the presence of an ignition source is referred to as the ______.
20. A chemical that affects the dermal layer of the body is referred to as ______.

Hazards of Liquids

Objectives

After studying this chapter, the student will be able to:

- Describe the physical and health hazards associated with liquids.
- Describe the physical and health hazards associated with solvents.
- Describe the safety precautions used when spray-painting is in progress.
- Describe the hazards associated with paints and adhesives.
- Compare and contrast acids and caustics.
- Explain the PPE used in handling acids and caustics.

Key Terms

- **Acute effect**—an immediate adverse effect on biological tissue.
- **Bulk containers**—have a rated design for liquids of 119 or more gallons, 882 pounds for solids, and a water capacity greater than 1,000 pounds for gases.
- **Chemtrec**—the Chemical Transportation Emergency Center provides information on chemicals around the clock. It is a service of the Chemical Manufacturers Association and can be reached by calling 1-800-424-9300.
- **Chronic effect**—a slow-developing adverse effect (cancer) that may take 10 to 20 years to appear; typically results after long-term exposure to a chemical.
- **Flammable liquid**—has a flashpoint below 100°F (37°C).
- **Flashpoint**—the lowest temperature at which a flammable liquid will produce a rich enough vapor concentration to ignite in the presence of an ignition source.
- **Hematopoietic system**—a term applied to the blood-forming system in the human body.
- **Hepatotoxin**—a chemical suspected of causing liver damage.
- **Liquefied petroleum gas (LPG)**—a liquid, gas mixture that has a gauge pressure of 40 psig at 70°F (21.1°C).
- **Nephrotoxin**—a chemical suspected of causing kidney damage.
- **Neurotoxin**—a chemical suspected of causing nerve damage; some links exist between this chemical type and behavioral and emotional abnormalities.
- **Non-bulk containers**—have rated capacities for liquids (119 gallons or less), solids (882 pounds or less or total capacity of 119 gallons), and gases (a water capacity of 1,000 pounds or less).
- **Primary hazard**—the hazard classification of the material with the greatest risk percentage component being shipped. See 172.101 table for CFR. (The Code of Federal Regulations [CFR] contains all of the permanent rules and regulations of OSHA and is produced in paperback format once a year.)
- **Subsidiary hazard**—material other than primary hazard.
- **Target organ toxin**—a chemical that selectively targets a specific organ in the body.

Introduction to the Hazards of Liquids

Safely handling, storing, and transporting chemicals are an essential element of a chemical technician's role. Process operations that include reactions, distillation, plastics, steam generation, extraction, and water treatment all use special procedures and equipment to limit the hazards associated with these liquids. The hazards associated with liquids are physical, chemical, or biological. *Physical hazards* of liquids are described as combustible liquid, compressed liquid, explosive, **flammable liquid,** organic peroxide, oxidizer, pyrophoric, unstable, and water reactive.

Chemical plants and refineries process hydrocarbons that will burn or explode violently. Chemical technicians work with vast chemical inventories that have characteristics that classify them as physical or chemical hazards. Process units use large amounts of water in cooling water systems, steam generation systems, fire-water systems, and domestic water. The hazards associated with water are woven into the many industrial applications associated with H_2O.

Chemical hazards of liquids include carcinogens, mutagens, teratogens, reproductive toxins, asphyxiation, anesthetic, neurotoxic, allergic response, irritants, sensitizers, corrosives, toxic, highly toxic, and those that target specific organs. Many process liquids have high and low pH levels that make them caustic or acidic. These liquids present various hazards to an unprotected technician. Acids and bases are chemical hazards that require special procedures, equipment, and personal protective equipment (PPE). Solvents can also be found in the inventory of chemical liquids used in a plant. Solvents, paints, and light-ends have characteristics that make them both chemical and physical hazards.

Biological hazards are associated with liquids that contain any living organism capable of causing disease in humans. These include insects, bacteria, fungi, and molds. Industrial water is one of the primary liquids where biological hazards can be found. Wastewater treatment, environmental control units, cooling water systems, and steam generation systems may expose a technician to biological hazards.

Handling, Storing, and Transporting Hazardous Chemicals Safely

- Transporting, storing, and handling chemicals requires that process technicians understand the systems, equipment, and technology they are working with, the physical hazards associated with chemicals in their facility, the health hazards associated with chemicals in their facility, chemical routes of entry into the human body, using the material safety data sheets (MSDSs), and proper labeling, signs, and tags usage. When materials are stored or transported the **primary hazard and subsidiary hazards are listed.** Primary hazard classification lists the material with the greatest risk percentage component being shipped. The **subsidiary hazard** is identified as the material other than primary hazard.

In order to safely store and transport a chemical, technicians should be familiar with the system they are working with. These systems vary from plant to plant, so it would be difficult to identify all of the industrial processes that are available in the petrochemical, gas processing, and refinery industries. Examples of these industrial processes include alkylation, isomerization, catalytic cracking, hydrodesulfurization, propylene, polyethylene, benzene, crude oil distillation, and over a hundred more. New technicians are carefully trained on these systems over a specified period of time. New technicians are carefully monitored and are not released to work alone until they have passed the rigorous requirements established by each plant. Site-specific training is a critical part of transporting, handling, and storing chemicals. Bulk and non-bulk containers have different capacities. For example, **non-bulk containers** have rated capacities for liquids (119 gallons or less), solids (882 pounds or less or total capacity of 119 gallons), and gases (a water capacity of 1,000 pounds or less).

During the apprentice training program, most technicians become familiar with the basic equipment and technology found in the petrochemical, gas processing, and refining industries. Although the processes in which this equipment is used may change, the equipment remains constant. Most apprentice training programs cover valves, piping, vessels, pumps, compressors, steam turbines, heat exchangers, cooling towers, fired heaters, boilers, distillation columns, reactors, extruders, flare systems, environmental awareness, safety training, math, chemistry, physics, roles and responsibilities of process technicians, working in self-directed work teams, quality control, basic process principles, process instrumentation, symbols and diagrams, industrial processes, basic hand tools, and minor maintenance training.

Physical Hazards Associated with Liquids

A physical hazard associated with liquids is a chemical that falls into one of the following categories: combustible liquid, flammable liquid, organic peroxide, corrosive, toxic, and unstable liquid. The characteristics associated with these fluids include the following:

- Combustible liquid—has a **flashpoint** between 100°F (38.8°C) and 200°F (93°C).
- Flammable liquid—has a flashpoint below 100°F (37°C).
- Organic peroxide—explodes when temperature exceeds a specified point.
- Pyrophoric—a chemical that ignites spontaneously with air at temperatures below 130°F (54.4°C).
- Unstable—a chemical that will react (condense, decompose, polymerize, or become self-reactive) when it is exposed to temperature, pressure, or shock.
- Water reactive—a chemical that reacts with water to form a flammable or hazardous gas.

Health Hazards Associated with Liquids

The health hazards associated with gases are similar to the health hazards associated with liquids. Liquids that fall under the following categories are listed on the MSDS, and safe handling procedures are clearly outlined:

- Carcinogens—liquids that are known cancer-causing substances.
- Mutagen—chemical that is suspected to have the properties required to change or alter the genetic structure of a living cell.
- Teratogen—liquid that is suspected to have an adverse effect on the development of a human fetus.
- Reproductive toxin—chemical that inhibits the ability of a person to have children.
- Asphyxiation—simple asphyxiation occurs when oxygen is removed or displaced by a chemical. For example, all hydrocarbon liquids are known as asphyxiants in the vapor phase.
- Anesthetic—liquid that dulls the senses. Example: alcohol.
- Neurotoxic—slows down the brain. Example: lead and mercury.
- Allergic response—a person may not have an allergic reaction immediately to a chemical. In some cases, it takes multiple exposures. Example: poison ivy.
- Irritants—a liquid that causes temporary discomfort when it comes into contact with human tissue.
- Sensitizers—a liquid that affects the nerves. Example: Phenol will be absorbed through the skin and will sensitize the affected area.
- Corrosives—chemical that causes severe damage to human tissue. Example: sulfuric acid.
- Toxic—chemical that has been determined to have an adverse health impact.
- Highly toxic—term applied to a chemical that requires only a small amount to be lethal.
- **Target organ toxin**—term applied to a chemical that contacts the body at a remote location and is transferred to another area of the body where it has an adverse effect on a specific organ.

Pressure and Pressurized Equipment

The primary variables a process technician works with are: (1) temperature, (2) flow, (3) level, (4) analytical, and (5) pressure. In 1990, a chemical plant in Houston, Texas exploded, killing 17 people. In the follow-up, it was determined that a small pressure gauge on the wastewater tank had been blocked in. Controlling pressure in process equipment is a critical aspect of a process technician's job. The pressure control loop in Figure 5-1 illustrates how pressure can be controlled automatically in an operating process.

Figure 5-1
Pressure Control Loop

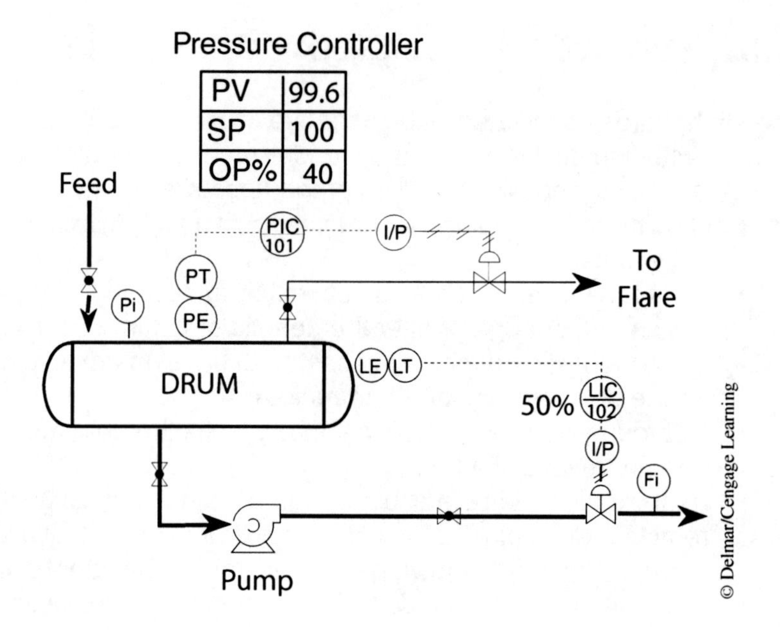

Process Systems

A process system is a collection of equipment that works together to perform a specific function. A distillation process is a collection of systems that are used together to separate the components in a mixture by their boiling points. Separating these components requires a variety of heating and cooling systems. One of the simplest systems is a pump-around that includes a tank, valves and piping, a pump, and critical instrumentation. Figure 5-2 shows how this system works as feed flows through the system. A small heat exchanger has been added to warm up the feed. A variety of control loops are used on this system, including: pressure, temperature, flow, and level. The instruments in the control loop make up a single system, the pre-heater equipment is another system, and the pump-round is another.

Flammable Liquid Storage

Small quantities of flammable liquids are often kept inside metal cabinets or in fire-resistant rooms. Figure 5-3 shows a typical metal cabinet. Frequently these cabinets have a separate ventilation system. Large quantities of flammable liquids are kept in tanks and vessels that are classified as open or closed systems. Flammable liquids include fuels, lubricants, solvents, paints, or other hydrocarbons. A number of operations require flammable liquids to be kept under pressure. An inert gas such as nitrogen is frequently used to increase the pressure in a tank. Closed tanks will respond to temperature variations. The vapor pressure exerted by a liquid

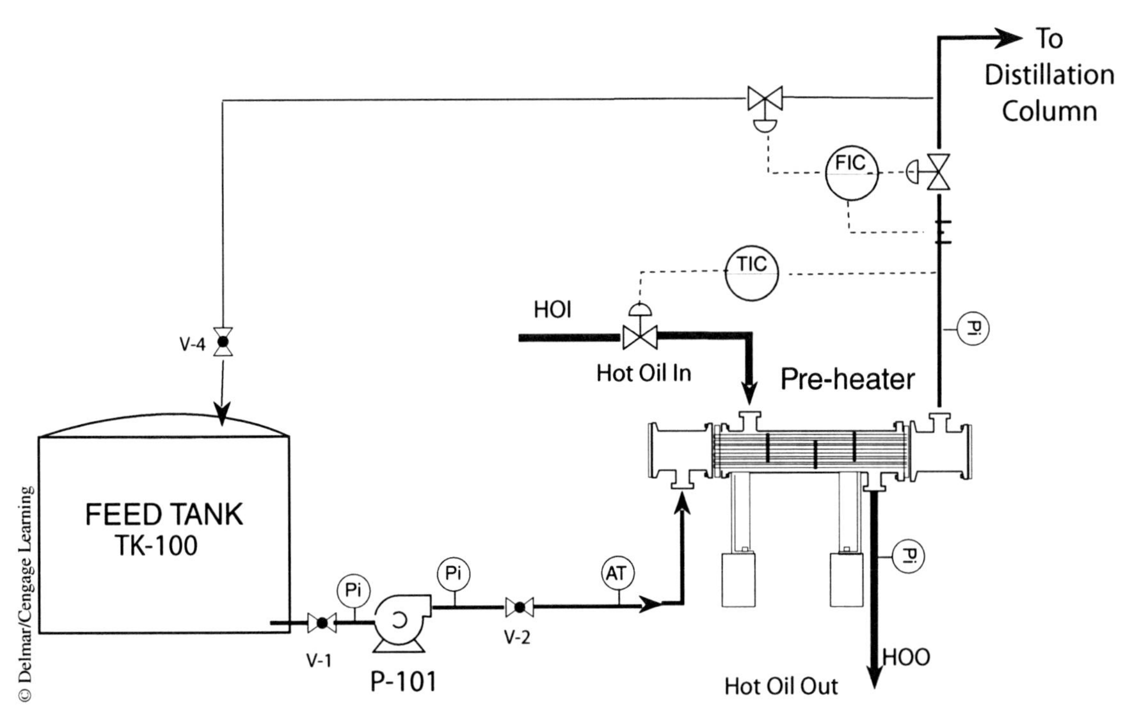

Figure 5-2 *Process Systems*

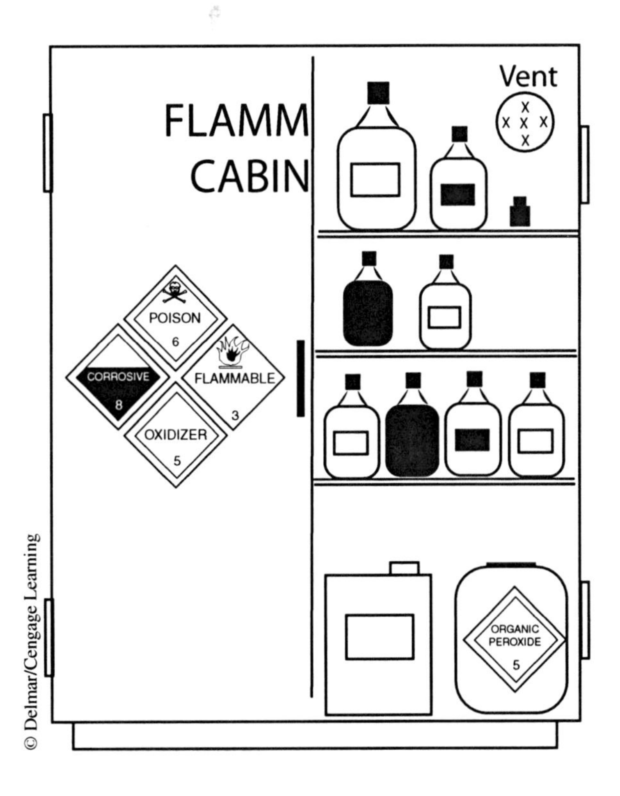

Figure 5-3
Metal Cabinets for Flammable Liquids

increases as the temperature rises. In an enclosed space, a flammable liquid will expand and the vapor pressure will increase as the temperature rises. Tanks that are open to the environment do not have pressure variations unless the level in the tank increases. Most operating plants are classified as non-smoking because of the risk associated with hydrocarbon technology. A single gallon of gasoline will explode with the force of 72 sticks of dynamite.

Spontaneous Combustion

Spontaneous combustion is the result of a slowly developing chemical reaction that produces its own heat. Oily rags and coal piles have the ability to spontaneously combust. In the case of oily rags, an oxidation reaction occurs that uses the fibers of the rags, the oil, and oxygen to start an exothermic reaction. This type of reaction gives off heat, whereas an endothermic reaction requires heat. Oily rags should be stored in metal containers. The three sides of the fire triangle are illustrated in Figure 5-4; however, another side can be added to the bottom of the triangle (making it a tetrahedron) that represents a chemical reaction. In coal pile fires, the slow oxidation reaction occurs when moisture is present in the coal. When fuel, air, and heat are combined in the correct proportions, combustion can occur. Spontaneous combustion uses slowly developing exothermic reactions to create the correct proportions required to initiate combustion.

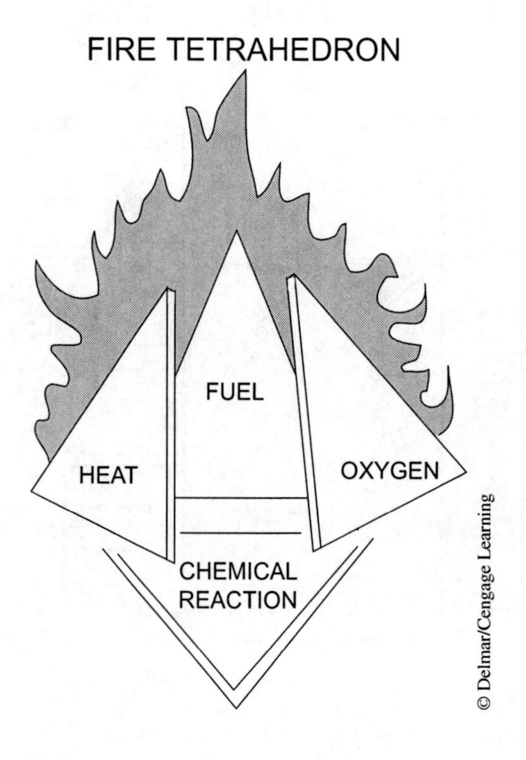

Figure 5-4
Fire Tetrahedron and Chemical Reactions

Oxidizers

An oxidizer is described as a substance that enhances the ability of oxygen to combine with fuel. When oxidization has occurred, little heat is needed to initiate a combustion reaction; in some cases room temperature is adequate. Common oxidizers include ammonium nitrate, chlorine, potassium nitrate, and pure oxygen. Common combustible items, such as petroleum grease products, and oils react violently, ignite, and occasionally explode when brought into contact with an oxidizer. It is difficult to fight a fire if an oxidizer is involved. A common mistake made by process technicians is to grease or lubricate the threads, gauges, and valves on an oxygen cylinder or to store oxygen cylinders with acetylene cylinders.

Hazards of Steam

Steam generation and application inside a chemical plant or refinery is the single largest utility. Boilers produce steam for a variety of uses including: warming process equipment; purging air and hydrocarbons from process equipment; direct injection into processes; running pumps, compressors, and generators; and vacuuming systems, heat exchangers, and furnaces. Steam is used in steam tracing to warm and protect equipment in the winter and to provide heat for buildings and warehouses. A typical steam generation system includes a boiler; a series of steam headers to process units; high-, medium-, and low-pressure steam; steam traps; a condensate return system; a de-aerator; a chemical additive system; and a make-up water system. Characterized by billowing white clouds of escaping water vapor, steam is the most visible utility used in the chemical processing industry.

The most common hazards associated with steam are physical exposure to live steam, contact with heated equipment, non-uniform heat transfer, and equipment over-pressuring. High-pressure steam discharged from a small opening has enough energy to cut through solid materials. Many exposed lines provide heat sources that can easily burn an unprotected technician.

Hazards of Water

The widespread use and application of water in the chemical processing industry creates a number of potential hazards. When water flashes to steam at 100°C, it expands 1,500 to 1,600 times its original volume. One gallon of water can produce 1,600 gallons of steam. Hazards associated with water are typically associated with uncontrolled mixing; however, a number of other causes can be identified that could be related to weather, equipment operations, and displacement procedures.

Uncontrolled Mixing

Uncontrolled mixing of water with hydrocarbons can generate high-pressure conditions that can damage systems that are not equipped with pressure relief devices. When water flashes to steam, it can rupture pipes and vessels. Thermal expansion and steam pressure can combine to be a destructive force. Heat transfer devices such as heat exchangers are typically designed with low or high point vents to remove water from the tubeside or the shellside of the device. Figure 5-5 illustrates the destructive forces associated with this device when water flashes to steam.

Foamover

Foamover occurs when water vaporizes under hot, heavy oil or asphalt. In many cases, the foam can expand to 20 or 30 times the original volume of the product pumped in. This violent foaming can overflow the tank and surrounding firewalls. In severe cases it can damage adjacent equipment. When asphalt is allowed to cool in a tank, it will need to be heated up slowly to prevent foamover. The term *delayed foamover* is used to describe a process in which hot asphalt does not initially mix with the water, allowing it to heat up enough to vaporize. To prevent this condition, process technicians should carefully inspect tanks, tank trucks, or tank cars before loading hot asphalt.

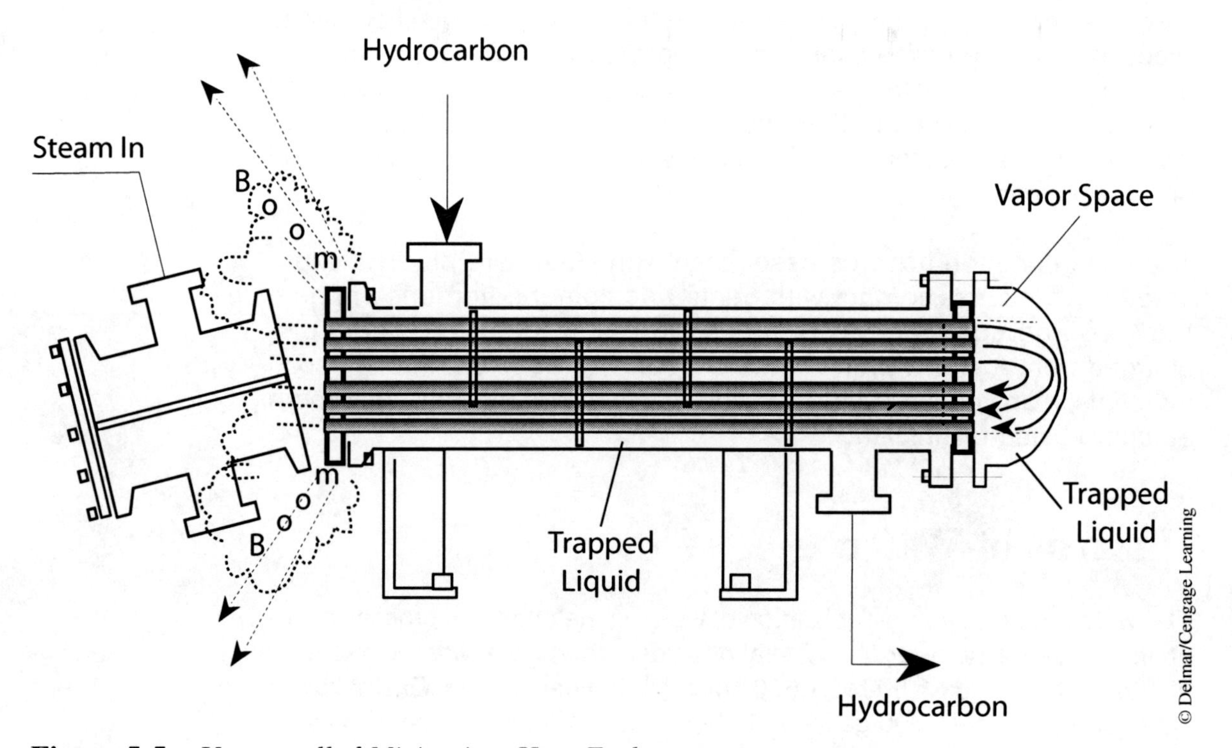

Figure 5-5 *Uncontrolled Mixing in a Heat Exchanger*

Water and Acid

Water and acid do not respond well to each other. In a variety of industrial processes in which acids may come into contact with water, it is important to remember this procedure: pour acid into water; never pour water into an acid. Large quantities of acid are found in refining and alkylation units. Sample containers should be handled carefully, and acid lines should be identified. Before performing typical housekeeping assignments, it is important to discuss with your trainer any special precautions.

Water Flashing

Water can accumulate in a wide variety of process equipment during extended shutdowns or under normal operational conditions. The hydrocarbon family has a wide range of boiling points when compared to water. Water and hydrocarbons do not mix well. Water in reflux drums should be drained and measured to prevent violent pressure surges in the column. Water in a steam turbine should be removed to prevent damage to the system. Water in compressor systems can seriously damage the unit. Tremendous pressures can rapidly be generated inside a heat exchanger unless it is properly shut down and started up. Pressure changes in a gas or vapor will respond differently inside a liquid. Open systems respond differently to rapid pressure surges than closed systems do. When water flashes or vaporizes to steam, it can expand to many times its original volume. Although water has a boiling point of 100°C, steam can achieve much higher temperatures. When a rupture occurs, escaping steam can artificially atomize a hydrocarbon, dispersing it into an explosive mixture.

Vacuum

Vacuum towers operate at pressures below atmospheric pressure. In these devices the boiling point of water is artificially lowered, creating a hazardous condition when water is present. The volume expansion is much greater in low-pressure situations. Water expands in volume 5,000 times at 160°F and 20 inches of mercury. Figure 5-6 illustrates how three different conditions affect the boiling point of water and expansion.

Static Generation

Electrostatic generation occurs when petroleum products are pumped through a pipeline. This process is associated closely with the velocity of the product being pumped and the amount of water present in the mixture. Under the right conditions, a static spark can ignite the mixture and rupture the line. Petroleum products should be kept free of water to avoid creating this hazardous condition.

Environmental Issues

Several environmental issues are closely associated with the hazards of water. For example, rapid weather changes cause temperature shifts that affect process instrumentation. Severe thunderstorms can quickly fill the sewer systems that connect the operating units to the water treatment system.

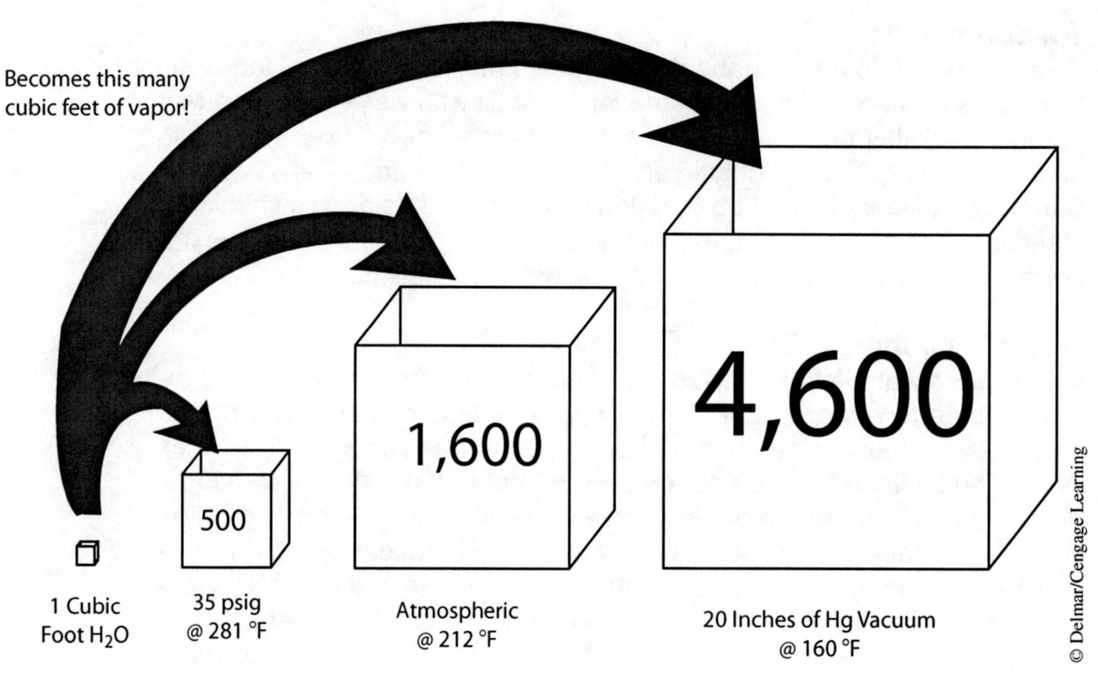

Figure 5-6 *Pressure Effects*

The hazards associated with this situation can place the local environment at risk from hazardous chemical contamination if the plant is unable to process the high volume of contaminated water. The weather can also produce ice, sleet, and hail, creating a series of different hazards. Cooling towers can produce thick banks of fog that can obscure visibility. Freezing conditions can cause pipes and equipment to burst. Hot water removes oxygen from local rivers and lakes if it is not carefully controlled. In the chemical processing industry, the environmental hazards associated with water cover a broad area.

Displacement

When water is used to flush process systems, oxygen can be pushed into hydrocarbon systems. As oxygen accumulates in vapor spaces, an explosive situation may be created. Displacement is often used to displace or remove hydrocarbons and to test for leaks. Because water is typically heavier than most hydrocarbons, engineering specifications should be followed closely before adding this additional weight to the unit. For each foot of water added to the unit, 0.433 psi should be added to the internal pressure.

Hazards of Light-Ends

Light-ends are hydrocarbons that are heavier than air and will quickly evaporate at room temperature and pressure. They typically have low boiling points and a Reid vapor pressure (RVP) of 18 psi or higher. For this reason, light-ends are often handled and stored as liquids. Examples of light-ends include methane, ethane, propane, butane, and pentane. A common liquid mixture

Propane	100°F	189 psi	remains liquid
Normal butane	100°F	52 psi	remains liquid
so-pentane	100°F	21 psi	remains liquid

Table 5.1
Pressure Characteristics

of propane and butane is often referred to as **liquefied petroleum gas, or LPG**. Light-ends are difficult to contain in pipelines and equipment because of their low viscosity and density characteristics. Because most of these hydrocarbons exist as a gas under normal conditions, pressure is added and in some cases temperatures are lowered to keep them in a liquid state. The explosive characteristics of light-ends are violent. Table 5-1 illustrates the pressure characteristics associated with light-ends, temperature, and pressure.

Viscosity

Light-ends under pressure can easily escape from enclosed systems, in many cases flowing through openings that heavier hydrocarbons could not escape. This relatively low viscosity ratio is a characteristic of light-ends. Viscosity is a measure of the flow characteristics of a liquid. For example, kerosene, a light-heavy hydrocarbon, is 15 times as viscous as propane and 6 times as viscous as pentane. Escaping light-ends are characterized by a white cloud similar to billowing fog or steam.

Frostbite

Another hazard associated with light-ends is the possibility of frostbite. When light-ends contact human skin, heat is quickly absorbed out of the tissue. The affected area should be flushed with water.

Thermal Expansion

Hazardous conditions associated with the term *thermal expansion* are also closely related to light-ends. A 20% vapor cavity is typically provided for sample containers and storage devices. Overheating is carefully avoided, and filled containers are refrigerated or kept in the shade.

Hazards of Air

Air is also eliminated from process equipment containing light-ends before startup. Steam or inert gas (nitrogen) is typically used to purge air out of the system. This reduces the possibility of an explosion. One gallon of propane transferred into a 10,000 gallon drum containing air at 60°F and at atmospheric pressure has the explosive energy equivalent to 13 pounds of trinitrotoluene (TNT).

Water Hazards

During the steaming process, water can be trapped in the system, for example in pump and compressor casings, low point bleeder valves, the bottom of tanks, and so on. This water can be drained from the system; however, ice formation can partially restrict flows and make valves appear to be closed. Ice in the valve can melt and allow light-ends to escape into the atmosphere.

Acids and Caustics

In 1884, a Swedish chemist named Svante Arrhenius (1859–1927) attempted to define the characteristics of an acid and a base. An acid produces H_3O (hydronium) ions in an aqueous solution, whereas a base produces OH (hydroxyl) ions in an aqueous solution. Modern chemists rarely taste acids and bases; however, in the past they did. This is how we know that an acid has a sour taste like a lemon, and a base is bitter. An acid typically has a pH value between 0 and 7.0, and a base or caustic has a value between 7.0 and 14.0. Table 5-2 lists acids and bases commonly used in the chemical processing industry and their chemical formulation. Figure 5-7 shows the values found on a pH scale. An acid will turn blue litmus paper red, and a base will turn red litmus paper blue.

Acids and caustics have a wide range of applications in the chemical processing industry. They are used in alkylation units, water treatment, storage batteries, chemical processing, and refinery operations. When acids and caustics come into contact with human tissue, they tend to dissolve or eat away any exposed area. Mists and vapors inhaled by the respiratory system will cause chemical burns and destroy soft throat and lung tissue. The proper protective equipment should always be used when handling these materials.

Although acids have several characteristics that occur during a reaction, the primary one is that the acid donates a proton to a base. For this reason, an acid is called a proton donor, and a base is called a proton acceptor.

Acids and caustics are typically referred to as chemical opposites because they are used to neutralize each other. This process is useful in water treatment for cooling towers, heat exchangers, boiler feed water, domestic water supplies, and wastewater treatment.

Table 5.2
Acids and Bases

Acid		Base	
Perchloric acid	$HClO_4$	Sodium hydroxide	NaOH
Hydroiodic acid	HI	Potassium hydroxide	KOH
Hydrobromic acid	HBr	Lithium hydroxide	LiOH
Sulfuric acid	H_2SO_4	Barium hydroxide	$Ba(OH)_2$
Hydrochloric acid	HCL	Ammonia	NH_3
Nitric acid	HNO_3	Magnesium hydroxide	$Mg(OH)_2$
Phosphoric acid	H_3PO_4		

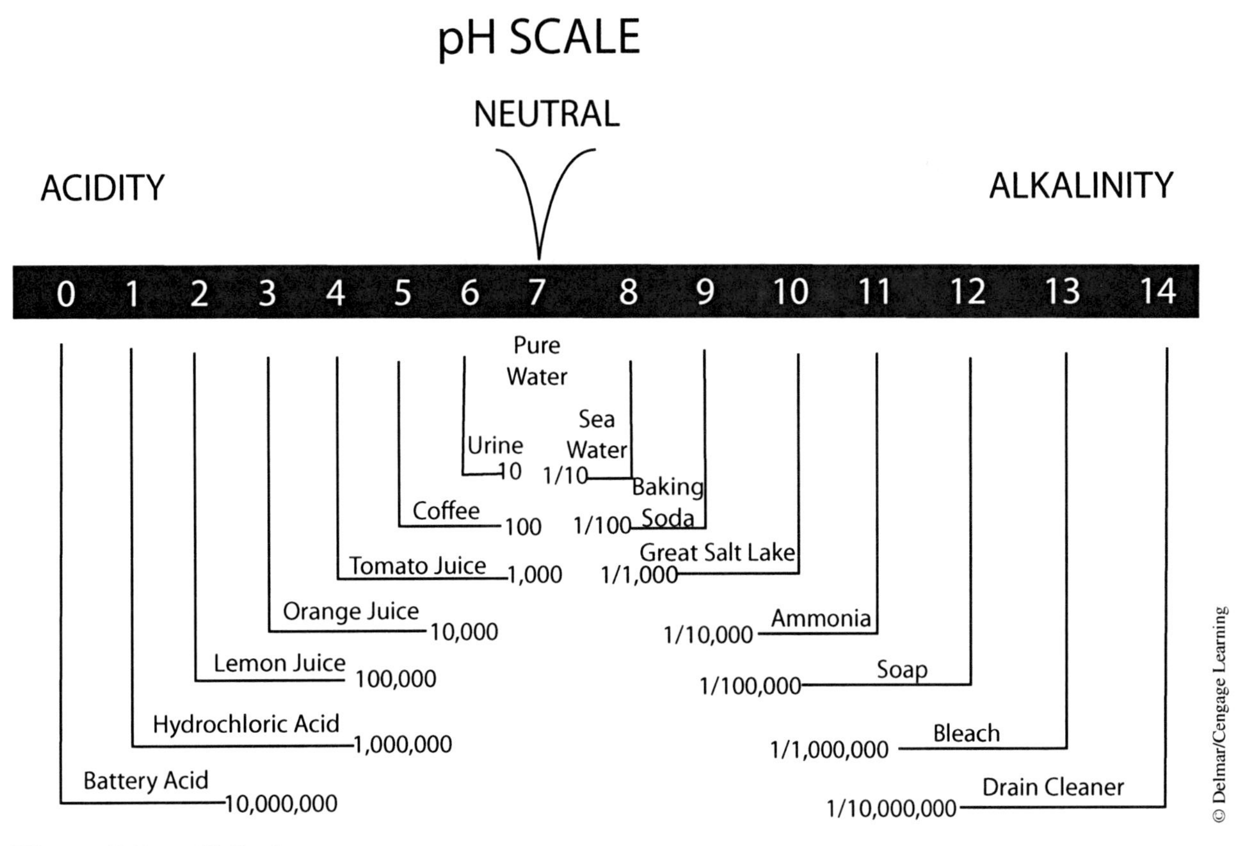

Figure 5-7 *pH Scale*

Solvents

Many industrial applications frequently use paints and adhesives in preventive maintenance programs and new construction. These products contain a hazardous substance called a solvent. Solvents are flammable, volatile, highly toxic, and produce large volumes of vapor. For a process technician, this presents a serious fire hazard. In addition to being a fire hazard, paints and adhesives containing solvents have harmful effects when inhaled, ingested, injected, or absorbed into the human system. Solvents should be handled carefully and stored properly.

Paints and Adhesives

Most industrial painting projects should be considered as a breathing hazard. Acrylic paints, oil-based paints, and epoxy paints contain a variety of hazardous components. Two important safety procedures include adequate ventilation and the use of respiratory equipment and PPE. Fine mists that are inhaled allow paint particles to build up on the inner surface of the lung, reducing respiration. If the paint contains a toxic substance, this is the most rapid method of acute or chronic poisoning. It should also be noted that the propellant used in many spray cans is toxic. Paints that are used for rust prevention contain chromates, an alkaline chemical that can lead to chronic poisoning. Lead is another product commonly found in paints that are used for rust prevention. The hardeners used in epoxy paints are also hazardous when absorbed through the skin and can produce an **acute** or **chronic effect.** Although most solvents have varying toxic effects, they each appear to attack specific organs in the human body and create injuries that are often irreversible. Adhesives contain chemicals such as toluene and xylene that can cause permanent damage when inhaled in sufficient quantities. Some adhesives, such as instant glues, contain cyanoacrylate, which is toxic when inhaled, injected, ingested, or absorbed through the skin.

Hematopoietic System Toxins, Hepatotoxic Agents and Other Harmful Agents

There are a number of toxic agents that act on different areas of the human body. **Hematopoietic system toxins** act on the blood cells or the bone marrow that produces blood cells. Examples of these toxins include Aniline, Benzene, Nitrites, Nitrobenzene, and Toluidine. **Hepatotoxic agents** like Carbon tetrachloride, Nitrosamines, and Tetrachloroethane cause liver damage.

The human nervous system is very sensitive to organometallic compounds and selected sulfide compounds. **Neurotoxic agents** can damage the nervous system. Examples of Neurotixic agents include Carbon disulfide, Methyl mercury, Manganese, Organic phosphorus insecticides, Tetraethyl lead, Thallium, and Tri-alkyl tin compounds.

Nephrotoxic agents like halogenated hydrocarbons and Uranium compounds damage the kidneys.

Pulmonary toxins target the soft pulmonary tissues in the lungs and trigger fibrotic changes or cause a restrictive disease called pneumoconiosis. Examples of these toxins include coal dust, cotton dust, and wood dust. Fibrotic changes can be triggered by asbestos and free silica.

Summary

The hazards associated with liquids are physical, chemical, and biological. *Physical hazards* of liquids are described as combustible liquid, compressed liquid, explosive, flammable liquid, organic peroxide, oxidizer, pyrophoric, unstable, and water reactive. Chemical plants and refineries process hydrocarbons that will burn or explode violently. *Chemical hazards* of liquids include carcinogens, mutagens, teratogens, reproductive toxins, asphyxiation, anesthetic, neurotoxic, allergic response, irritants, sensitizers, corrosives, toxic, highly toxic, and those that target specific organs. Acids and bases are chemical hazards that require special procedures, equipment, and personal protective equipment. Solvents can also be found in the inventory of chemical liquids used in a plant. Solvents, paints, and light-ends have characteristics that make them both chemical and physical hazards.

Biological hazards are associated with liquids that contain any living organism capable of causing disease in humans; this includes insects, bacteria, fungi, and molds. Industrial water is one of the primary liquids where biological hazards can be found. Wastewater treatment, environmental control units, cooling water systems, and steam generation systems may expose a technician to biological hazards.

Steam generation and application inside a chemical plant or refinery is the single largest utility. The most common hazards associated with steam include physical exposure to live steam, contact with heated equipment, nonuniform heat transfer, and equipment over-pressuring. High-pressure steam discharged from a small opening has enough energy to cut through solid materials. Many exposed lines provide heat sources that can easily burn an unprotected technician. When water flashes to steam at 100°C, it expands 1,500 to 1,600 times its original volume. One gallon of water can produce 1,600 gallons of steam.

Hazards associated with water are typically due to uncontrolled mixing; however, a number of other areas can be identified. This would include environmental topics such as the weather, equipment operations, and displacement procedures. *Light-ends* are hydrocarbons that are heavier than air and will quickly evaporate at room temperature and pressure. Examples of light-ends include methane, ethane, propane, butane, and pentane. A common liquid mixture of propane and butane is often referred to as liquefied petroleum gas, or LPG. Light-ends are difficult to contain in pipelines and equipment because of their low viscosity and density characteristics. Because most of these hydrocarbons exist as a gas under normal conditions, pressure is added, and in some cases temperatures are lowered to keep them in a liquid state. The explosive characteristics of light-ends are violent. Light-ends under pressure can easily escape from enclosed systems, in many cases flowing through openings that heavier hydrocarbons cannot escape. This relatively low viscosity ratio is a characteristic of light-ends. Other hazards associated with light-ends are the possibility of frostbite and thermal expansion. One gallon of propane transferred into a 10,000-gallon drum containing air at 60°F and at atmospheric pressure has the explosive energy equivalent to 13 pounds of TNT.

Acids and caustics have a wide range of applications in the chemical processing industry. An acid produces H_3O (hydronium) ions in an aqueous solution, whereas a base produces OH (hydroxyl) ions in an aqueous solution. An acid typically has a pH value between 0 and 7.0, and a base or caustic has a value between 7.0 and 14.0. An acid will turn blue litmus paper red, and a base will turn red litmus paper blue. When acids and caustics come into contact with human tissue, they tend to dissolve or eat away any exposed area. Mists and vapors inhaled by the respiratory system will cause chemical burns and destroy soft throat and lung tissue. The proper protective equipment should always be used when handling these materials. Although acids have several characteristics that occur during a reaction, the primary one is that the acid donates a proton to a base. For this reason, an acid is called a proton donor, and a base is called a proton acceptor.

Many industrial applications frequently use paints and adhesives in preventive maintenance programs and new construction. These products contain a hazardous substance called a solvent. Solvents are flammable, volatile, highly toxic, and produce large volumes of vapor. Most industrial painting projects should be considered breathing hazards. Acrylic paints, oil-based paints, and epoxy paints contain a variety of hazardous components. Two important safety procedures include adequate ventilation and the use of respiratory and personal protective equipment.

Review Questions

1. Describe the hazards associated with paints and adhesives.
2. Compare and contrast the differences between acids and bases.
3. List and describe the hazards associated with light-ends.
4. List the specific hazards associated with solvents.
5. Describe the effects of static charge.
6. Describe the hazards associated with water.
7. List the hazards associated with steam.
8. List and describe common oxidizers.
9. Explain the scientific principles associated with spontaneous combustion.
10. Describe flammable liquid storage.
11. Describe the key elements of a process system.
12. List the primary variables a process technician works with.
13. Define the term *mutagen*.
14. Describe and list physical hazards.
15. Describe and list chemical hazards.
16. Describe and list biological hazards.
17. Compare and contrast the terms *nephrotoxin* and *neurotoxin*.
18. Define the term *acute effect*.
19. Compare and contrast the terms *flashpoint* and *flammable liquid*.
20. Explain the best way to safely handle, store, and transport hazardous chemicals.

Hazardous Chemical Identification: Hazcom, Toxicology, and DOT

Objectives

After studying this chapter, the student will be able to:

- Describe the Hazard Communication Standard.
- Identify the physical properties and hazards associated with handling, storing, and transporting chemicals.
- Describe physical and health hazards associated with exposure to chemicals.
- Describe the key elements of a material safety data sheet.
- Describe a hazardous chemical inventory list.
- Explain the purpose of a written hazard communication program.
- Identify methods used to protect process technicians from hazardous chemicals.
- Identify safety signs, tags, and warning labels utilized by process technicians.
- Describe toxicology and the terms associated with it.
- Describe the material classification system for the Department of Transportation (DOT).
- Describe the Hazardous Materials Identification System.
- Describe the National Fire Prevention Association labeling system.

Key Terms

- **DOT**—U.S. Department of Transportation, which regulates shipments of hazardous materials. These regulations contain specific information on how hazardous materials are identified, placarded, documented, labeled, marked, and packaged.
- **Hazardous chemical**—a chemical that has been determined to be a physical hazard to humans.
- **HAZCOM standard**—Hazard Communication Standard (HCS), known as the workers' right-to-know. Hazard Communication 29 CFR 1910.1200 training is required upon initial assignment.
- **HMIS**—Hazardous Materials Identification System is another labeling system frequently used by industrial manufacturers to identify the hazards associated with chemicals.
- **Lethal Concentration 50 (LC_{50})**—when the concentration of an airborne chemical reaches the level at which it kills 50% of the test animals.
- **MSDS**—a typical material safety data sheet has ten sections: Product Identification and Emergency Information, Hazardous Ingredients, Health Information and Protection or Hazards Identification, Fire and Explosion Hazard, Data and Chemical Properties, Spill Control Procedure, Regulatory Information, Reactivity Data, Storage and Handling, and Personal Protective Equipment.
- **Toxicology**—the science that studies the noxious or harmful effects of chemicals on living organisms.

Introduction to Hazardous Chemical Identification

Chemical manufacturers are required by the **HAZCOM standard** to: (1) analyze and assess the hazards associated with chemicals and develop written procedures for evaluating chemicals, (2) document hazards and develop material safety data sheets **(MSDSs)** and warning labels, (3) disseminate the information to affected individuals, and (4) label, tag, and attach warning documentation to chemicals leaving the workplace. Industry employers of process technicians are responsible for the development of a written hazard communication program, a **hazardous chemical** inventory list, and associated material safety data sheets.

The Hazard Communication Program—The Workers' Right-to-Know Act

Government Mandate: Hazard Communication 29 CFR 1910.1200 training is required for workers upon initial assignment.

The purpose of the Hazard Communication Standard (HCS) is to ensure that the hazards associated with the handling, transport, and storage of chemicals in a plant are evaluated and transmitted to affected personnel. A fundamental principle of the chemical Hazard Communication Standard (HCS) is that informed technicians are less likely to be injured by chemicals and chemical processes than uninformed technicians. According to the standard, all of the chemical inventories and processes within a chemical plant or refinery must be evaluated for potential hazards and risks. Where a risk is found, essential information and training is required for the existing workforce and new employees.

Requirements of the Standard

Because the chemical processing industry manufactures chemicals and employs technicians, it is responsible for both sections of the OSHA standard that address (1) chemical manufacturers' and users' responsibilities and (2) employers' requirements. Chemical manufacturers are required by the HAZCOM standard to perform the following functions:

- Analyze and assess the hazards associated with chemicals
- Develop written procedures for evaluating chemicals
- Document hazards and develop MSDSs and warning labels
- Disseminate the information to affected individuals
- Label, tag, and attach warning documentation to chemicals leaving the workplace

Employers are responsible for the following functions:

- Preparing a written HAZCOM program
- Organizing a hazardous chemical inventory list
- Designing individual MSDSs

This written program should be designed so that it is given to the new employee upon initial assignment. The materials should be site-specific, readily accessible by plant personnel, and include an evergreen feature that will keep it up to date. Employers are also required to provide training to employees on the hazards associated with chemicals they will be working with, how to read an MSDS, how to select and use personal protective equipment (PPE), and how to read and use one of the three standard labeling systems: **DOT**—Department of Transportation, **HMIS**—Hazardous Materials Identification System, and **NFPA**—National Fire Protection Association.

Delivery of the Standard to Employees

The HAZCOM standard requires employers to provide information or training to employees about their plant's HAZCOM program. (See Figure 6-1.) Fundamental information that must be provided to a process technician includes the key elements of the HAZCOM standard, the plant's written hazard communication program, a detailed hazardous chemical inventory list, and associated MSDSs along with warning labels, tags, and signs. Information should be included on how to access the HAZCOM system, chemical inventory list, and MSDSs 24 hours a day, 7 days a week. Employers are required by law to provide open access to HAZCOM materials.

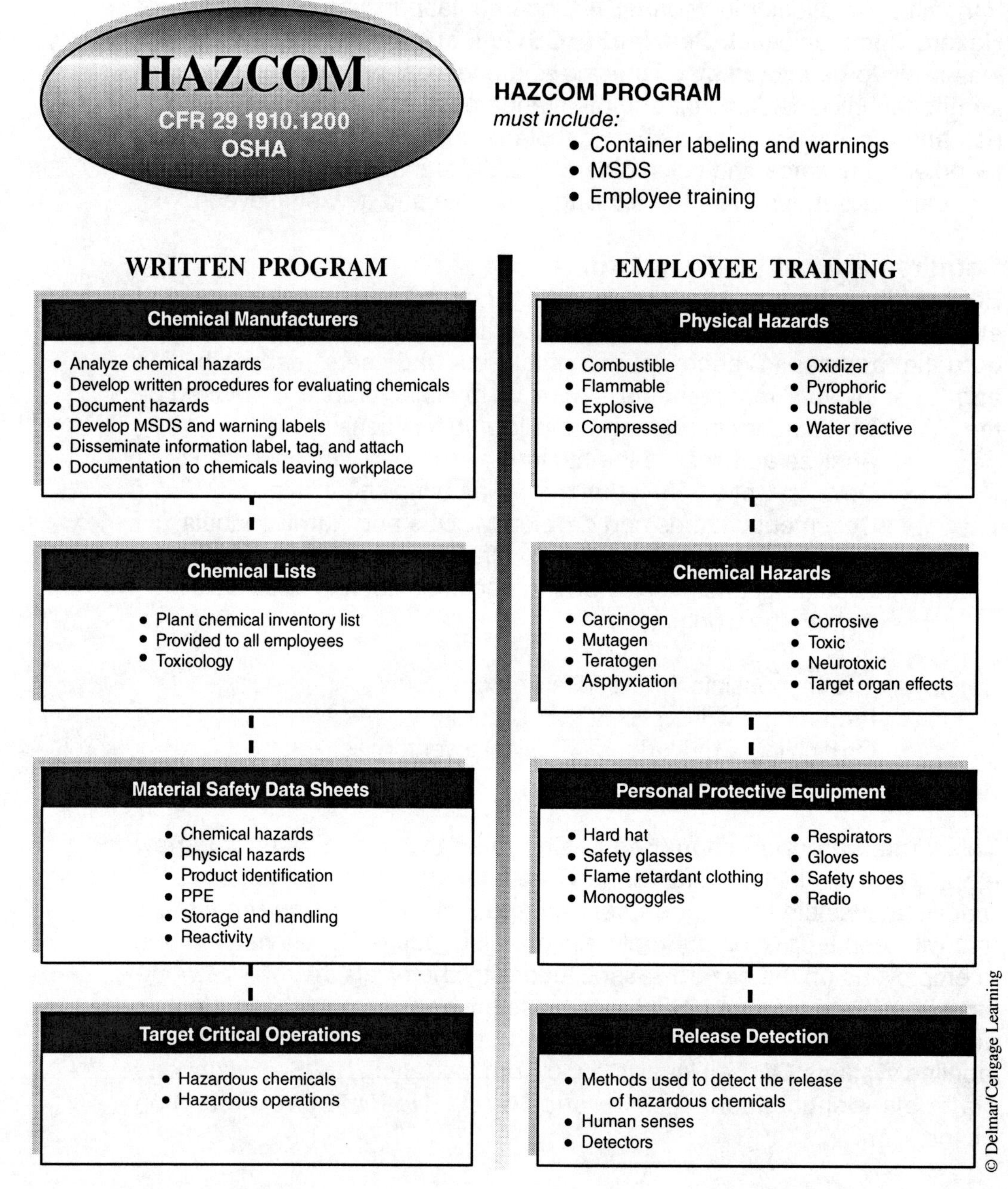

Figure 6-1 *HAZCOM*

This is why the HAZCOM standard is frequently referred to as the Workers' Right-to-Know Act.

The chemical processing industry initiates the delivery of HAZCOM training upon initial assignment of an employee to the plant. Training focuses on the physical and health hazards associated with exposure to chemicals. Additional information is provided on the **toxicology**, physical properties, and hazards associated with handling, storing, and transporting chemicals. New technicians are required to review company procedures used to protect employees from hazardous chemicals, and specific operations are identified that may expose an employee to a chemical. The training section also includes the selection and use of PPE and the methods and observations utilized to detect the release of hazardous chemicals.

Material Safety Data Sheet (MSDS)

It has been estimated that one out of every four workers in the United States handles a chemical. Although most chemicals are not harmful in small doses, caution should be exercised when the dose (amount) exceeds the recommended exposure (time) limits. The development of the MSDS is the responsibility of the chemical's manufacturer. A sample MSDS is given in Figure 6-2. A typical MSDS has the following ten sections:

- Section 1, Product Identification and Emergency Information—Chemical name, chemical family, formula, trade name, company code, manufacturer address, and emergency telephone number available 24 hours a day.
- Section 2, Hazardous Ingredients—Chemical component percentages and OSHA hazard.
- Section 3, Health Information and Protection or Hazard Identification—Nature of hazard and first aid; eye and skin contact, inhalation, and ingestion; exposure limits, precautions, and personal protective equipment.
- Section 4, Fire and Explosion Hazard—Flashpoint, flammable limits, auto-ignition temperature, general hazards, firefighting procedures, and hazardous combustion products.
- Section 5, Physical Data and Chemical Properties—Vapor pressure, boiling point, evaporation rate, appearance and odor, specific gravity, solubility in water, percent volatile, density, freezing point, and melting point.
- Section 6, Spill Control Procedure—Land spill procedure, water spill procedure and vapor release procedure. Methods to control and to protect yourself.
- Section 7, Regulatory Information—Department of Transportation shipping description; flashpoint; TSCA information; Comprehensive Environmental Response, Compensation and Liability Act (CERCLA) information; and Superfund Amendments and Reauthorization Act (SARA) information.
- Section 8, Reactivity Data—Stability of material, conditions to avoid instability, materials and conditions to avoid incompatibility, hazardous decomposition of product, and hazardous polymerization.

MATERIAL SAFETY DATA SHEET

IDENTIFY ***(As used on label and list)***

Section 1 **Manufacturer's name, address**	**Section 5** **Reactivity Data**
Emergency telephone: Information telephone: Date prepared: Signature of preparer:	Incompatibility: Hazardous decomposition or byproducts: Hazardous polymerization:

Section 2 **Hazardous Ingredients**	**Section 6** **Health and Hazard Data**
Hazardous components: Chemical name: Exposure limits: - OSHA PEL, permissible exposure limit - ACGIH TLV, threshold limit value - OEL, occupational exposure limit % of composition:	Routes of entry: Health hazards: Carcinogenicity: Signs and symptoms of exposure: Medical conditions generally aggravated by exposure: Emergency and first-aid procedures:

Section 3 **Physical / Chemical Characteristics**	**Section 7** **Precautions for Safe Handling and Use**
Boiling point: Vapor pressure: Vapor density: Solubility in water: Specific gravity: Melting point: Evaporation rate: Appearance and odor:	Steps to be taken if material is released or spilled: Waste disposal methods: Precautions to be taken in handling and storing: Other precautions:

Section 4 **Fire and Explosion Hazard Data**	**Section 8** **Control Measures**
Flashpoint: Flammable limits: Extinguishing media: Special firefighting procedures: Unusual fire and explosion hazards:	Respiratory protection: Ventilation: Protective gloves: Eye protection: Other protective clothing or equipment: Work / hygienic practices:

Figure 6-2 *Sample MSDS*

- Section 9, Storage and Handling—Electrostatic accumulation hazard, storage temperature, storage, transport pressure, loading temperature, unloading temperature, and viscosity at loading.
- Section 10, Personal Protective Equipment—Respiratory protection, clothing and gloves, face shield, goggles, ventilation, and so on.

Toxicology

Toxicology is the science that studies the noxious or harmful effects of chemicals on living organisms. The fundamentals of toxicology include a relationship between dose and response. Dose is defined as the amount of chemical entering or being administered to a subject; response is defined as the toxic effect the dose has upon the subject.

During a scientific study, a graph is developed that plots dose and response. Statistical analysis is used to determine the lethal dose (LD_{50}).

The LD_{50} is the expected dosage that will kill 50% of the lab animals in a controlled experiment. The term **LC_{50} (lethal concentration 50)** is used when the dosage is applied through inhalation. Both the LC_{50} and LD_{50} are indicators of the toxicity of a specific substance. For example, toxicity levels can range as follows:

LC_{50} .350 grams	Least toxic
LC_{50} .150 grams	
LC_{50} .005 grams	Most toxic

The degree of toxicity of a substance also has a relationship with dose and response. As toxicity increases, the required dosage decreases. Some substances are eight million times more toxic than others. Chemicals have four possible routes of entry into the human body: skin absorption, inhalation, ingestion, and injection. When phenol is absorbed through the skin, it damages the nerves. The degree of damage depends on the dosage. A small dose may result in a tingling sensation or numbness, a medium dose results in partial paralysis, and a large dose results in death. When benzene is inhaled into the lungs, it is carried by the blood cells throughout the system. Small quantities of benzene are harmless; a high LC_{50} would result in bone marrow damage and an inability to produce red blood cells. When potassium cyanide is ingested, it destroys living cells.

Exposure limits to various materials are designed to allow a process technician to work safely with a chemical over an extended period of time. Exposure limits are numerical values from four different exposure limit categories as follows:

1. Threshold limit value (TLV)—established standard.
   - Time-weighted average (TLV-TWA)—based on an 8-hour day, 40 hours a week without harmful side effects
   - Short-term exposure limit (TLV-STEL)—maximum 15-minute exposure, 4 times a day, 60-minute intervals
   - Ceiling (TLV-C)—maximum limit (Do not exceed!)
2. Occupational exposure limits (OELs)—developed by OSHA and industrial hygiene to protect workers from hazards when a TLV does not exist

3. Permissible exposure limits (PEL)—OSHA-mandated regulatory limits
4. Supplier recommended limits

There are three ways to detect the release or presence of hazardous chemicals in the workplace: bodily senses, monitoring equipment, and industrial hygiene. The first way is by smell, sight, sound, taste, or touch. Technicians should always be alert to unusual vapor releases, strange smells, abnormal sounds and vibrations, and chemical irritants that affect the eyes, ears, and nose. The second way to detect a release is with portable and continuously operated detectors. This process provides a safer layer of protection than relying on the bodily senses alone. The third way to detect chemicals is through industrial hygiene. Each plant should have an industrial hygienist who is trained in the use of detection equipment appropriate for the chemicals used in the facility.

Safety Signs, Tags, and Warning Labels

Petrochemical facilities, gas processing facilities, and refineries all store, transport, and move large quantities of chemicals. Each of these chemicals must be properly labeled and identified. Appropriate safety signs must be displayed so that everyone can see them. These signs must be located in the places where they are most needed. Tags should be used where needed and a well-defined permit system should be established. Training on each of these areas must be given to all employees. Examples of warning labels, safety signs, and tags include the following:

- Warning labels—chemical information: identify the chemical, hazards, and manufacturer
- Safety signs—chemical hazard: benzene, H_2S, chlorine, eye wash, safety stations, speed limits, and so on
- Tags—cold work permits, opening or blinding permits, hot work permits, permits to enter, lock-out, tag-out, energy isolation, and unplugging

The chemical processing industry uses three widely recognized hazard communication labeling systems: Department of Transportation (DOT), Hazardous Materials Identification System (HMIS), and National Fire Protection Association (NFPA).

Department of Transportation Labeling System

The DOT labeling system—Government Mandate: DOT 49 CFR 171-177—utilizes a diamond-shaped pattern with colors, symbols, and numbers that correspond to specific hazards. These signs are typically used as placards for tank cars and trucks. The DOT numbering system does not indicate the severity of the hazard but rather the hazard's division or classification, as shown in Figure 6-3.

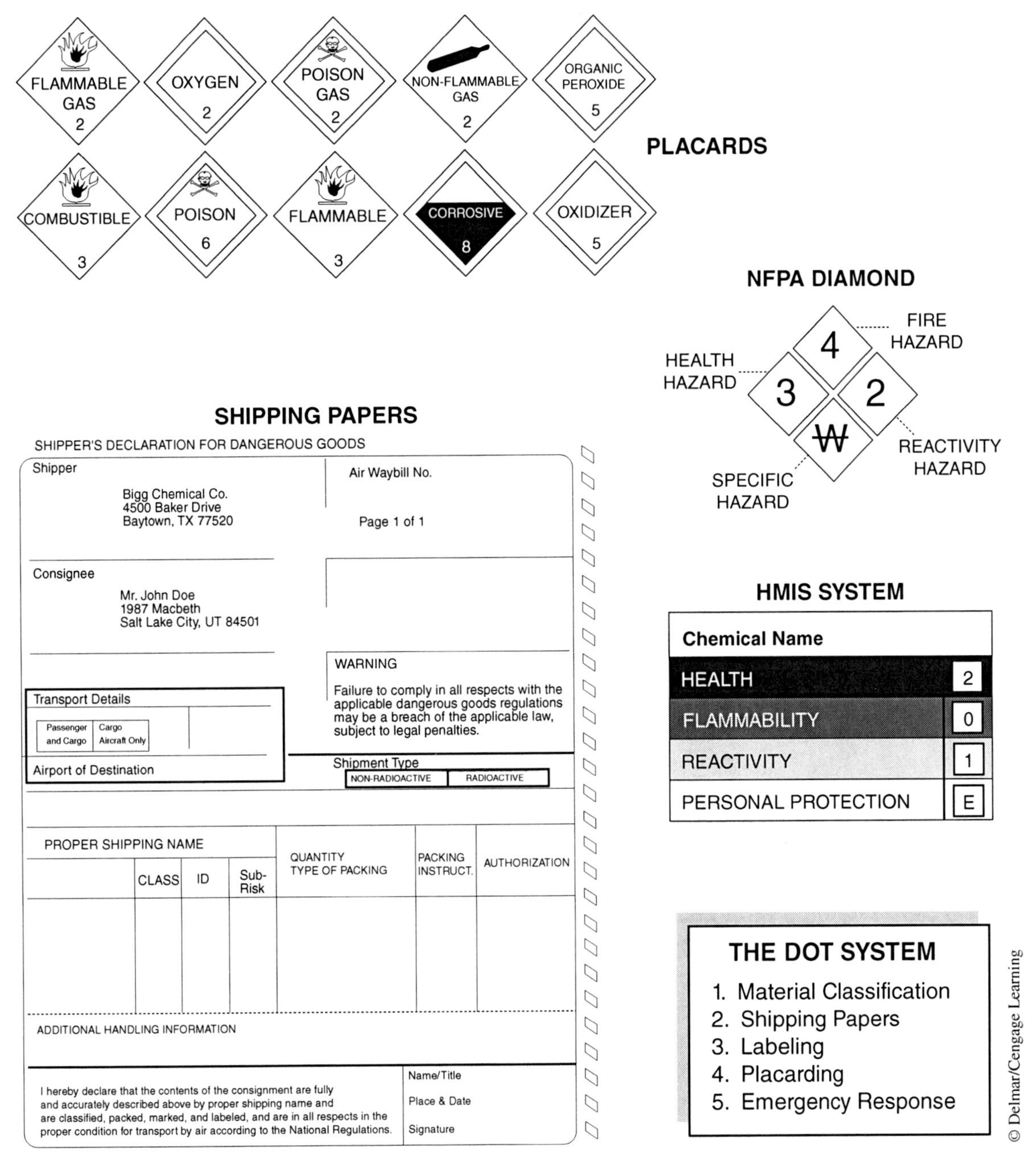

Figure 6-3 *DOT System*

DOT Material Classification

Shipments of hazardous materials are regulated by the DOT. These regulations contain specific information on how hazardous materials are identified, placarded, documented, labeled, marked, and packaged. Hazardous material shipments that are not in compliance with federal regulations will be delayed and can result in severe penalties. In civil cases, marking the wrong name on a container can result in fines of up to $25,000 per violation. In criminal cases, fines of up to $500,000 and five years in jail can be imposed for intentionally shipping a hazardous chemical and attaching the wrong MSDS.

Materials to be transported are classified into ten different categories. They are:

- Explosive
- Gases
- Flammable Liquids
- Self-Reactive
- Oxides
- Poisonous and Infectious Materials
- Radioactive Materials
- Corrosive Materials
- Miscellaneous Hazardous Materials
- Peroxides

DOT Shipping Papers

The shipping papers for a chemical must contain information that will allow easy access and understanding to the material being transported. The information should be in English and may be handwritten. Codes and abbreviations cannot be used in the official document. (See Figure 6-4 for an example of a DOT shipping form.) The following information should be included in the shipping papers:

- Basic description of material
- Name and address of shipper
- CHEMTREC number or company 24-hour hotline
- Proper shipping name of material
- Hazard class or division (immediate hazards to health)
- UN or NA material classification number
- Packing group of material
- Basic description: N.O.S., flammable liquid, and so on (risk of fire or explosion)
- Quantity, volume
- Precautions to be taken in the event of an accident
- Methods for handling fires, spills, leaks
- First aid procedures

Note: Additional information is needed for radioactive material. An MSDS should be included with any material being shipped.

DOT Marking, Labeling, and Placarding

The marking and labeling procedure for chemicals that are being shipped must follow specific guidelines. Labels, signs, and tags should be in an unobscured, durable format, and in English. In most cases this information is contrasted against an eye-catching background. Package orientation markings are designed to ensure that the product is stored in the safest possible way.

SHIPPING PAPERS

SHIPPER'S DECLARATION FOR DANGEROUS GOODS

Shipper

Bigg Chemical Co.
4500 Baker Drive
Baytown, TX 77520

Air Waybill No.

Page 1 of 1

Consignee

Mr. John Doe
1987 Macbeth
Salt Lake City, UT 84501

WARNING

Failure to comply in all respects with the applicable dangerous goods regulations may be a breach of the applicable law, subject to legal penalties.

Transport Details

Passenger and Cargo	Cargo Aircraft Only

Airport of Destination

Shipment Type

NON-RADIOACTIVE	RADIOACTIVE

PROPER SHIPPING NAME	CLASS	ID	Sub-Risk	QUANTITY TYPE OF PACKING	PACKING INSTRUCT.	AUTHORIZATION

ADDITIONAL HANDLING INFORMATION

I hereby declare that the contents of the consignment are fully and accurately described above by proper shipping name and are classified, packed, marked, and labeled, and are in all respects in the proper condition for transport by air according to the National Regulations.

Name/Title

Place & Date

Signature

Figure 6-4 *Typical Shipping Form*

Hazardous materials must be labeled with primary and subsidiary hazard labels. The primary hazard will take precedence over the subsidiary, with the hazard class number being removed from the secondary hazard label. Aircraft cargo must be labeled with an Authorized Cargo Aircraft Only label. Labels should be placed near the proper shipping name and not on the bottom of a container where it is not easily seen.

Placards are used to display the proper shipping name and classification of a chemical. Placards typically contain the following information:

- Proper shipping name
- Material classification number at the bottom of placard
- Material identification number at the center of placard

Process technicians place placards on all chemicals leaving the physical boundaries of the plant except for infectious substances and combustible liquids in non-bulk packages. Cargo tanks carrying hazardous materials should be placarded on all visible sides. Placards are typically the responsibility of the shipper and are displayed in the diamond pattern. (See Figure 6-5.)

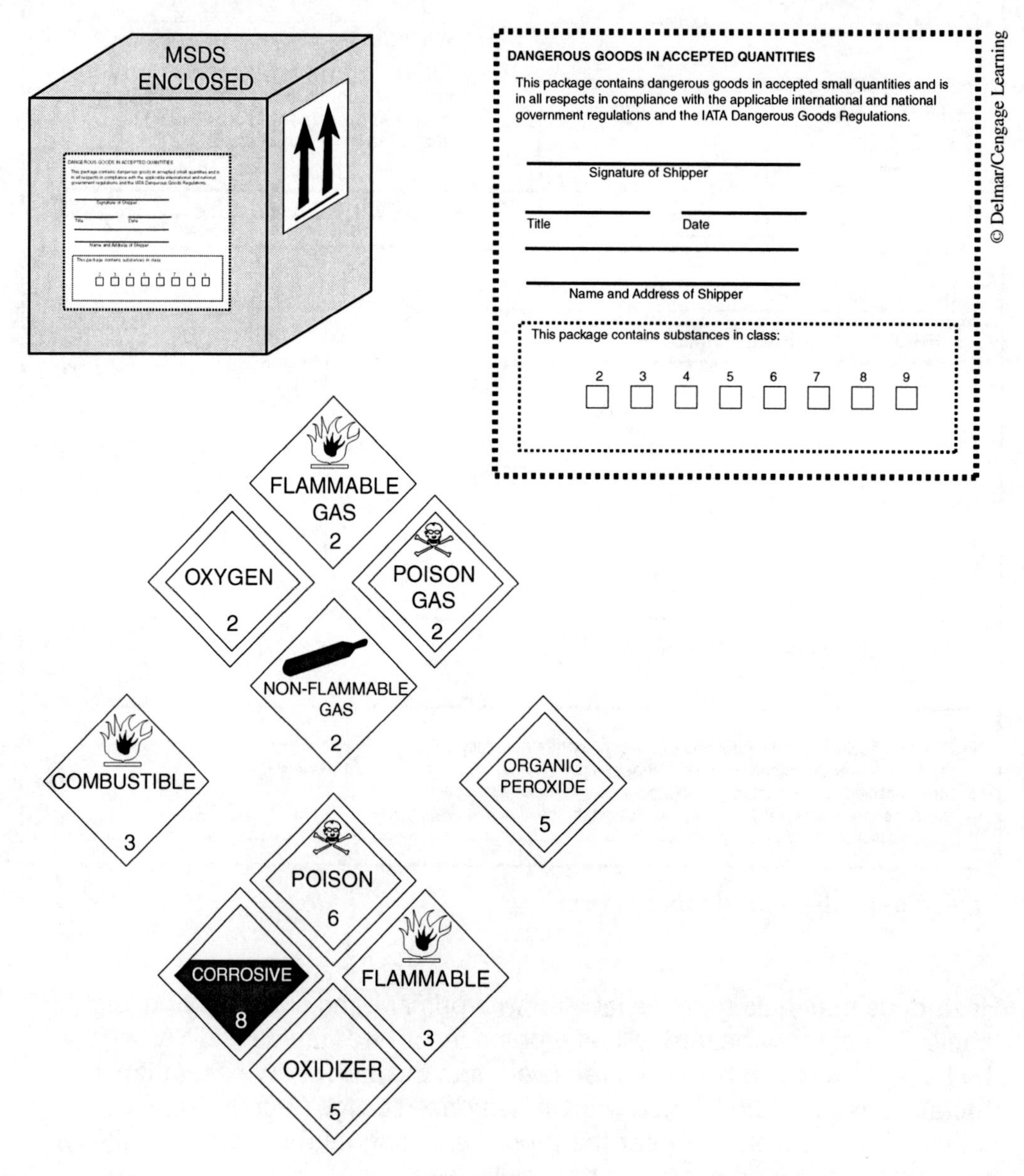

Figure 6-5 *Marking, Labeling, and Placarding*

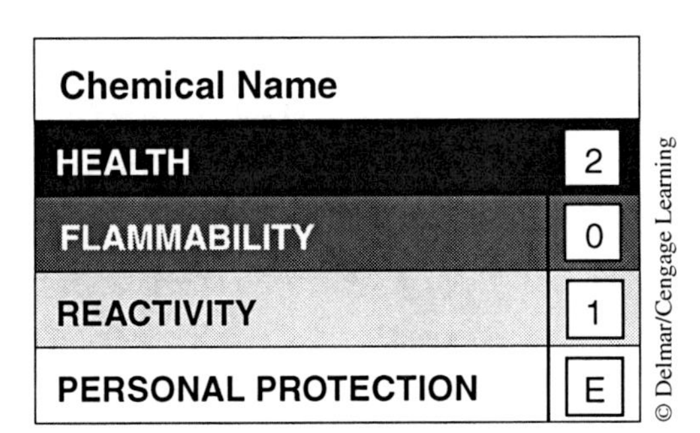

Figure 6-6
HMIS System

Hazardous Materials Identification System (HMIS)

HMIS is another labeling system frequently used by industrial manufacturers to identify the hazards associated with chemicals. The HMIS communicates its information to process technicians by listing five essential topics: the chemical name, health hazard, flammability hazard, reactivity hazard, and PPE requirements (as shown in Figure 6-6). Each topic block is color coded and receives a severity ranking from 1 to 4. The only exception to this rule falls under the PPE topic block, in which a letter that corresponds to an index is displayed.

National Fire Protection Association (NFPA)

A third hazard communication labeling system is the NFPA. This system utilizes the four small diamonds arranged so they form one large diamond. (See Figure 6-7.) Each small diamond is color coded and has its own severity rating from 1 to 4.The four color-coded topic blocks are: blue for health hazards, red for fire hazards, yellow for reactivity hazards, and white for specific hazards.

Any unmarked vessel that contains a chemical must be identified. Portable samples should also have the appropriate labels. The law requires that any vessel storing a chemical be properly marked. Some flexibility exists in this area so permanent signs may not be required. Some plants use electronic systems to identify the contents of a vessel by its number. The only exception to the labeling rule is if a small quantity of chemical is being moved to another location where it will be immediately used. This location must be within the same facility. When a chemical leaves a plant, it falls under the hazard communication program, must be properly labeled, and, in most cases, it must be accompanied by an MSDS.

Pipeline transfers fall under the HAZCOM guidelines, and the appropriate paperwork is electronically shipped during the transfer process. Chemicals transported in trucks fall under HAZCOM and the DOT. Truck drivers are

HEALTH HAZARD
- 4 - Deadly
- 3 - Extremely hazardous
- 2 - Hazardous
- 1 - Slightly hazardous
- 0 - Normal material

FIRE HAZARD (flash points)
- 4 - Below 73°F
- 3 - Below 100°F
- 2 - Below 200°F
- 1 - Above 200°F
- 0 - Will not burn

REACTIVITY HAZARD
- 4 - May detonate
- 3 - May detonate with heat or shock
- 2 - Violent chemical change
- 1 - Not stable if heated
- 0 - Stable

SPECIFIC HAZARD

Oxidizer	OXY
Acid	ACID
Alkali	ALK
Corrosive	COR
Use no water	W̶
Radiation Hazard	☢

NFPA DIAMOND

FIRE HAZARD 4
HEALTH HAZARD 3
REACTIVITY HAZARD 2
SPECIFIC HAZARD W̶

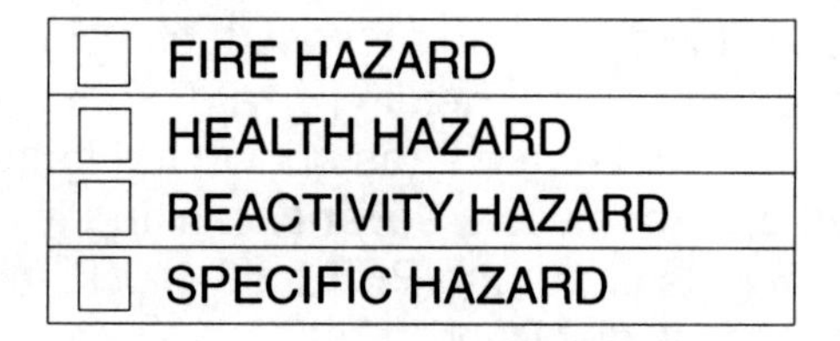

RATING EXPLANATION GUIDE

HEALTH		FLAMMABLE		REACTIVE	
Recommended Protection		**Susceptibility to Burning**		**Susceptibility to Energy Release**	
4	Full protection suit & breathing apparatus	4	Very flammable	4	May detonate under normal conditions
3	Full protection suit & breathing apparatus	3	Ignites under normal temperature conditions	3	May detonate with shock or heat
2	Breathing apparatus with full face mask	2	Ignites with moderate heating	2	Violent chemical change but does not detonate
1	Breathing apparatus may be worn	1	Ignites when preheated	1	Not stable if heated; use precautions
0	No precautions	0	Will not ignite	0	Normally stable

Figure 6-7 *NFPA System*

given an MSDS and must display the appropriate placards, labels, and signs.

When equipment is removed from service, dismantled, and placed on a truck, an MSDS for each chemical trace left in the system must accompany the truck driver. The law requires that the equipment be flushed, cleared, and tagged before it can be placed on the truck. If a hazardous chemical has been used in the system, a red tag stating DANGER—THIS EQUIPMENT HAS CONTAINED A HAZARDOUS MATERIAL must be attached. The plant's radioactive safety office is responsible for low-level radioactive material, and all questions should be deferred to this office.

Summary

The purpose of the Hazard Communication Standard (HCS) is to ensure that the hazards associated with the handling, transport, and storage of chemicals in a plant are evaluated and transmitted to affected personnel. According to the Hazard Communication Standard (HCS), chemical manufacturers and employers are required to analyze and assess the hazards associated with the chemical and develop written procedures for evaluating chemicals; document the hazard and develop material safety data sheets and warning labels; disseminate the information to affected individuals; and label, tag, and attach warning documentation to chemicals leaving the workplace.

Employers are responsible for the following:

- The development of a written hazard communication program
- A hazardous chemical inventory list and associated material safety data sheets
- Providing training to employees on the hazards of the chemicals they will be working with
- Providing instructions on how to read an MSDS
- Providing instructions on how to select and use personal protective equipment
- Using one of the three standard labeling systems: DOT, HMIS, or NFPA

Physical hazards associated with chemicals include the following:

- Combustible liquids
- Combustible gases
- Explosives
- Flammable gases
- Flammable liquids
- Organic peroxides
- Oxidizers
- Pyrophorics
- Unstables
- Water reactives

Health hazards associated with chemicals include the following:

- Carcinogens
- Mutagen
- Teratogens
- Reproductive toxins
- Asphyxiation
- Anesthetics
- Neurotoxic
- Allergic responses
- Irritants
- Sensitizers
- Corrosives
- Toxic
- Highly toxic
- Target organ effects

Review Questions

1. How do hazardous chemicals enter the human body?
2. Define the term *toxicology*.
3. Describe a hazardous chemical.
4. What is a health hazard?
5. In civil cases, marking the wrong chemical name on a container can result in ____________?
6. List the typical personal protective equipment (PPE) worn by a process technician.
7. What is a fundamental principle of the chemical hazard communication program?
8. What is a physical hazard?
9. What are chemical manufacturers required by the HAZCOM standard to do?
10. What are employers required by the HAZCOM standard to do?
11. Describe the components of a chemical hazard communication program.
12. Who develops the material safety data sheet (MSDS)?
13. Describe the Hazard Communication Standard.
14. Identify the physical properties and hazards associated with handling, storing, and transporting chemicals.
15. Describe the key elements of an MSDS.
16. Describe a hazardous chemical inventory list.
17. Explain the purpose of a written hazard communication program.
18. Identify methods used to protect process technicians from hazardous chemicals.
19. Describe toxicology and the terms associated with it.
20. Explain the following terms: HMIS, DOT, and NFPA.

Fire and Explosion

Objectives

After studying this chapter, the student will be able to:

- Describe the principles of fire prevention, protection, and control.
- Review the chemistry of fire.
- Describe the fire classification system.
- Evaluate the different types of fire extinguishers.
- Analyze the different fire stages.
- Identify the various types of firefighting equipment.
- Respond to a fire emergency.
- Describe flammable and explosive materials.
- Describe the 1947 Monsanto chemical plant explosion in Texas City, Texas.
- Evaluate the impact of the 1989 Phillips vapor release and explosion in Houston, Texas, and the 1990 ARCO explosion in Houston, Texas, on the development of the process safety management standard.

Key Terms

- **Absorbed heat, effects of**—increase in volume, temperature, change of state, chemical change, and electrical transfer.
- **Aqueous**—water-based solvent system. Examples include acids, alkalis, and detergents. Properties include low vapor pressure at ambient temperatures and low system toxicity.
- **Burnable materials**—have flashpoints above 100°F and are referred to as combustibles. This would include: kerosene, no. 6 fuel, brake fluid, and antifreeze.
- **Extinguishers, types of**—Halon, CO_2, dry chemical, water, and foam.
- **Fire hydrants**—located throughout the plant to provide water for fire trucks.
- **Fire monitors**—used to cool down exposed facilities and equipment and limit the spread of the fire. The fire monitor is equipped with an adjustable 500-gallon-per-minute nozzle that has three distinct spray patterns: fog, straight stream, and power cone.
- **Fire tetrahedron**—the four sides of the tetrahedron are represented by fuel, oxygen, chemical reaction, and a source of ignition. The removal of any one of these components will extinguish the fire.
- **Fire, elements of**—fuel, oxygen, and heat source.
- **Fire, reporting**—give (1) name and phone number, (2) fire location and extent of fire, and (3) products involved in fire.
- **Fire, stages of**—the Incipient Stage (no smoke or flame, little heat, combustion begins), Smoldering Stage (increased combustion, smoke, no visible flame), Flame Stage (flames become visible), and Heat Stage (excessive heat, flame, smoke, toxic gases).
- **Fires, classes of**—Class A (wood, paper), Class B (gas, LPG), Class C (electrical), and Class D (combustible metals).
- **Flammable substances**—chemicals that have flashpoints below 100°F (38°C). Examples include: benzene, toluene, gasoline, ethyl alcohol, and ethyl ether.
- **Gas**—typically described as a formless fluid that assumes the shape of the vessel it is in, exerting pressure equally in all directions at normal pressures, temperatures, and conditions. Examples include: ammonia, argon, helium, hydrogen, methane, nitrogen, and oxygen.
- **Heat transfer**—the three modes of heat transfer are radiant, conduction, and convection. Radiant heat transfers energy through space by means of electromagnetic waves. Conduction occurs as the molecules that make up a solid begin to vibrate and transfer energy across their matrices. Convection occurs as the molecules that make up a liquid or a gas speed up as heat energy is added and move naturally from hot areas to cold areas.
- **Heat**—a form of energy that comes in a variety of forms: sensible, latent heat of vaporization, latent heat of fusion, latent heat of condensation, and specific heat; measured in BTUs or calories.
- **Hose reel**—designed to take the hose to the fire. Most hose reels have an adjustable nozzle and approximately 100 feet of 1½-inch hose.

- **Ignition temperature**—the lowest temperature at which a substance will automatically ignite.
- **Radiation sickness**—the breakdown of cells in the body due to radiation exposure.
- **Solvent**—used to dissolve another material; includes aqueous and non-aqueous.
- **Temperature**—the hotness or coldness of a substance; measured in degrees Fahrenheit or Celsius.
- **Vapor**—a gas that is formed when a chemical vaporizes.

Fire, Explosion, and Detonation

There are many sources of ignition in a chemical plant or refinery. These mixtures combine readily with oxygen when heated. The lowest **temperature** at which an air-hydrocarbon mixture will explode or burn is called the **ignition temperature.** A variety of ignition sources can be found in the chemical processing industry. Examples include cutting, catalyst effect, furnaces, electrical sparks, static charges, and welding. New research conducted at the University of Utah suggests that a mixture of oxygen and hydrogen could ignite under a cascading catalytic effect if exposed to a fresh metallic surface inside a valve body being opened or closed.

Most chemicals manufactured in a refinery or chemical plant will burn and detonate under the right conditions. Detonations are a special type of explosion that are extremely destructive because they travel at high velocities, through plant pipe work and process equipment, and produce high internal pressures. A detonation wave for air-hydrocarbon mixtures expands at a high velocity: 4,000 to 8,000 feet per second. The pressures created in a detonation wave can be more destructive than normal explosions. Figure 7-1 shows how a typical detonation develops. Most chemical manufacturers have procedures designed to avoid mixtures of hydrocarbons and oxygen.

Chemical Explosions

A chemical explosion can be described as a rapidly building, self-contained fire that generates gases and **vapors** capable of exceeding the pressure ratings of the equipment it is in. Chemical explosions can be described five different ways: thermal explosions, combustion explosions, condensed phase explosions, physical explosions, and nuclear explosions. Other terms used to describe this type of phenomenon include detonation, fireball, overpressure damage, and unconfined vapor cloud explosions.

A thermal explosion is the result of two or more chemical compounds combining and reacting violently. This type of explosion has been identified as the most common in the chemical processing industry. Combustion

Figure 7-1
Evolution of a Detonation

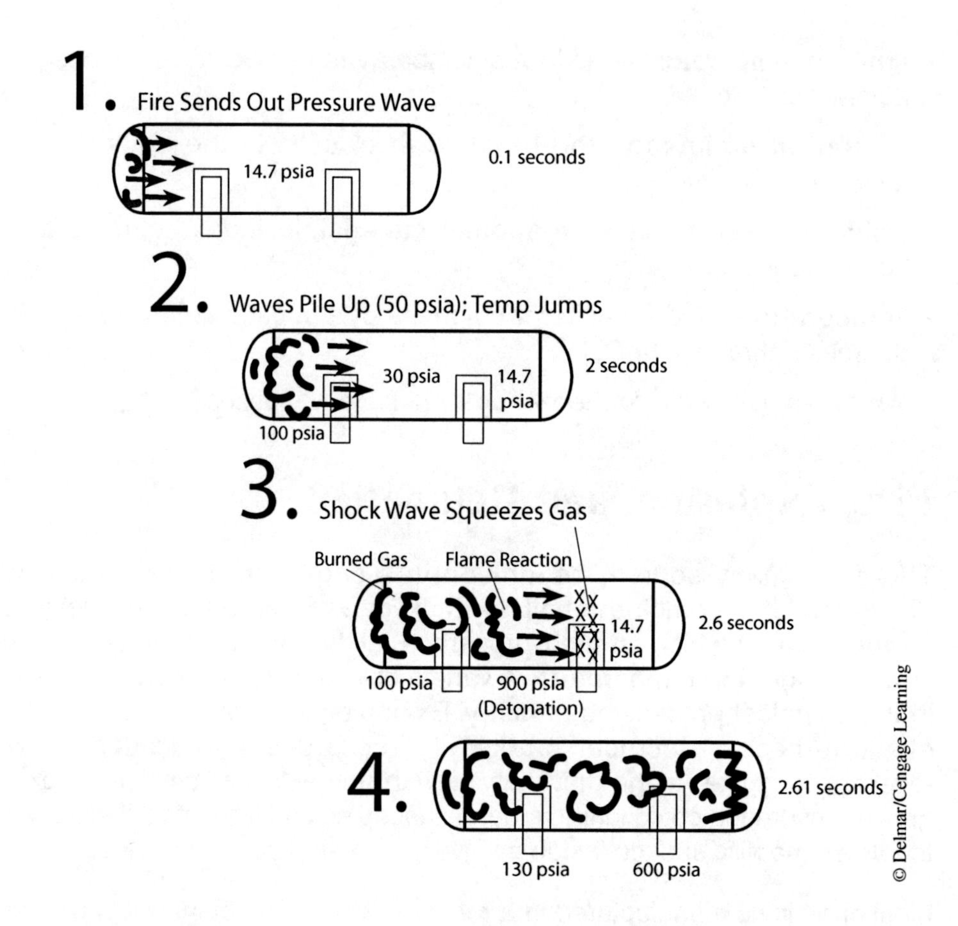

explosions include unconfined vapor cloud explosions (UVCE), **gas**, dust, mists, and backdrafts. Condensed phase explosions are rapid chemical component reactions that occur between liquids and solids. The term nuclear explosion is used to describe fission and fusion of matter. Nuclear power is used in electrical power generation facilities.

Texas City Fire and Explosion

On a cool spring morning, shortly after 8:00 a.m. on April 16, 1947, Texas City residents experienced one of the worst industrial accidents to occur in the United States. The *S.S. Grandcamp*, a French ship, was anchored in Galveston Bay, beside the new Monsanto plant. Around 8:00 a.m. the crew noticed that a small fire had started in the number four hold. More than likely a carelessly discarded cigarette was responsible for igniting the fire. Inside the hold, longshoremen had been loading a crystalline powder, ammonium nitrate. Ammonium nitrate was used as a fertilizer for crops or as an additive for TNT. Ammonium nitrate has about half the explosive capacity as TNT, but many of the hazards associated with this chemical were not generally known.

The crew responded to the emergency immediately but were unable to control the fire. The captain was alerted, and the crew was ordered to seal the hatch and start the ship's fire-smothering steam system. The steam system was designed to keep the cargo from being water damaged. This decision would end up being the worst thing possible the captain and crew could have done. Instead of smothering the fire, the steam created the following conditions within the hold of the ship:

- The increased pressure and temperature from the fire and steam were trapped inside the hold.
- This initiated thermal decomposition of the ammonium nitrate.
- Combustible gases were given off by the smoldering fertilizer.
- The hatch covers blew off as hot gases belched out of the hold.

At 8:45 a.m., the Texas City volunteer fire department started spraying water on the fire. (See Figure 7-2a.) The 27 members of the team fought furiously to control the fire, but the ship was so hot that the water vaporized upon contact with the ship (as shown in Figure 7-2b). The captain gave the order to abandon ship.

The fire gave off a brilliant orange color that predictably drew a crowd. Explosions and fires were common in the boomtown of Texas City. Many of the 18,000 residents had spent hours watching the various chemical company fires. Kids from several local high schools cut class and headed to the wharves to watch the fire. The fire also attracted many of the local chemical plant workers. The Republic Oil Company began shuttling people down to watch the fire. The plant manager, assistant manager, lab manager, and transportation manager were among this group. At the nearby Monsanto Chemical Company, employees began to gather to watch the fire from a large drafting room that afforded an excellent view of the burning ship. To most of the spectators, this was an exciting event. None of the chemical plant employees, bystanders, schoolchildren, ship's crew, or airplane pilots circling the area expected the nightmare that unfolded next. (See Figure 7-3.)

Figure 7-2a
Texas City Volunteer Firefighters Before Explosion

Figure 7-2b
Texas City—S.S. Grandcamp Shortly Before the Explosion

Figure 7-3
Texas City—S.S. Grandcamp at the Port of Houston

At 9:12 a.m., the *S.S. Grandcamp* unexpectedly and violently exploded. The force of the blast caused the following damage, some of which is shown in Figure 7-4:

- It killed 581 bystanders, the ship's crew, and volunteer firefighters.
- It knocked two airplanes out of the air, killing their pilots and injuring 3,500 other people.

Figure 7-4
Texas City Harbor After the Explosion

- It was heard over 150 miles away and sent a mushroom cloud up 2,000 feet.
- The blast's force launched the *S.S. Grandcamp's* 1.5 ton anchor two miles and sent a tidal wave that propelled vessels and debris 100 to 200 feet inland.
- $700 million worth of property and equipment were destroyed. Everything within 1,500 feet of the blast was totally destroyed.
- Burning fragments of the *Grandcamp* were scattered all over the area, setting fire to the Monsanto plant.
- Another ship, the *High Flyer*, was blown from its moorings, caught fire, and drifted into another ship, the *Wilson B. Keene*. (See Figure 7-5a.)
- Like the *S.S. Grandcamp,* the *High Flyer* was loaded with ammonium nitrate. At 1:11 a.m., the *High Flyer* exploded and sunk the *Wilson B. Keene.* Casualties were light because the area had been evacuated.

Approximately 880 tons of ammonium nitrate had exploded in the hold of the *S.S. Grandcamp*. (See Figure 7-6.) This amount can be contrasted with the 2.5 tons of ammonium nitrate that was used in the tragic 1995 Oklahoma City blast. The Texas City blast was 300 times more powerful than the Oklahoma City bomb. Nearly 60 years later, the residents of Texas City still remember running north toward the blue sky and away from the billowing black eruptions that rocked Galveston Bay. (In Figure 7-5b, the *S.S. Grandcamp's* propeller pin, ripped from the ship during the explosion, shows the magnitude of the damage caused by this disaster.) As the city was rebuilt from the ashes of the explosion, a new era of chemical awareness was ushered in.

Figure 7-5a *Texas City—The Wilson B. Keene Was Destroyed When the S.S. Grandcamp Exploded 600 Feet Away*

Figure 7-5b *Texas City—Propeller Pin from the S.S. Grandcamp*

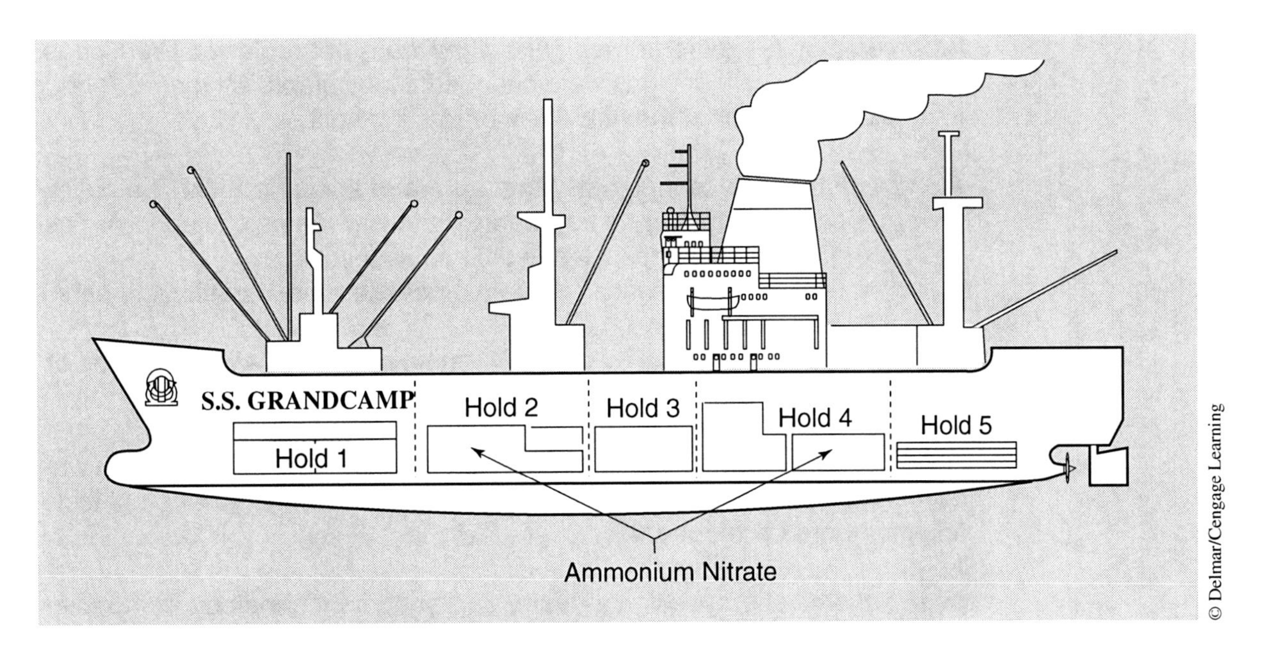

Figure 7-6 *S.S. Grandcamp*

It should be noted that many of the companies that were heavily damaged retained most of their employees and were back up and operational within two years. This spirit of cooperation and communication between industry and the community has lead directly to the development of Community Awareness and Emergency Response (CAER) programs and allowed Texas City to rebound economically.

As consequences of the Texas City explosion, wartime studies on the conditions that could destabilize ammonium nitrate were released, and emergency response procedures and communication between industry and community were improved. In addition, a new awareness of chemical hazards was ushered in, CAER was established, and a catastrophic model existed for future reference.

Houston's Phillips and ARCO Explosions

Working in the chemical processing industry is typically characterized as safe and responsible. Chemical manufacturers have for the most part established a safety-conscious atmosphere. In 1980, the chemical processing industry (CPI) as a whole began the process of restructuring. New and innovative technologies were brought into the plants. The rapid assimilation of this new technology was a difficult process for most facilities. In some ways it compromised the traditional safety systems as bigger and better toys were brought into the plants. Older workers had some difficulty in adjusting to this new age of computer automation.

Because of this new technology, many companies could produce much more with fewer people. Instead of laying off the old work force, most chemical manufacturers allowed their numbers to be reduced through attrition.

When employees retired or resigned, they were not replaced. Workloads, schedules, and shift designs were changed. Many plants left the old 8-hour rotating shift in favor of the new 12-hour rotating shift.

Along with this new technology came an era of specialization. The old refinery processes had clearly fragmented into new chemical technologies. The new structure was most clearly defined as follows:

- Refinery processes (19 main processes with variations of each)
- Gas processes
- Petrochemical processes (41 main processes with variations of each)

Research into synthetic fuel alternatives and petrochemical processes was accelerated during the late 1970s and early 1980s. Many new patent technologies were developed.

As the chemical processing industry upgraded its technology, companies scrambled to put in place adequate safety systems. The term *adequate* does not necessarily mean good. When asked to rank safety, quality, and production by their company's priorities, most technicians ranked them in this order:

1. Production!
2. Quality
3. Safety

Chemical manufacturers were attempting to keep safety near the top of their companies' priority list; however, the necessary synergy was not there to put it at the top. Because of this a number of work practices existed in the industry, which set the stage for catastrophic events. Some of these work practices included the following:

- Industry use of unskilled, unqualified contract labor to perform jobs in the plant
- Industry chemical hazards were not regulated
- Prevention of catastrophic releases was not regulated

Phillips Chemical Company in Houston produced high-density polyethylene for use in a number of plastic-related specialties. During the early 1980s, many seasoned employees had retired from Phillips. The knowledge base and experience levels dropped significantly, and a mechanism did not exist for capturing the technical expertise of this group and disseminating it into the existing workforce. Few of the people who retired were replaced by the company, as work assignments were shifted and redistributed to the remaining employees. Fewer people were doing much more.

In August of 1989, two workers were accidentally killed inside the plant. A sense of malaise remained over the Phillips complex as the company attempted to uncover what they needed to do to prevent this from happening again. Unfortunately, the system that was in place at Phillips and at many other companies was responsible for the high-risk environment. On October 22, 1989, company employees at Phillips were asked to remove a polyethylene blockage to a loop reactor. The employees refused the assignment because hydrocarbon vapors were heavy in the area, and they believed a valve was leaking through. Phillips employees were willing to perform the job if the company would shut down the operation and secure the area. The company made a decision to bring in contract workers and have them remove the polyethylene restriction while the unit was kept on line.

The difficulty of the operation became apparent as the contract crew attempted to complete their assignment. After working on the assignment for many hours, the decision was made to bypass one of the unit's safety features in order to speed up the work. A critical valve in the procedure had two high-pressure hoses connected to it in order to help position the valve. Unknown to the contract employees, the high pressure hoses were installed backwards, and the valve could not be closed without switching the hoses. When the valve was opened, 85,000 pounds of high-pressure gases dumped from the reactor system. A mixture of ethylene, hexane, hydrogen, and isobutane dumped to the concrete mat. Within a few minutes, a deadly vapor cloud engulfed the plant.

Contract and maintenance workers fled the site as the vapor cloud grew. The Phillips operational staff responded immediately to the emergency. As nonessential personnel fled, they remained in their area frantically attempting to stop the release. When the explosion occurred, 23 members of the operational staff were killed, and the entire plant was destroyed. The explosion registered 3.5 on the Richter scale as debris was scattered over a six-mile radius, and damage estimates ranged from $750 to $800 million. OSHA cited Phillips for 566 willful safety violations.

Eight months later, on July 5, 1990, the ARCO chemical plant in Sheldon, Texas, was having trouble with a safety system on a wastewater tank. The problem had been going on for several months and seemed to have finally reached its limits. When the tank exploded, it leveled the immediate area and killed 17 men. It was later discovered that two gauges that could have prevented the accident were not working. One was broken, the other was blocked out.

Most process technicians view OSHA as a reactive rather than proactive organization. Government officials usually show up after someone has been killed or a complaint has been filed. Over a ten-month period, 42 men had been killed within a six-mile radius. National attention had been focused on the Houston area. OSHA and the EPA were determined to investigate the incidents, identify the problems, and put a plan in place to resolve them. The final conclusion of the EPA and OSHA was to put in place a standard designed to prevent catastrophic releases of toxic, reactive, or flammable chemicals.

Outcomes

The Phillips and Arco incidents led directly to the development of the process safety management (PSM) standard. The PSM standard poured new life into existing programs such as Hazard Communication (HAZCOM) and Hazardous Waste Operations and Emergency Response (HAZWOPER). OSHA and the EPA called into question industries' use of contract employees. One of the largest influences of the PSM standard was the creation of industrial partnerships between education, government, and industry. This alliance formed the foundation upon which the community college and university process technology program was built.

Polymers and Fire

Understanding how polymers react when exposed to fire is essential in working as a process technician. Polymers are described as natural or synthetic compounds linked together in large chains. These molecules are often of high molecular weight consisting of millions of sequential linked units. Human-made polymers include polyvinyl chloride (PVC), polyethylene, and polypropylene. PVC has a wide range of applications, including automobile upholstery, apparel fabric, house siding, and piping. Polyethylene is used in trash bags, toys, packaging, and plastic bottles. Polypropylene is used in plastic washing machine parts, automobiles, and plastic baby diapers. These products have thousands of practical applications and are found in virtually every home in the United States. Naturally occurring polymers include cotton, leather, natural rubber, paper, wood, and wool. All polymers are composed of the element carbon. A single carbon atom bonded to 4 hydrogen atoms is called methane, 2 carbon atoms with its complement of hydrogen atoms is called ethane, and 10 carbon atoms is called decane. When 22 carbon atoms combine with 42 hydrogen atoms, a paraffin wax is produced. When carbon atoms bond to other carbon atoms in a long chain, the result is a polymer. Polyethylene contains hydrocarbon chains with over 100,000 carbon atoms.

It is important to know what happens when polymers burn. The primary hazards are temperature, oxygen levels, and toxic fumes. Temperatures in an enclosed room will reach over 1,500°F, hot enough to melt brass. Within minutes of the start of the fire, toxic fumes fill the room and carbon monoxide levels jump from 0.1% to 1.4%, high enough to cause unconsciousness and death. Smoke can contain levels of carbon monoxide above 7% in a polymer fire. Oxygen levels in a polymer fire can quickly drop to below 10%. Polymer fires react differently depending on the arrangement of atoms in the polymer. Chlorine and PVC produce hydrogen chloride (HCl). Nitrogen and wool produce hydrogen cyanide (HCN). The best protection in this type of situation is a self-contained breathing apparatus (SCBA). Process technicians should not hesitate to don this emergency equipment during a fire.

Flammable, Explosive, and Radioactive Hazards

Working in the chemical processing industry goes hand in hand with working with flammable, explosive, and radioactive hazards. Most of these hazards are carefully controlled, and new technicians are exposed early to all of the precautions and information surrounding the handling, storage, operations, and sampling of these materials. Explosive and flammable materials are fire hazards. These materials are also considered to be toxic hazards, and precautions are taken to limit their poisonous effects on people and the environment. The hazards associated with radioactive materials are linked to the emission of harmful radiation that tends to break down human tissue. This material is also considered to be highly toxic if accidentally released into the environment.

Flammable and Explosive Materials

Given the right condition, any flammable or combustible mixture can be explosive. This process relates to the increase of surface area of the flammable material in relationship to the correct percentage of air. Many fuels are atomized to increase their burning efficiency. Fire is a chemical process that requires fuel, **heat**, and oxygen in the correct proportions to chemically react together. In Chapter 5, Figure 5-4 actually illustrates the key elements of the Fire Tetrahedron that includes in addition to fuel, heat and oxygen, the term chemical reaction. This is typically associated with spontaneous combustion that utilizes a slowly developing exothermic reaction to generate combustion. The combination of these is illustrated in the fire triangle. The chemical reaction in a fire can be stopped by removing one of the three required components. Each flammable substance has a unique set of characteristics that separate it from the others. The term flashpoint is used to describe the lowest temperature at which a flammable liquid will produce a rich enough vapor concentration to ignite in the presence of an ignition source. Chemicals with flashpoints below 100°F (38°C) are called **flammable substances**. Examples of these are benzene, toluene, gasoline, ethyl alcohol, and ethyl ether. **Burnable materials** that have flashpoints above 100°F are referred to as combustibles. This would include kerosene, no. 6 fuel, brake fluid, antifreeze, and many others. Table 7-1 shows the flashpoint for a variety of chemicals.

Radiation Hazards

The most common form of radiation exposure in industry comes from uranium oxides used in nuclear fuels, iridium-191, and other gamma rays. **Radiation sickness** occurs when a person is exposed to harmful radiation, and the individual cells in the body begin to break down. Radioactive materials are highly toxic and may cause genetic damage or cancer in later

Table 7-1
Flashpoints of Common Chemicals

Chemical	Flashpoint
Ethyl ether	−49°F
Gasoline	−45°F
Benzene	12°F
Toluene	40°F
Methyl alcohol	52°F
Dioxane	54°F
Ethyl alcohol	55°F
Amyl acetate	77°F
Hydrazine	100°F
Kerosene	110°–162°F
Fuel oil no. 2	126°–204°F
Fuel oil no. 6	150°–270°F
Ethylene glycol	232°F

years. Metallic uranium and plutonium are extremely toxic poisons and will cause serious damage if they are inhaled or swallowed.

A process technician can be exposed to radioactive hazards by inhaling or ingesting airborne radioactive particles, or being exposed to a radiation source. Exposure can be internal or external depending on the situation. Distance, time, and shielding are three factors that need to be considered when working with radioactive materials. The term *distance* refers to the drop-off of radiation, because radiation drops off with the square of the distance of one foot from the source. If 20 units can be measured at a distance of one foot from the source, at two feet the reading will decrease by a factor of 4, or read 5 units. Radiation is measured in units per hour. Time is an important factor in calculating exposure. Radiation shielding is made of lead. Lead is an element that can absorb radiation. Industry uses three devices to monitor exposure: a film badge, a whole-body counter, and a dosimeter. The dosimeter has a visible scale that can be read; however, the film badge must be developed. The dosimeter and the film badge measure external radiation. The whole-body counter is similar to a Geiger counter and measures internal and external exposure. Extensive training is provided in areas where a process technician could be exposed to radioactive hazards.

Fundamentals of Fire Prevention, Protection, and Control

The fundamentals of fire prevention, protection, and control include a variety of complex and technical standards. Most of these standards can be found in *The Handbook of Industrial Loss Prevention* by Factory Mutual Engineering

Corporation and *The Handbook of Fire Protection* of the National Fire Protection Association. A summary of these standards includes the following:

- Analyze the physical layout of the plant or facility and surrounding structures.
- Ensure that firefighting equipment is available and strategically placed.
- Design engineering systems that allow for the safe handling, storage, and shipment of flammable and combustible materials.
- Establish written procedures and emergency response plans.
- Develop structured safety training programs.
- Design and implement early fire detection systems.
- Prevent the outbreak and spread of fires.

The Chemistry of Fire

Fire is a chemical reaction between oxygen, fuel, and heat. Fire may be defined as rapid oxidation with the evolution of light and heat. In the fire triangle, *fuel* must be present in vapor form, in the flammable range, and there must be *oxygen* or an oxidizing agent. In addition, *heat* must be present in the form of high environmental temperatures, hot surfaces, open flames, friction, and so on. If any one of these three elements is removed the fire will go out. Figure 7-7 shows the three sides of a fire triangle.

One of the greatest dangers to personnel and facilities in an industrial environment is from fire and explosions. Most potential hazards are engineered out; however, the nature of the work requires routine handling of highly flammable materials. Therefore, it is important to be constantly aware of the following:

- Ingredients of a fire
- Fire prevention and control measures
- Properties of flammable and combustible materials

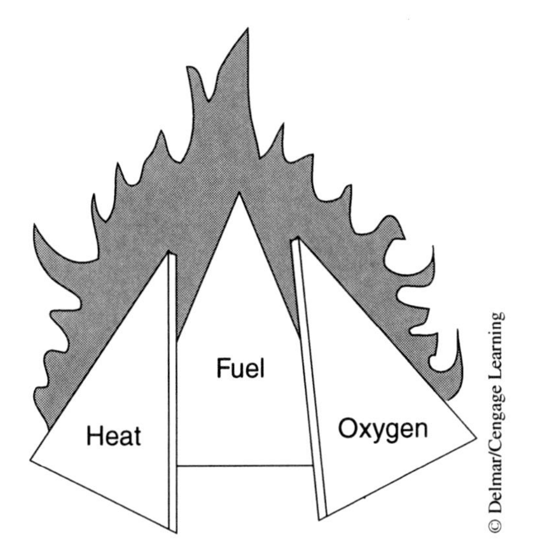

Figure 7-7
The Fire Triangle

When all of the elements of the fire triangle combine and combustion occurs, the flame both radiates and convects heat to the surrounding area. If flame impingement occurs on a metal vessel or tank, conduction carries heat from the flame to the contents of the vessel or tank. Radiant heat, conduction, and convection contribute equally after combustion takes place. The heat source in the fire triangle can be delivered by any of these three **heat transfer** methods. **The effects of absorbed heat** include an increase in volume, temperature, change of state, chemical change, and electrical transfer.

Radiant heat provides the energy feedback from the flame to the burning material. This affects the rate of burning. Flame spread during a fire occurs primarily through radiant heat transfer. Radiant heat may be so intense that a process technician cannot approach the flame. When technicians locate and report a fire in the facility, a number of systematic steps are required. **Correct steps in reporting a fire**: give (1) name and phone number, (2) fire location and extent of fire, and (3) products involved in fire.

The Hazard of Air

The atmosphere that surrounds the earth is composed of 78.1% nitrogen, 20.9% oxygen, and 0.9% argon, and the rest is carbon dioxide and other gases. The pressure exerted by this air is equal to 14.7 psi at sea level.

Oxygen in the atmosphere is consumed by most living things on the earth. The hazards associated with air are limited when compared to the benefits; however, a good understanding of these hazards is important. Some of these hazards include the following:

- Combustion—Oxygen is essential in the fire triangle.
- Compression—High pressure discharges of air can be dangerous.
- Absorption—Oxygen can be absorbed and then released at the wrong time from a mixture.
- Concentrating oxygen—Propane and butane are lighter hydrocarbons that pick up more oxygen than heavier hydrocarbons do.
- Separation—Oxygen can separate from a mixture and does not easily return back into a solution, creating an oxygen-rich vapor above the liquid phase.
- Mixing—Oxygen mixes with hydrocarbons to create a hazardous condition. (In a furnace, increased O_2 can cause more fuel to be used; in a vessel with a small amount of hydrocarbon present, the air creates an explosive condition.)
- Oxidation—Oxygen is aggressive and corrosive. (Example: OxyClean®)

Fire Stages

The development of a fire can be traced through four distinct stages:

1. Incipient Stage—no smoke or flame, little heat, combustion begins
2. Smoldering Stage—increased combustion, smoke, no visible flame
3. Flame Stage—flames become visible
4. Heat Stage—excessive heat, flame, smoke, toxic gases

Flashpoint, Flammable Limits, and Ignition Temperature

As discussed earlier, the term flashpoint is used to describe *the lowest temperature at which a flammable liquid will produce a rich enough vapor concentration to ignite in the presence of an ignition source*. It is well known that hydrocarbons must be combined in the vapor phase with the correct proportions of air in order to burn. For example, propane-air mixture will burn when propane is between 2.1% and 9.5%. Propane and air mixtures above 9.5% are too rich to burn, and those that are below 2.1% are too lean. The concentration of hydrocarbons in the air has a direct relationship to the flammability limits as shown in Table 7-2. In an enclosed system it is possible for a combustible mixture to form if air is accidentally allowed to enter. A combustible mixture will not burn until it reaches its unique ignition temperature. The ignition temperature is described as the lowest temperature at which a substance will automatically ignite. This process varies for each hydrocarbon and does not require a spark or flame to occur. If the mixture is within its flammability limits and has reached its ignition temperature, it will ignite.

Table 7-2
Flammability Limits and Ignition Temperatures

Hydrocarbon	Flammability Limits % in Air		Ignition Temperature °F
	Min.	Max	
Methane	5	15	1,200°F
Ethane	2.9	13	970°F
Propane	2.1	9.5	915°F
N-Butane	1.4	8.4	850°F
Gasoline	1.1	7.5	900°F
Kerosene	N/A	N/A	565°F
Crude Oil	N/A	N/A	600°F

*Not applicable (N/A)

Fire Classification System

The fire classification system is designed to simplify the selection of fire-fighting techniques and equipment:

- Class A fires involve the burning of combustible materials such as wood, paper, plastic, cloth fibers, and rubber.
- Class B fires involve combustible and flammable gases and liquids and grease.
- Class C fires are categorized as electrical fires. These involve energized equipment and Class A, B, and D materials that are located near the fire.
- Class D fires cover combustible metals such as aluminum, magnesium, potassium, sodium, titanium, and zirconium.

Types of Fire Extinguishers

Portable, hand-held fire extinguishers are carefully placed around the process unit. These devices are a fundamental part of the fire prevention, protection, and control standard. Because most fires require split-second reactions, the portable fire extinguisher is an excellent first-line defense. Portable fire extinguishers can be used to stop a fire before it gets out of control, provide an escape route through a large fire, and contain a fire until help or the fire team arrives.

Fire extinguishers come in a variety of shapes and designs. The selection and use of a fire extinguisher depends on the type of fire being fought. Process technicians need to be familiar with the classification of fire system in order to select the proper extinguisher in an emergency situation.

The *carbon dioxide extinguisher* (*CO_2 extinguisher*) is composed of a cylinder filled with compressed carbon dioxide. (See Figure 7-8.) A dip tube penetrates the top of the cylinder and continues to the bottom. A carrying handle and operating lever is located at the top of the extinguisher. A discharge hose is attached to the dip tube on the cylinder side and an oversized discharge horn at the exit. CO_2 extinguishers are effective on Class B and C fires because they displace oxygen. Process technicians prefer CO_2 extinguishers because they are inexpensive and do not leave a chemical residue. Limitations of this type of extinguisher include poor performance in windy conditions, asphyxiation in enclosed spaces, and frostbite if the discharge touches human tissue. A fully charged CO_2 extinguisher weighs about 45 pounds and has an effective range of about 10 feet and a 20-second lifetime.

Dry chemical fire extinguishers are composed of a cylinder, dip tube, pressure gauge, hose and nozzle, BC or ABC dry chemical agent, carrying handle, operating lever, locking pin, and a compressed nitrogen or carbon dioxide

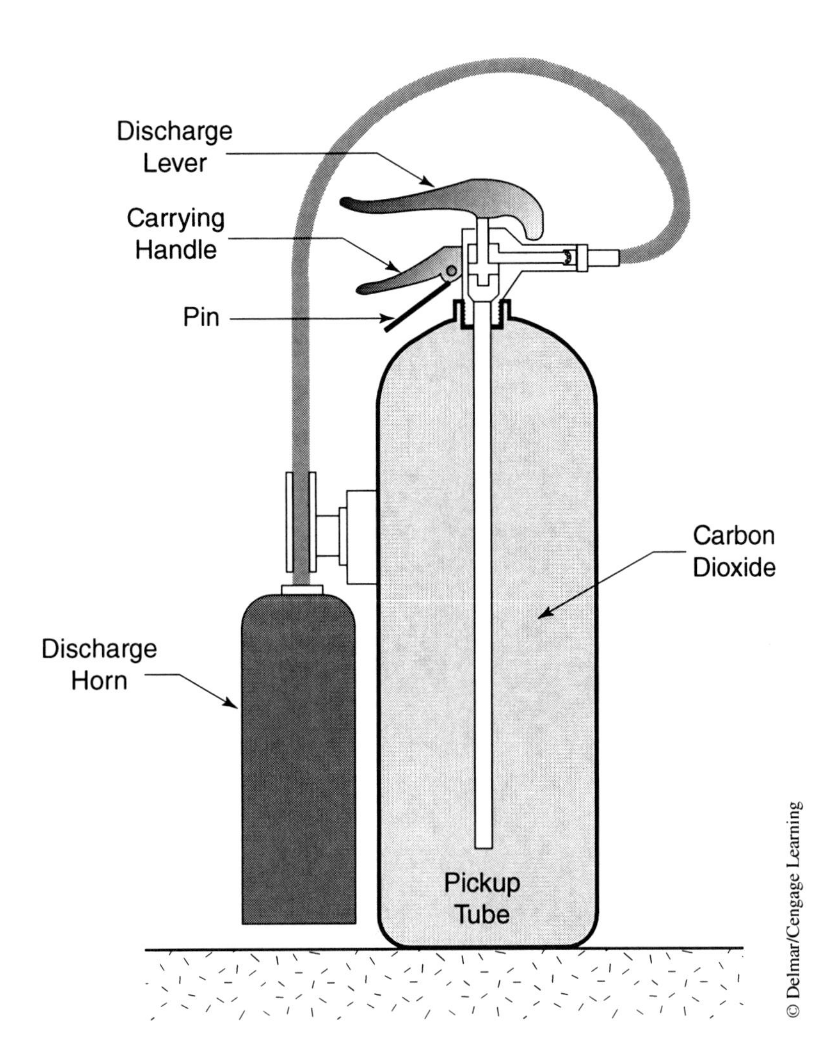

Figure 7-8
Carbon Dioxide Fire Extinguisher

cartridge. (See Figure 7-9.) The extinguisher needs to be operated in the upright position in order to pick up the extinguishing agent from the bottom of the dip tube. The dry chemical fire extinguisher can be used on Class A, B, or C fires. This type of portable extinguisher leaves a residue that can damage computer equipment and sensitive electronics. Using this device in a confined space is dusty and messy. When charging the cylinder with an 800 psi CO_2 cartridge, the extinguisher should be pointed away from the upper body and head. A fully charged dry chemical cylinder has about 200 pounds of pressure. It has an effective range of 30 feet and weighs about 40 pounds. It is approved for confined space work and available in 150- or 250-pound, wheeled units.

Foam fire extinguishers are used to control flammable liquid fires. The foam forms an effective barrier between the flammable liquid and the oxygen needed for combustion. Many of the large open top tanks used to store hydrocarbons in industry are equipped with foam depression systems. Portable, hand-held, foam fire extinguishers are equipped with a cylinder that has two sections. One section contains a water and aluminum sulfate mixture, and the other section contains water, foam stabilizer, and sodium bicarbonate. (See Figure 7-10.) In order to use this type of fire extinguisher,

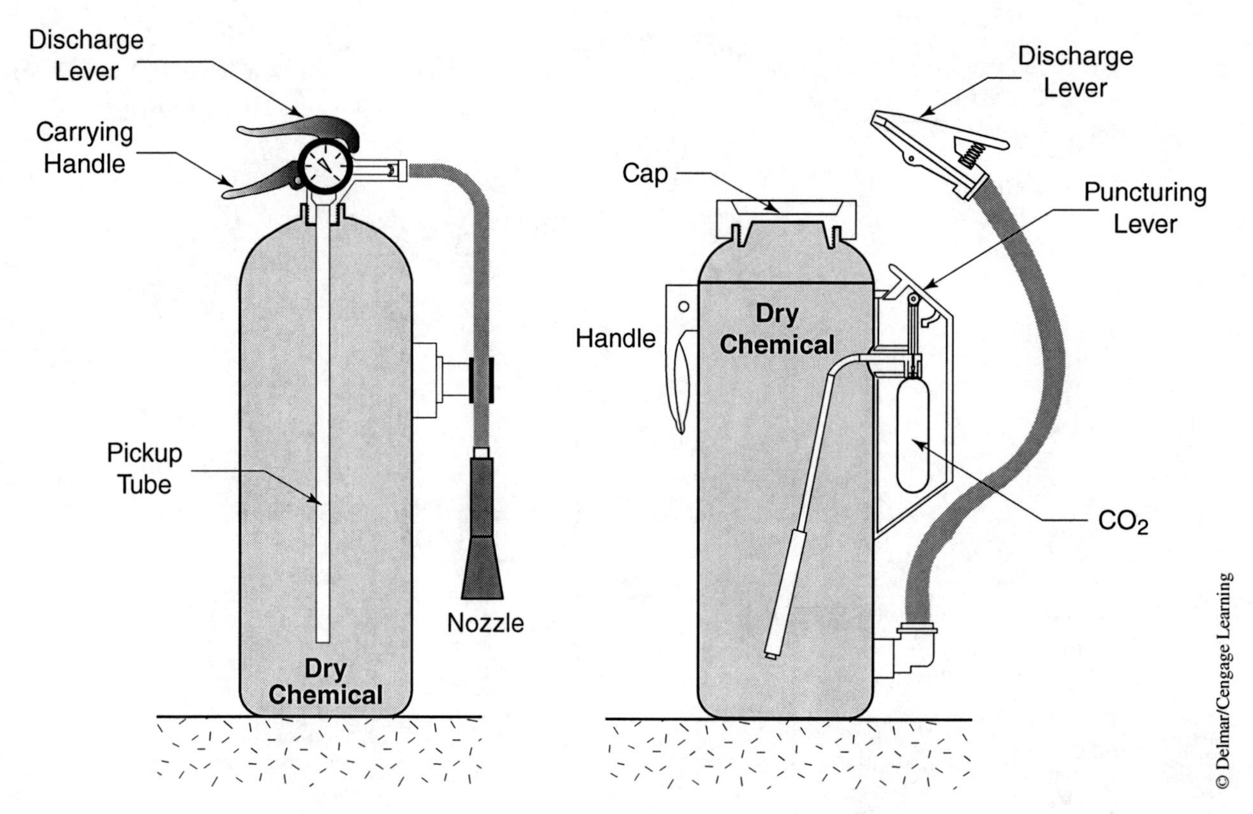

Figure 7-9 *Dry Chemical Fire Extinguishers*

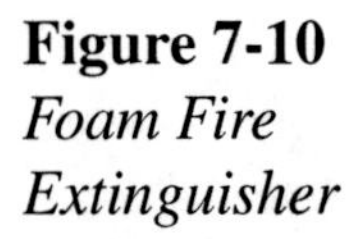

Figure 7-10
Foam Fire Extinguisher

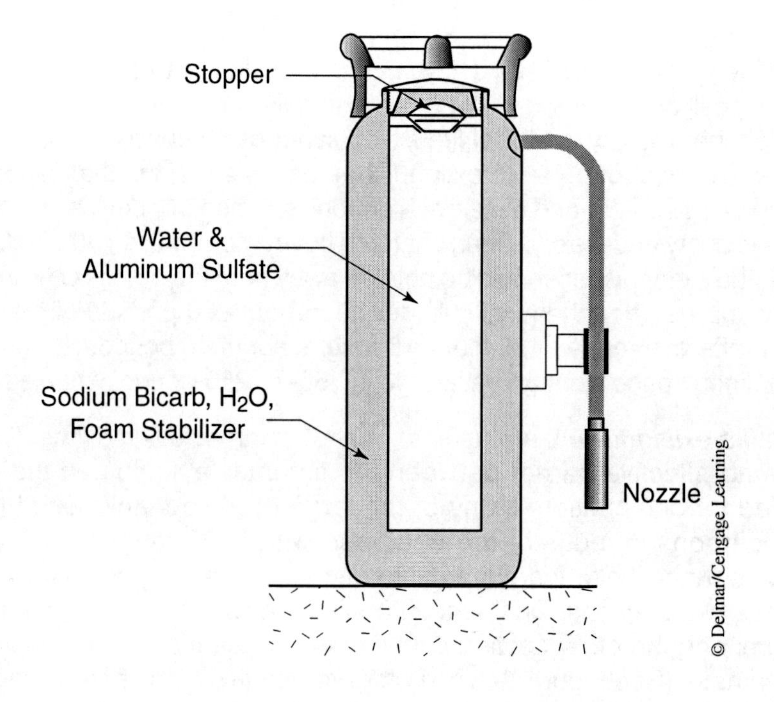

the device needs to be inverted. This allows a lead stopper to fall out and the two chambers to combine. The combination of these two chambers forms a compressed foam. Foam extinguishers are effective on Class A and B fires.

Halon fire extinguishers are composed of a cylinder filled with high-pressure halon gas, dip tube, hose, carrying handle and operating lever, pressure gauge, and locking pin. (See Figure 7-11.) This type of extinguisher is considered to be superior to a dry chemical extinguisher because it does not leave a corrosive residue on sensitive electronics or computers. Halon extinguishers are designed for use on Class A, B, and C fires. They have an effective range of 20 feet and a lifetime of 30 seconds. Halon extinguishers should never be used in an enclosed space because the halon gas and fire produce chemical vapors that can damage the respiratory tract.

Water fire extinguishers are composed of a cylinder, dip tube, pressure gauge, carrying handle, locking pin, operating lever, and overfill tube. (See Figure 7-12.) Portable, hand-held water-filled fire extinguishers are designed for use on Class A fires only.

- **Fire hydrants** are located throughout the plant to provide water for fire trucks. Water pressure to the fire water header is maintained by a series of pumps and back-up pumps. **Hose reels** are also connected to the fire water header and are

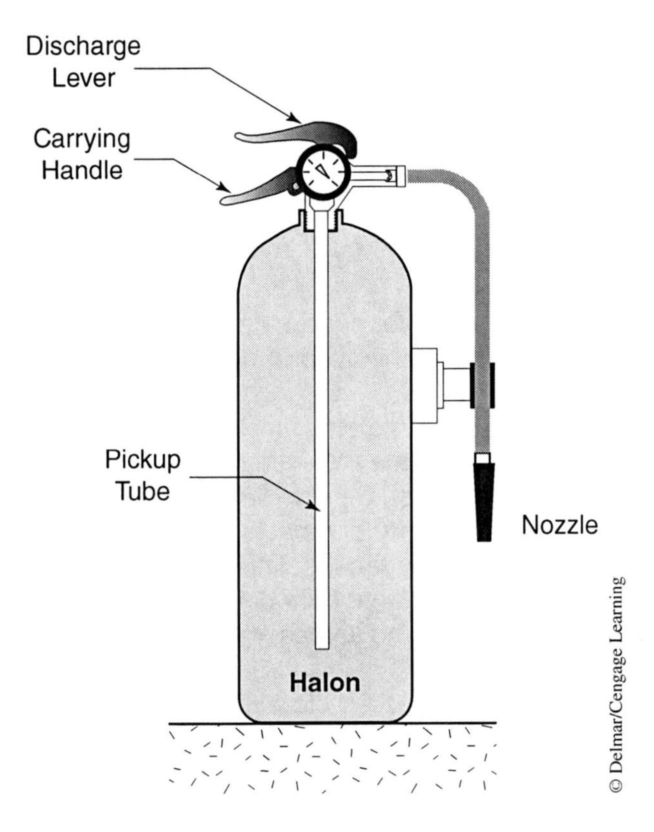

Figure 7-11
Halon Fire Extinguisher

Figure 7-12
Water Fire Extinguisher

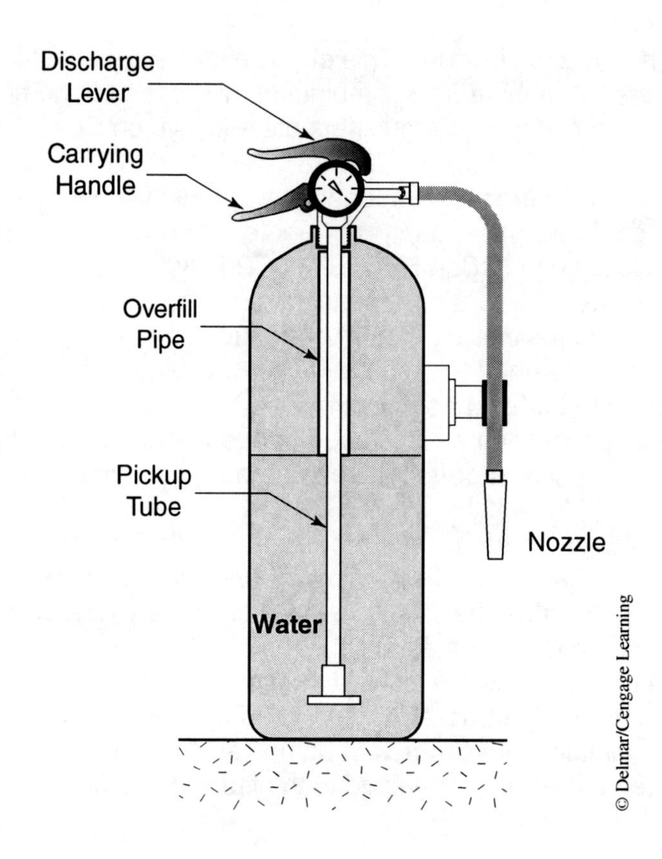

designed to take the hose to the fire. Most hose reels have an adjustable nozzle and approximately 100 feet of 1½-inch hose. Technicians are given specific training in operating this equipment.

Fire Extinguisher Use

In the proper hands, a portable, hand-held fire extinguisher can be an effective tool. The basic rules involved in using a fire extinguisher are as follows:

- Identify the type of material involved in the fire.
- Select the appropriate fire extinguisher.
- Test the extinguisher. Severe burns could result if you get to the fire and find an empty or defective extinguisher in your hand.
- Approach upwind of the fire. This will keep the heat, smoke, and extinguishing agent from blowing back on you.
- Aim at the base of the flames and use a sweeping motion to extinguish the fire.
- Back away from the fire before the extinguisher is emptied. Never turn your back on a fire.

Fighting Fires

Some of the safety equipment used to prevent the spread of fire include **fire monitors**, fire extinguishers, fire hose stations, sprinkler systems, CO_2, and Halotron systems.

Liquid fires depend on a gas and air mixture above the product. The liquid itself does not burn, but the vapor-rich air above the liquid ignites and burns. Heat increases the vaporization process of the product by stimulating the activity of the molecules. The higher the temperature, the higher the flammable vapor concentrations will be above the liquid. Most liquid fires are pool type. Pool-type configurations can be found in a tank, diked area, unconfined spill (river, lake, and land), or leaking mechanism. In contrast to a gas fire, a liquid fire's burning rate is dictated by how fast the flame can transfer heat back to the liquid.

In solid fuel fires, heat is fed back to the solid fuel and a complex gasification process takes place. The individual characteristics of the solid fuel complicates heat release estimates. Solid fuels tend to thermally decompose rather than evaporate. The heat of gasification for solids includes the heating of the material, melting, and the heat of decomposition. Firefighters use the same method for calculating pool fire burning rates on solid fuel fires.

As mentioned previously, solid and liquid fires must burn in the gas phase. Liquid and solid fires are limited in burn rate by heat transfer to the source. Gas fires are usually jet flames. Jet flames usually result when gas escapes from its special containment. Gas release rates and heat release rates are determined by the chemical characteristics of the gas. The only difference between liquid, solid, and gas fires is the *method and rate of fuel gas evolution*.

Chemical explosions take place when high pressure gas is discharged from the generation of chemical reactions. There are five basic types of chemical explosions:

- Combustion explosions—includes dust, gas, mists, and backdrafts (An Unconfined Vapor Cloud *Explosion* [UVCE] is included in this group.)
- Thermal explosions—two or more chemical compounds combine to rapidly react and explode; common to petrochemical and refinery operations
- Condensed phase explosions—rapid chemical component reactions between solids or liquids; associated with high explosives
- Nuclear explosions—fission and fusion of matter
- BLEVE—a physical explosion involving a *boiling liquid expanding vapor explosion, involving the release and ignition of an expanding vapor*

When a fire starts in a leak-tight room, substantial pressures are created by expanding gases. Fire growth and relative tightness of the room determine total pressures. Heating and expansions are normally linked. Usually this maximum pressure is never achieved because excessive pressure escapes easily from normal structures. Compartment fire pressures are closely linked to buoyancy forces and not expansion of hot gases.

After a fire has started, heat is fed back to the fuel, and the fire begins to spread. As a smoke plume forms, thermal radiation continues to heat the compartment. The warm smoke plume entrains cool air, which increases upward flow rates. As the plume strikes the ceiling, it forms a ceiling jet. The ceiling jet is distributed equally across the ceiling and flows until it contacts a wall. At this point the room begins to fill with smoke from the ceiling down. Total plume gas flow depends on fire size and height:

- Zone 1 ceiling—hot uniform
- Zone 2 floor—cool uniform
- Zone 3—gas temperature uniform from 2.1 meters to 1 meter

As the smoke descends, it will select the first available release or vent to the outside (door, window, and so on). Flow rate depends on the smoke layer size, smoke temperature, and vent size. It should be noted that ambient pressure increases from the top of the room to the floor. The weight of the air increases with the column depth similar to the hydrostatic pressure of the water in an elevated tank. The higher the column of water, the higher the pressure at the base of the tank.

Fire growth is characterized by five stages:

1. Ignition—Fire starts.
2. Growth—Fire size increases by using up initial combustible material.
   - Radiation stage: vigorous burning.
   - Enclosure stage: gas layer at ceiling, other heated objects reradiate substantial heat back toward the fuel surface.
   - Ceiling stage: flames reach ceiling, and heat is reradiated back.
3. Flashover—Temperature is 500°C to 600°C; the entire room is on fire.
4. Fully developed fire—There is a steady burning of combustibles.
5. Decay—Fuel begins to be consumed; temperature declines consistently.

The fire plume is located directly above the fire. As the fire burns, hot gases are released into the atmosphere. As these gases rise they entrain cooler air, which speeds upward velocity. Fire plumes constantly change during the life of the fire in size, temperature, and upward flow rate. The type of fire also determines the characteristics of the plume: turbulent, hydrocarbon, wood, and so on.

Fire growth and shape initially depend on the materials that made up the fire during the ignition stage. At the start of the growth stage, the fire begins to burn vigorously. At this point combustible materials in the room are exposed to excessive heat. Low density, thin, vertical materials will be consumed easily. If the fuel in the fire is a flammable gas or liquid, then the growth rate will be enhanced. Sustained temperature increases or decreases will come from the physical properties of the combustible material.

A liquid hydrocarbon fire is characterized by high temperatures and emissivity rates. Because this type of fire has a higher heat release rate than a typical wood fire, its plume is taller and more turbulent, and the flames reach much higher. Hydrocarbon fires are also more luminous because of the amount of soot they have entrained in the flames.

The shape of the plume depends upon plume velocity and entrainment. The flow entrained is proportional to the plume velocity at a particular elevation. Flame growth and shape is affected by the following:

- Sudden wind changes, which increase flame exposure to combustible materials, causing concurrent flame spread
- Dry conditions
- Variety in chemical components consumed by the fire
- Type of fire: gas, liquid, and solid

Summary

The fundamentals of fire prevention, protection, and control include: analyze the physical layout of the plant or facility and surrounding structures; ensure firefighting equipment is available and strategically placed; and design engineering systems that allow for the safe handling, storage, and shipment of flammable and combustible materials. Written procedures and emergency response plans, structured safety training programs, and early fire detection systems will help prevent the outbreak and spread of fires.

The development of a fire can be traced through the incipient stage, the smoldering stage, the flame stage, and the heat stage. Fire growth is characterized by five stages: ignition, growth, flashover, fully developed fire, and decay.

Fire is a chemical reaction between oxygen, fuel, and heat. In the fire triangle, fuel must be present in vapor form, in the flammable range, and there must be oxygen or an oxidizing agent and heat in the form of high environmental temperatures, hot surfaces, open flames, friction, and so on. If any one of these three elements is removed, the fire will go out.

The fire classification system is designed to simplify the selection of fire-fighting techniques and equipment:

- Class A fires involve the burning of combustible materials such as wood, paper, plastic, cloth fibers, and rubber.
- Class B fires involve combustible and flammable gases and liquids, and grease.
- Class C fires are categorized as electrical fires. These involve energized equipment and Class A, B, and D materials that are located near the fire.
- Class D fires cover combustible metals such as aluminum, magnesium, potassium, sodium, titanium, and zirconium.

The five most common fire extinguishers found in the chemical processing industry include: CO_2 extinguishers, effective on Class B and C fires; dry chemical fire extinguishers, effective on Class A, B, and C fires; foam fire extinguishers, used to control flammable liquid fires, effective on Class A and B fires; Halon fire extinguishers, designed for use on Class A, B, and C fires; and water fire extinguishers, designed for use on Class A fires only.

The basic rules involved in using a fire extinguisher are: identify the type of material involved in the fire, select the appropriate fire extinguisher, test the extinguisher, approach upwind of the fire, aim at the base of the flames and use a sweeping motion to extinguish the fire, and back away from the fire before the extinguisher is emptied.

There are five basic types of chemical explosions: combustion explosions, thermal explosions, condensed phase explosions, nuclear explosions, and BLEVEs.

Review Questions

1. In your own words, describe the explosion of the *S.S. Grandcamp*. What could have been done to prevent this accident before it happened?
2. Why were so many people killed during the *S.S. Grandcamp* explosion?
3. Describe the hazards of handling, storing, and transporting ammonium nitrate.
4. Describe the special conditions, described in chapter 4, that are required to safely handle methyl isocyanate.
5. What were the major causes of the Phillips explosion?
6. What were the major causes of the ARCO explosion?
7. List the mistakes that occurred during the Bhopal (chapter 4), Texas City, Phillips, and ARCO incidents. What is common between them?
8. List the three sides of the fire triangle.
9. Your computer console has just caught on fire. List the steps you would take. Identify the type of extinguisher you would use.
10. The motor on pump 102 has just caught on fire. List the steps you would take. Identify the type of extinguisher you would use.
11. A liquid spill from tank 102 has just caught on fire. Tank 102 is empty. List the steps you would take. Identify the type of extinguisher you would use.
12. Fire monitors are used to ___ exposed facilities and equipment and limit the spread of the fire. The fire monitor is equipped with an adjustable ___ gallon per minute.
13. The nozzle on a fire monitor has three settings. Name them.
14. List the four stages a fire goes through.
15. When reporting a fire, what information should you provide?
16. Describe three principles of fire prevention, protection, and control.
17. In two or three paragraphs, describe the fire classification system.
18. Describe the hazards associated with fighting a fire that has the potential to produce a BLEVE.
19. List the five most common fire extinguishers used in the chemical processing industry.
20. Describe the evolution of detonation and chemical explosions.

Electrical, Noise, Heat, Radiation, Ergonomic, and Biological Hazards

OBJECTIVES

After studying this chapter, the student will be able to:

- Describe the hazards associated with electricity.
- List the hazards associated with the operation and maintenance of electrical equipment.
- Describe the hazards of bonding and grounding.
- Describe the hazards of working with noise, heat, and radiation.
- List the ergonomic hazards found in the chemical processing industry.
- Analyze the hazards connected to confined space entry.
- List the hazards associated with industrial lifting.
- Identify the biological hazards found in a chemical facility.
- Describe the primary concerns about blood-borne pathogens (Blood-borne Pathogens—CFR 1910.1030).
- Describe the hearing conservation standard (Occupational Noise Exposure—CFR 29 1910.95).
- Explain the effects of noise on hearing.
- Review the purpose of hearing protection devices.
- Describe the selection, use, fit, and care of hearing protection.
- Explain the purpose of audiometric testing.

Key Terms

- **Administrative noise abatement**—reduction of employee exposure time to industrial noise.
- **Anacusis**—total hearing loss.
- **Audiometry**—the science of testing, measuring, and recording hearing ability.
- **Biological hazards**—include any living organism capable of causing disease in humans.
- **Bonding**—is described as physically connecting two objects together with a copper wire.
- **Electrical shock**—causes ventricular fibrillation and paralysis of the respiratory system.
- **Engineering Noise Abatement**—reducing noise through new equipment design and innovation.
- **Ergonomics**—the science of how people interact with their work environment.
- **Grounding**—a procedure designed to connect an object to the earth with a copper wire and a grounding rod.
- **HCP**—hearing conservation program.
- **Heat exhaustion**—dizziness, weak pulse, cool, moist skin, and painful heat cramps.
- **Heatstroke**—a dangerous condition in which the body is no longer able to cool itself. When this happens the internal body temperature rises sharply and constitutes a real medical emergency.
- **Legionnaires' disease**—this biological hazard is a form of pneumonia caused by inhaling Legionellae bacteria. This strain of bacteria has been found in cooling towers and heat exchangers.
- **Lightning**—an electrical discharge equivalent to 1,000,000,000,000,000 hp.
- **Noise**—loosely defined as valueless or unwanted sound.
- **OSHA regulation CFR 29 1910.95**—requires all process technicians who are exposed to over 85 decibels (dB) over an eight-hour period be placed in a hearing conservation program.
- **Reducing noise at its source**—a technique that utilizes engineering controls and limits employee exposure to high noise areas.
- **Spark**—occurs when electricity jumps a gap. Sparks and arcs occur during the normal operation of electrical equipment.
- **TWA**—time weighted average.

Plant-Specific Hazards

A new process technician should consider the various hazards found in a plant before starting the initial job rotation. Most of these hazards are covered by this text; however, on-the-job training should fill in any gaps that are specific to your organization. A key part of operating safely is recognizing and understanding the equipment and technology you will be working with. This chapter will specifically focus on the following hazards:

- Electricity and electrical equipment
- **Bonding** and grounding
- Noise, heat, and radiation
- **Ergonomics**
- Confined space entry
- Lifting
- Biological and blood-borne pathogens

Electricity

Process technicians are typically not asked to connect a three-phase motor, replace a breaker, or pull new wiring. Qualified electricians are always assigned this specific task, and working with equipment that utilizes high voltage electricity is an important part of a technician's job assignment. The hazards of working with electricity include: (1) sparks and arcs, (2) static electricity, (3) **lightning**, (4) stray currents, (5) energized equipment, and (6) electric shock.

When electricity jumps a gap, it is called a **spark**. Sparks and arcs occur during the normal operation of electrical equipment. The insulation on older equipment breaks down over a long time period. A spark can be as hot as 1,500°F and may ignite a flammable mixture. Modern facilities are pre-engineered with explosion-proof equipment that does not allow arcs or sparks out of a confined area. Static electricity was discussed in earlier chapters and can lead to a deadly situation. Examples of this problem include pumping, filtering, or moving flammable materials from one place to another. An accumulated static charge has enough voltage to ignite a flammable mixture. To prevent this problem, a procedure called **grounding** is used. Lightning strikes are difficult to predict; however, tall vessels in an electrical storm can attract this phenomenon. Lightning can ignite flammable and combustible materials in tanks and vessels. Lightning has an electrical discharge equivalent to 1,000,000,000,000,000 hp. Most industrial equipment is grounded to prevent static charges and lightning strikes from damaging equipment and personnel. Stray currents in a plant are considered to be power line leakage or battery action of soils and metals. One hazard associated with stray currents is current flowing through grounded piping, causing the metal to corrode. Typical electrical equipment found in the industry include motors, switches, heaters, lighting, breakers, motor control centers (MCCs), ovens, control panels, alarms, thermocouples, and so on. Electric shock is always a hazard when working with electrical equipment because some equipment is exposed to the elements, and often older equipment has not been replaced.

Over 1,000 people die each year as a result of **electrical shock**. The direct cause of death appears to be ventricular fibrillation and paralysis of the

Table 8-1
Effect of Various currents

Current	Result
1–8 mA	Tingling or slight shock sensation
8–15 mA	Painful shock
15–20 mA	Painful shock and loss of muscle control
20–50 mA	Painful shock and loss of respiration
over 50 mA	Fatality, in most cases

respiratory system. If the amount of the current and the current path are sufficient, a person's breathing will stop and death results from asphyxiation. In accidental electrocutions, the amount of current can vary and is determined by voltage, contact area, and body resistance. Dry surface areas provide high resistance; however, if a process technician is wet and standing in water, the contact area is increased and body resistance is lowered. This is how most accidental electrocutions occur. The amount of current a process technician can tolerate is low and is typically measured in milliampere (mA). Table 8-1 shows the effect of various currents. The best way to prevent electrical shock is to do the following:

- Ensure that the equipment is grounded or unplugged.
- Keep electrical devices away from bathtubs.
- Avoid work when you are physically tired or fatigued.

Shift-work extracts a serious price and can make it difficult to physically and mentally recover. Because overtime is fairly abundant in the plant, each technician should carefully consider his or her individual limits, as most accidental electrocutions occur during this time.

Electrical Equipment

Electrical equipment is typically classified into two groups: (1) equipment that produces arcs and sparks under normal use, and (2) equipment that produces arcs and sparks at the instant of failure. A simple household light switch is a good example of group 1, and a common induction motor is a good example of group 2. Motors are used to operate compressors, pumps, and fans, and will generally operate for years without failure. In process plants where the environment is hazardous, explosion-proof housings are used. The National Electrical Code (NEC) classifies a typical refinery or chemical plant as Class 1, Group D. A Class 1 area has flammable gases or vapors in high enough concentrations to produce explosive or ignitable mixtures. Class 1 facilities are also divided into Division 1 or Division 2 depending on if they are always hazardous (Division 1) or rarely hazardous (Division 2). Atmospheres containing alcohol, benzene, butane, gasoline, hexane, lacquer solvent vapors, naphtha, or natural gas are classified as Group D. Most refineries and chemical plants are classified as Class 1, Group D, Division 2. The NEC requires all Class 1, Group D, Division 1 sites to have explosion-proof housings on all energized equipment.

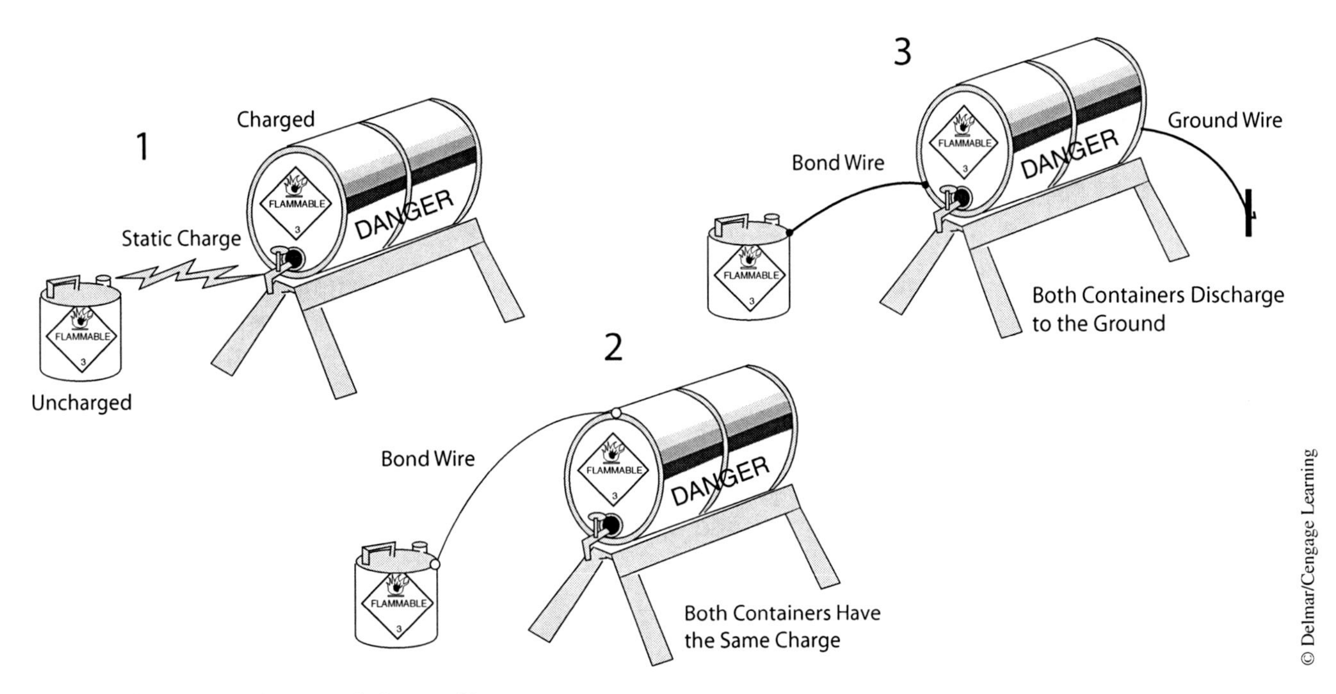

Figure 8-1 *Bonding and Grounding*

Bonding and Grounding

A static electric spark can be an ignition source between a charged body and an uncharged body. At a certain distance, an electric spark is produced and discharged into the uncharged body. Flammable liquid containers can build up static charges as the material is pumped in. Fluid movement of any type can produce a similar effect. The chemical industry has found two methods to prevent this hazard from occurring: bonding and grounding. **Bonding** is described as physically connecting two objects together with a copper wire. Grounding is described as a procedure designed to connect an object to the earth with a copper wire and a grounding rod. In some cases, underground water pipes are used. This procedure provides an alternate path for the electricity to flow. When two objects are connected, a spark cannot jump between them. A grounding rod connects the object to the earth. This process is illustrated in Figure 8-1.

Heat and Radiation

Summer heat and relative humidity in the Gulf Coast area stays in the 90°F plus range in the afternoons and above 90% humidity during the early morning hours. **Heat exhaustion** is a primary concern for chemical and refinery workers in these areas. Industrial equipment is typically hot and noisy. Temperature extremes seriously affect how much work a person can do. The human body produces heat that must be removed in order to maintain a temperature range between 96°F and 99°F. Sweating is the primary way the body removes heat. Body sweat is composed of salt and water and accounts for around 1 liter per day under resting conditions. Heavy work schedules can push this to over 4 liters within a four-hour work period. Both the water and salt must be replaced in order to ensure good health.

Radiant heat from the sun produces electromagnetic waves that are absorbed into objects and human tissue. Radiant heat transfer is line of sight and can be prevented only by staying indoors or wearing heat-reflective clothing. The effects of high temperatures include: heat exhaustion; dizziness; weak pulse; cool, moist skin; and painful heat cramps. High temperatures can result in **heatstroke**, a dangerous condition in which the body is no longer able to cool itself. When this happens, the internal body temperature rises sharply and constitutes a real medical emergency.

Hearing Conservation and Industrial Noise

When OSHA was enacted in 1970, federal regulations for controlling noise in the workplace were implemented. The Occupational Safety Health Administration adopted the regulations contained in the Walsh-Healy Act and applied them to all employees. These new standards were brief and had far-reaching impact that could be categorized as having two major components (as shown in Figure 8-2):

1. Maximum noise exposure
2. Actions that employers must take if the limits are exceeded include the following:
    - Reduce noise by using engineering and administrative controls
    - Provide hearing protection for employees
    - Implement a **hearing conservation program (HCP):**
        - Monitor sound levels
        - Conduct audiometric tests
        - Provide hearing protection
        - Provide training

Reducing noise at its source is a technique that utilizes engineering controls and limits employee exposure to high noise areas. **OSHA regulation CFR 29 1910.95** requires all process technicians who are exposed to over 85 decibels (dB) over an eight-hour, **time weighted average (TWA)** period be placed in a hearing conservation program. A decibel is not a linear unit; it is more closely identified as points on a sharply rising curve. The standard requires that each process technician receive initial training and refresher training on the following:

- Effects of noise on hearing
- Purpose of hearing protection devices

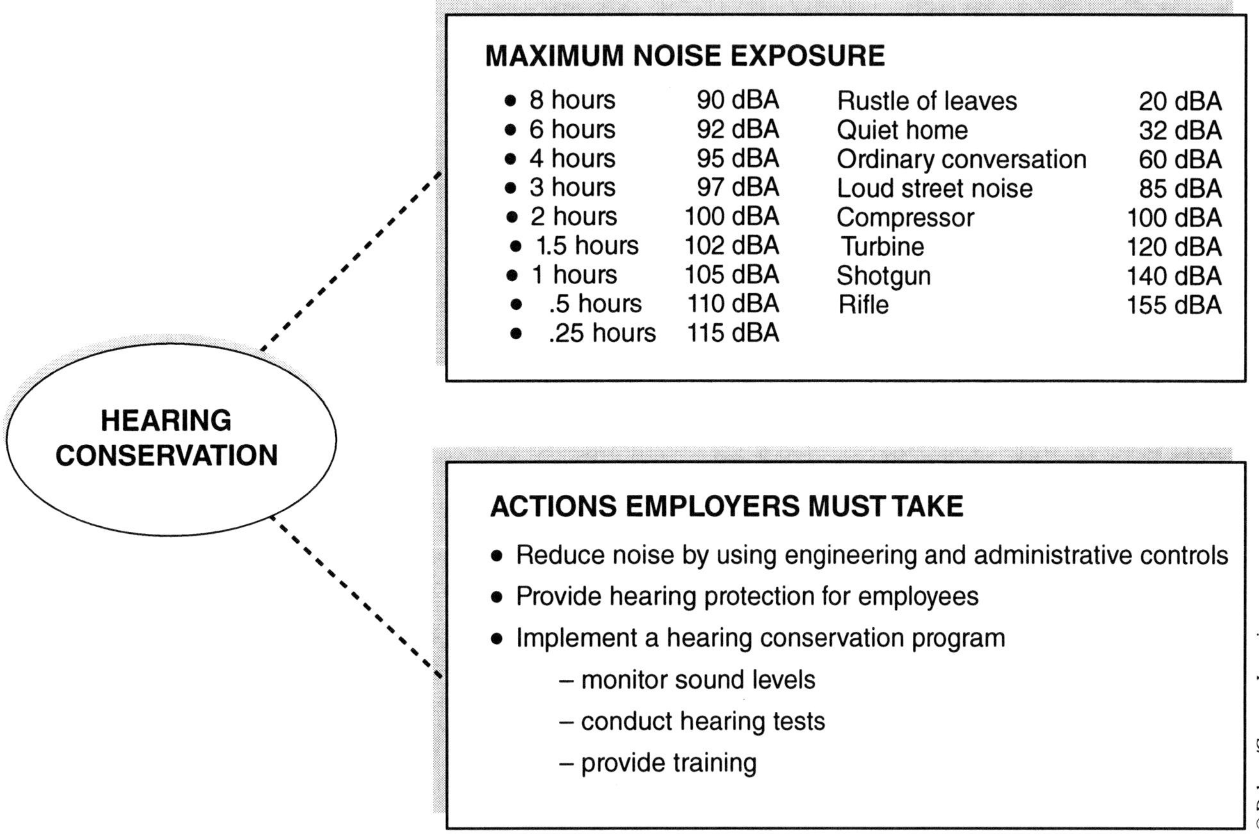

Figure 8-2 *Hearing Conservation*

- Selection, use, fit, and care of hearing protection
- Purpose of audiometric testing

Noise is loosely defined as valueless or unwanted sound. Ambient noise is defined as the composite of all-encompassing noise found in an environment. The physiological effects of noise include hearing loss, fatigue, speech problems, nausea, circulatory problems (prolonged exposure), heart attack (prolonged exposure), aural pain, tinnitus (ringing of ears), nerve damage, and structural damage to the eardrum. Hearing loss occurs slowly and painlessly. After process technicians have experienced hearing loss due to industrial noise, it is unlikely they will ever recover. **Anacusis** is a term used to describe total hearing loss. **Audiometry** is another term used to describe the science of testing, measuring, and recording hearing ability. Approximately 10 million U.S. and Canadian workers suffer from noise-induced hearing loss. The psychological effects of noise include being easily startled or annoyed and the inability to concentrate, relax, sleep, or communicate. It can also interfere with job performance and safety.

When a compressor or pump is started, it initiates a disturbance in the air called sound. Sound is generated in two ways:

1. Air flows over a sharp edge and becomes turbulent and creates sound waves.
2. A vibrating surface sends pressure waves through a fluid or solid medium. Process equipment typically falls into this group.

As illustrated in Figure 8-3, sound waves travel from the source of the noise outward, downward, and upward. These pressure waves roughly

OSHA TABLE D-2

Permissible Exposure Limits		Noise Comparison	
Hours	dB(A)		
8	90	Rustle of Leaves	20
6	92	Quiet Home	32
4	95	Ordinary Conversation	60
3	97	Loud Street Noise	85
2	100	Compressor	100
1.5	102	Turbine	120
1	105	Shotgun	140
.5	110	Rifle	155
.25	115		

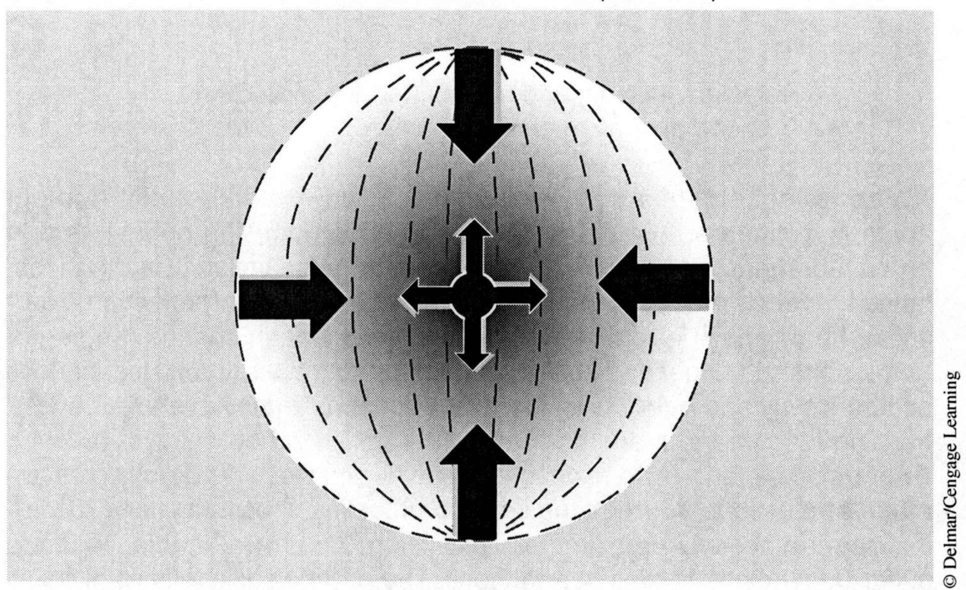

Figure 8-3 *Sound Waves Sphere*

resemble the circular ripples created by a stone thrown into a still pool of water. The waves can easily be seen traveling outward in a ring. The medium that the sound travels through—air, fluid, solid—does not move. Sound energy is transferred from one atom to the next through vibration. Sound waves form a three-dimensional ring that closely represents a sphere.

Noises over 140 decibels can cause permanent hearing loss in a single exposure. Hearing loss occurs through four major categories of noise exposure:

- Overall noise level
- Intensity of noise
- Duration—workday
- Duration—lifetime

Audiometric testing is performed on each employee annually. When a process technician is first hired, a baseline audiometric test is run. Hearing in each ear is tested and recorded. Annual hearing tests are then performed and compared to the baseline results for the rest of the technician's career.

Hearing protection can be classified in three ways:

- **Engineered noise abatement**—reducing noise through new equipment design and innovation
- **Administrative noise abatement**—reducing employee exposure time to industrial noise
- Personal hearing protection—earplugs and earmuffs. Expandable earplugs are commonly used by process technicians. The earplug must be rolled between the thumb and forefinger and compressed before it can be inserted into the ear canal. The foam ear plugs will expand and form a seal between the noise and the eardrum. Earmuffs provide the same protection as earplugs. The sealing area must be kept clear of hair, earrings, and glasses in order to provide protection. Earmuffs and earplugs (110 dBA) can be used together to provide maximum protection from noise. Hearing protection is typically selected by availability, intensity of noise, and personal comfort.

Ergonomic Hazards

Ergonomics is the science of how people interact with their work environment. The term ergonomics is derived from the Greek words ergon, meaning human work and strength, and nomos, meaning law or rule. This term was developed in the 1950s by a group of scientists attempting to design work tasks and equipment to fit workers. The primary goal is to minimize illnesses and accidents due to chronic physical and psychological stresses while maximizing efficiency, quality, and productivity. Ergonomic stresses or hazards can be initiated by the following:

- Repetitive or forceful work
- Improper work techniques
- Poorly designed tools and workplaces

Process-related hazards include lifting, material handling, sitting at a computer console for long periods of time, shift work, work teams, using the wrong tools for the job, working at heights, and working in confined spaces.

Hazards of Confined Spaces

Occasionally process technicians are required to work in confined spaces. This may include excavations, opened vessels and equipment, tank cars, distillation columns, reactors, or large tanks. Confined space ergonomic stressors include repetitive movement, exposure to potentially hazardous environments, and fear of enclosed or tight spaces. A variety of visible symptoms include anxiety, muscle pain, sweating, nausea, dizziness, loss of consciousness, vertigo, inability to communicate, and inability to communicate effectively.

Hazards of Lifting

Few problems are as serious to an active process technician as a bad back. Protecting the structural integrity of this part of the body is fundamental in having a productive work career. Proper lifting techniques are important in lifting 55-gallon drums, additive drums, bags, and other materials. Proper lifting techniques require a technician to keep the load close to the body:

- Bend at the knees and lift with the legs, keeping the heels on the ground.
- Keep the feet about shoulder-width and turned out.
- Avoid twisting by turning the entire body in the direction you want to move.
- Bend knees and squat down carefully when setting down loads.

Older employees should recognize the effect of age on their physical abilities and limit some of the activities they used to be able to do.

Biological Hazards

The field of biological safety was started during the 1950s at Fort Detrick in Frederick, Maryland. Special containment procedures were needed to prevent the release and exposure of infectious agents in the event of biological warfare. **Biological hazards** include any living organism capable of causing disease in humans. The purpose of biological safety is to minimize exposure to biological hazards through anticipation, control, evaluation, and recognition. Biological hazards are found in microorganisms, insects, allergens and toxins, and protein allergens from vertebrate animals. This type of exposure is rare in the chemical industry; however, with the advent of blood-borne pathogens such as hepatitis B (HBV) and human immunodeficiency virus (HIV), exposure is a possibility. Other process-related illnesses include **Legionnaires' disease**. This biological hazard is a form of pneumonia caused by inhaling Legionellae bacteria. This strain of bacteria has been found in cooling towers and heat exchangers. Another form of biological hazard comes from the spores found in pigeon droppings and another, meningitis, is carried by mosquitoes.

Blood-Borne Pathogens

The blood-borne pathogens standard became effective in March of 1992. The standard is designed to protect employees who come into contact with blood or infectious materials that are suspected of containing HIV and HBV. The Occupational Safety and Health Administration implemented the standard to eliminate or reduce exposure through engineering controls, administrative guidelines, PPE, medical vaccinations and surveillance, employee training, and signs and labels.

HIV causes acquired immunodeficiency syndrome (AIDS). The virus can survive only in the blood or body fluid environment. Transmission of HIV is through physical contact with these mediums. Process technicians can come into contact with this infected medium through sexual contact, puncture wounds, blood, or body fluid splash onto open injuries or the lips, mouth, nose, or eyes.

The hepatitis B virus causes inflammation of the liver and is a major contributor of acute and chronic hepatitis. Over 300,000 people are infected with HBV. Transmission of HBV and its flu-like symptoms is the same as for HIV.

Process technicians should use recommended personal protective equipment (PPE) when they are in situations in which exposure could occur. Examples of PPE used in this case are safety glasses, chemically resistant gloves, pocket masks, and lab coats. Decontamination of soiled PPE is typically through careful and cautious disposal.

Summary

A variety of other plant-specific hazards include working with: (1) electricity and electrical equipment, (2) bonding and grounding, (3) noise, heat, and radiation, (4) ergonomic hazards, (5) confined space entry, (6) lifting, and (7) biological and blood-borne pathogens.

The hazards of working with electricity include: (1) sparks and arcs, (2) static electricity, (3) lightning, (4) stray currents, (5) energized equipment, and (6) electric shock. Over 1,000 people die each year as a result of electrical shock. The direct cause of death appears to be ventricular fibrillation and paralysis of the respiratory system. Electrical equipment is typically classified into two groups: (1) produces arcs and sparks under normal use and (2) produces arcs and sparks at the instant of failure. The National Electrical Code (NEC) classifies a typical refinery or chemical plant as Class 1, Group D.

A static electric spark can be an ignition source between a charged body and an uncharged body. The chemical industry has found two methods to prevent this hazard from occurring: bonding and grounding. Bonding is described as physically connecting two objects together with a copper wire. **Grounding** is described as a procedure designed to connect an object to the earth with a copper wire and a grounding rod.

Ergonomics is the science that studies how people interact with their work environment. The term ergonomics is derived from the Greek words ergon meaning human work and strength and nomos meaning law or rule.

OSHA regulation CFR 29 1910.95 requires all process technicians who are exposed to over 85 decibels (dB) over an eight-hour TWA period be placed in a hearing conservation program. A decibel is not a linear unit; it is more closely identified as points on a sharply rising curve. Noise is loosely defined as valueless or unwanted sound. The physiological effects of noise include: hearing loss, fatigue, speech problems, nausea, circulatory problems (prolonged exposure), heart attack (prolonged exposure), aural pain, tinnitus (ringing of ears), nerve damage, and structural damage to the eardrum. Hearing loss occurs slowly and painlessly.

Biological hazards include any living organism capable of causing disease in humans. The purpose of biological safety is to minimize exposure to biological hazards through anticipation, control, evaluation, and recognition. Biological hazards are found in microorganisms, insects, allergens and toxins, and protein allergens from vertebrate animals. Other blood-borne pathogens include hepatitis B, human immunodeficiency virus (HIV), and Legionnaires' disease.

Review Questions

1. Describe the various hazards associated with working with electricity.
2. Compare and contrast the terms bonding and grounding.
3. Explain how the National Electrical Code (NEC) classifies a chemical plant or refinery.
4. Describe the key features of OSHA regulation CFR 29 1910.95.
5. List the physiological effects of industrial noise.
6. Explain how hepatitis B, human immunodeficiency virus (HIV), and Legionnaires' disease can be transmitted in the plant.
7. Describe the ergonomic hazards found in a refinery or chemical plant.
8. What are the hazards associated with operating electrical equipment?
9. Describe the symptoms of heat exhaustion.
10. Describe the correct procedure for lifting an object.
11. What are the ergonomic symptoms of working in confined spaces?
12. Describe the three ways hearing protection is classified.
13. Describe the hazards associated with working with noise, heat, and radiation.
14. Describe the hazards connected with confined space entry.
15. Identify the biological hazards found in the chemical processing industry.
16. Explain the primary concerns associated with blood-borne pathogens.
17. List the key elements of the hearing conservation standard.
18. Describe the term *audiometric testing*.
19. Define the term *anacusis*.
20. Compare and contrast the terms *heat exhaustion* and *heat stroke*.

Safety Permit Systems

Objectives

After studying this chapter, the student will be able to:

- Describe the different types of permit systems found in the chemical processing industry. Describe hot work permits (CFR 29 1910.119).
- Describe energy isolation and lockout-tagout (CFR 29 1910.147).
- Analyze the confined space entry procedure (CFR 29 1910.146).
- Review opening and blinding permits.
- Utilize a permit to enter.
- Describe an energy isolation permit.
- Recognize an unplugging permit.
- Describe cold work or routine maintenance permits.

Key Terms

- **CFR**—Code of Federal Regulations.
- **Degradation**—the breakdown or loss of physical properties.
- **Designated equipment owner**—the process technician who operates a piece of equipment or process.
- **Electrical lockout log**—includes the name of the person locking out the equipment, the date, time, and location. This same information is recorded when the lock is removed.
- **Enclosed space**—any space that has restricted entry.
- **Mechanical person**—the person who performs the work.
- **Permit system**—a regulated system that uses a variety of permits for various applications. The more common applications are cold work, hot work, confined space entry, opening or blinding, permit to enter, lockout/tagout.
- **Routine hot work area**—mechanical or fabrication shops.
- **Standby**—a technician who is certified and trained to support and warn technicians who have entered a confined space.
- **Vapor pressure**—is the pressure exerted in a confined space by the vapor above its own liquid.

Types of Permits

Describing the different permitting systems used by the chemical processing industry is difficult. Variations between plants are enormous with only a few common **permit systems**. Each plant has its own permitting systems that address routine work and maintenance, hot work, confined space entry, isolation of hazardous energy or lockout/tagout, opening piping, blinding lines, removing chemical waste, and unplugging lines. There are several permit systems that will not be discussed in this chapter. Refer to your plant's specific program for this training.

Three permit systems are common in the industry: the control of hazardous energy (lockout/tagout), confined space entry, and hot work. These three permitting systems are common within the CPI because they are government-mandated programs. Examples of a permitting system could include the following:

- hot work permit—any maintenance procedure that produces a spark or excessive heat, or requires welding or burning.
- energy isolation procedure, lockout/tagout—isolates potentially hazardous forms of energy: electricity, pressurized gases and liquids, gravity, and spring tension (The standard is also designed to shut down a piece of equipment at the local start/stop switch, turn the main breaker off, attach lockout adapter and process padlock, try to start the equipment, and tagout and record in lockout logbook.)

- confined space entry, permit to enter—designed to protect employees from oxygen-deficient atmospheres, hazardous conditions, power-driven equipment, and toxic and flammable materials.
- opening/blinding permit—removing blinds, installing blinds, or opening vessels, lines, and equipment.
- routine maintenance permit or "cold work permit—general maintenance and mechanical work that does not involve hot work or opening up a vessel.
- unplugging permit—barricades area, clears lines for unplugging, informs personnel, issues opening blinding permit, and issues unplugging permit.

Hot Work Permits

CFR 29 1910.119 Process Safety Management Standard Hot Work. The purpose of a hot work permit is to protect personnel and equipment from explosions and fires that might occur from hot work performed in an operational area. Hot work is defined as any maintenance procedure that produces a spark or excessive heat, or requires welding or burning. Examples of hot work include portable grinders, open fires, welding equipment, energized electrical circuits, internal combustion engines, electric motors, hot plates, Coriolis meters, turbine meters, concrete busters, soldering irons, dry sandblasting, and so on. The hot work permit has multiple layers of protection. One large chemical processing company involves the following people during the issue of a hot work permit:

- Process technician
- First-line supervisor
- Person performing the work
- Mechanical supervisor
- Safety permit inspector
- **Standby**

The process technician is responsible for securing the process to be worked on. If the lines are full of hydrocarbons or toxic materials they must be drained, inerted (explosive or hazardous vapors removed), and blinded. All sewers and manholes must be covered and appropriate firefighting equipment placed at the worksite. Before calling the safety inspector, the area should be cleaned up if needed. Good housekeeping is an essential element of the hot work permit procedure. If the hot work will be conducted inside a confined space, then all of the procedures for confined space entry come into play and must be completed before the hot work permit can be filled out.

The hot work permit, shown in Figure 9-1, must be filled out and ready for the appropriate signatures before the inspector is called. When the system is ready the mechanical supervisor and safety inspector will show up to inspect the area. Chemical concentrations and potential hazards are accessed. The need for a standby will be determined by the mechanical supervisor and the basic equipment operator. If everything is in order, the safety inspector will sign the card, and the work can be started.

Figure 9-1 *Example of a Hot Work Permit*

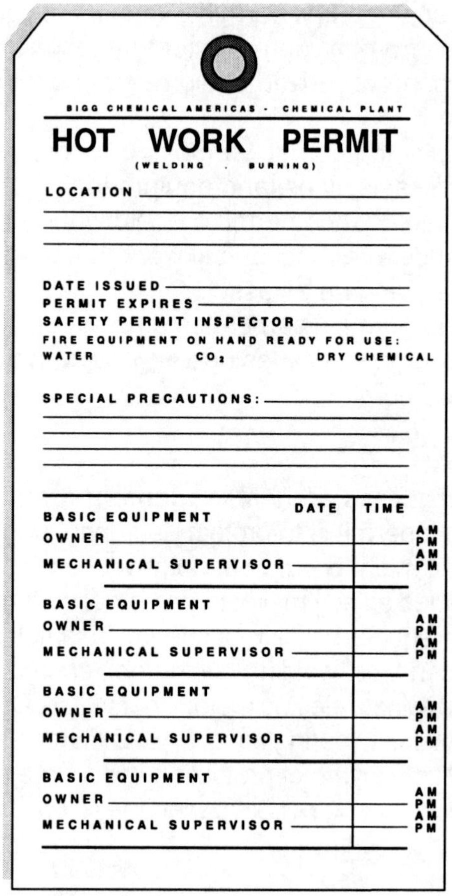

BIGG CHEMICAL AMERICAS - CHEMICAL PLANT

HOT WORK PERMIT

(WELDING . BURNING)

LOCATION

DATE ISSUED
PERMIT EXPIRES
SAFETY PERMIT INSPECTOR
FIRE EQUIPMENT ON HAND READY FOR USE:
WATER CO_2 DRY CHEMICAL

SPECIAL PRECAUTIONS:

	DATE	TIME
BASIC EQUIPMENT		
OWNER		AM PM
MECHANICAL SUPERVISOR		AM PM
BASIC EQUIPMENT		
OWNER		AM PM
MECHANICAL SUPERVISOR		AM PM
BASIC EQUIPMENT		
OWNER		AM PM
MECHANICAL SUPERVISOR		AM PM
BASIC EQUIPMENT		
OWNER		AM PM
MECHANICAL SUPERVISOR		AM PM

The following list outlines an *example* of the typical roles and responsibilities of people during the issue of a hot work permit:

- Process technician:
  - Inspects area and ensures housekeeping
  - Blinds, isolates, and clears equipment, vessels, tanks, and piping
  - Immobilizes power-driven equipment (lockout/tagout procedure)
  - Determines personal protective equipment (PPE) required
  - Fills out permit, posts at job site, and signs
- First-line supervisor:
  - Delegates responsibilities to process technician and ensures all established procedures are completed
- Person performing the work:
  - Inspects job site
  - Gathers information from process representative and mechanical supervisor about potential hazards, special procedures, or conditions
  - Selects and dons appropriate safety equipment
  - Performs work (other permits may be required)

- Mechanical supervisor:
    - Inspects area and ensures it is ready for safety inspector
    - Ensures equipment, vessels, and piping are cleared
    - Ensures safety equipment is located near job site
    - Reviews procedure with person performing the work
    - Confirms PPE required
    - Signs permit
- Safety permit inspector:
    - Inspects area and ensures it is safe
    - Performs gas test, determines oxygen level
    - Ensures equipment, vessels, and piping are cleared
    - Confirms PPE required
    - Signs permit and sets time limit
- Standby:
    - Ensures the person performing the work is safe
    - Has received special standby training
    - Wears PPE required to perform the job
    - Warns the person performing the work if a hazardous condition develops (Klaxon horn, sign language, radio)

Note: The standby does not go into the confined space until help arrives.

According to the CFR 29 1910.119 PSM standard, hot work employers are required to have a permit for hot work operations. The standard defines hot work as involving welding, cutting, brazing, or similar spark-producing operations. The permit must include the following topic blocks:

- Fire prevention and protection measures in place before hot work was initiated
- Date(s) permit is approved for
- Location, equipment, and item where hot work is performed
- Fire watch posted and in place during procedure and 30 minutes after work is complete

The permit must be displayed at the work site until the hot work operation is complete.

Confined Space Entry

The Permit Required Confined Spaces for General Industry is part of the OSHA CFR 1910.146 standard. The chemical processing industry has a variety of areas where confined space entry routinely takes place. Confined space awareness is the ability to do the following:

- Define confined space entry.
- Describe the hazards associated with confined space entry.
- List those involved in the confined space procedure:
    - Process representative
    - Mechanical supervisor

    - Person performing the work
    - Standby
- Identify hazards.
- Describe testing and monitoring equipment, ventilation equipment, radio equipment, and retrieval equipment.

Restricted or limited access into or out of confined spaces is essential. Entry occurs when the technician's body breaks the plane of the opening. A confined space is defined as a space large enough so a person can enter, has restricted entry, and is not designed for continuous occupancy. This includes the following: excavations, sewers, pits, reactors, boilers, furnaces, distillation columns, strippers, absorbers, vessels, tanks, silos, blenders, drums, piping, pumps, compressors, heat exchangers, extruders, or any space that restricts entry or allows the head of the worker to go below the top of the confined space. **Vapor pressure** from chemicals is often generated inside enclosed spaces. These vapors can displace oxygen and form hazardous conditions. Vapor pressure has a direct correlation with temperature.

- A confined space may contain a hazardous atmosphere, chemical, or asphyxiate. Inside a refinery or chemical plant, a variety of confined spaces, vessels, piping, excavations, power-driven equipment, and so on exist. The permit to enter, shown in Figure 9-2, is designed to protect employees from oxygen-deficient atmospheres, hazardous conditions, power-driven equipment, and toxic and flammable materials. Often the chemicals inside a vessel will break down over extended periods of time. This process is called **degradation**, or the breakdown or loss of physical properties.

The permit to enter must be completed before any **enclosed space** can be entered. The permit covers confined space entry only. Additional permits are required for cold work, hot work, opening/blinding, or energy isolation inside the enclosed space. Confined space entry is a complex procedure that requires a number of individuals to work together as a team. Several large industrial manufacturers utilize the following team approach when dealing with a confined space entry. This is only an example!

- Process technician
- First-line supervisor
- Person performing the work
- Mechanical supervisor
- Safety permit inspector
- Standby

Upon initial assignment to the unit, process technicians are given confined space awareness training. This training identifies what a confined space is and the specific hazards it presents. Hazards from a confined space include asphyxiation, engulfment, heat stress, electrocution, moving equipment, getting wedged in a narrow space, falling, and so on. These hazards can be removed when the steps for confined space entry are followed.

CONFINED SPACE ENTRY PERMIT 012345

Date and time issued: ____________________
Job site/space I.D.: ____________________
Equipment to be worked on: ____________________

Date and time expires: ____________________
Job supervisor ____________________
Work to be performed: ____________________

Stand-by personnel: ____________________ ____________________ ____________________

1. Atmospheric checks: Time ________
   Oxygen ________ %
   Explosive ________ % L.F.L.
   Toxic ________ PPM

2. Tester's signature:

3. Source isolation (no entry):

	N/A	Yes	No
Pumps or lines blinded,	( )	( )	( )
disconnected, or blocked	( )	( )	( )

4. Ventilation modification:

N/A	Yes	No
( )	( )	( )
( )	( )	( )

5. Atmospheric check after isolation and ventilation:
   Oxygen ________ %
   Explosive ________ % L.F.L.
   Toxic ________ PPM
   Time ________
   Tester's signature ____________________

6. Communication procedures: ____________________
   ____

7. Rescue procedures: ____________________
   ____
   ____

Permit prepared by: ____________________ (supervisor)

Approved by: ____________________ (Unit supervisor)

Reviewed by : ____________________ (CS Operations Personnel)

8. Entry, stand-by, and back-up persons:

(Successfully completed required training?)	Yes	No
Is it current?	( )	( )

9. Equipment:

	N/A	Yes	No
Direct reading gas monitor tested	( )	( )	( )
Safety harness & life for entry and stand-by persons	( )	( )	( )
Hoisting equipment	( )	( )	( )
Powered communications	( )	( )	( )
SCBAs for entry & stand-by persons	( )	( )	( )
Protective clothing	( )	( )	( )
All electric equipment listed Class 1, Div. 1 Group D and non-sparking tools	( )	( )	( )

10. Periodic atmospheric tests:
   Oxygen ________ % Time ________
   Oxygen ________ % Time ________
   Explosive ________ % Time ________
   Explosive ________ % Time ________
   Toxic ________ % Time ________
   Toxic ________ % Time ________

Figure 9-2 *Example of a Permit to Enter*

Employers are required to develop a written confined space entry program, identify all permit places in their plant, post warning signs, prevent unauthorized entry into confined spaces, develop procedures, and provide training in all of these areas and the use of personal protective equipment. Process technicians should be aware that conditions within a confined space can change and should be monitored carefully. Welding or cutting can expose hazardous chemicals that have been absorbed by the metal, use up oxygen, and concentrate noxious fumes. Contractors may also accidentally mix chemicals inside a confined space that they have brought in for specific work. It is during this time period that the role of standbys becomes critical as they are required to monitor work being performed in the confined space. An example of a confined space entry can be observed in Figure 9-3.

A typical confined space entry permit should include the following:

- Work to be performed
- Location
- Date(s) and authorization period
- Atmospheric testing results and entry conditions
- Entrant's name
- Entry supervisor and standby names
- Hazards and how to control
- Isolation and control methods
- Communication procedures
- Other active permits—hot work
- rescue and emergency procedures
- equipment to be used

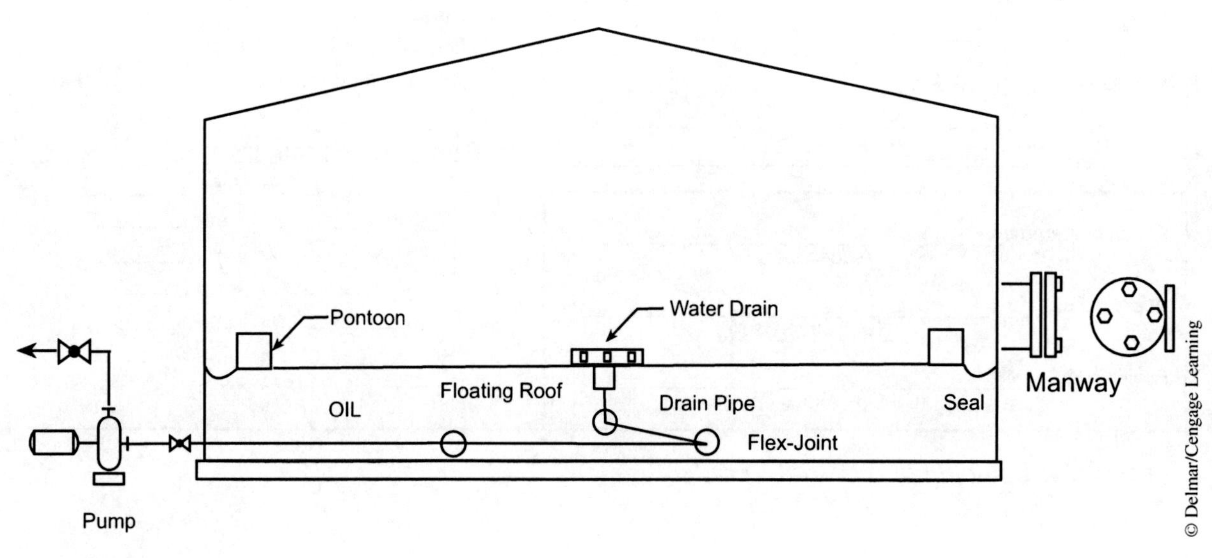

Figure 9-3 *Confined Space Entry*

Prepare for Confined Space Entry

To prepare a confined space for entry, the process technician should carefully follow the plant's procedure. Because each facility is unique, careful consideration of each step-by-step task should be given. Typically the process operator will secure the vessel or area that is being isolated. The term *isolated* refers to all points of entry into the confined space. For example, all pipes entering and exiting the confined space should be isolated or blinded. This procedure will stop chemicals from being accidentally admitted into the confined space during work activities. Any chemical remaining in the system must be removed prior to the confined space entry taking place. This procedure may require water washing, purging, or inerting. All electrical equipment in the confined space should be de-energized, locked-out, and tagged prior to entry. Adequate ventilation should be provided to the confined area. This procedure might require the use of fans or blowers. Back-up fans and blowers should be available in case of failure. Operators should post signs, erect barricades, and pull barricade tape during the preparation phase. Housekeeping is also important prior to having work performed on your unit. After the area is clear to be worked on, the atmosphere will be tested. Oxygen concentrations must be between 19.5% and 23.5%. Flammable gas concentrations must be less than 10% of their lower flammable limit. Airborne contaminates must fall into correct guidelines or work cannot be performed. The safety inspector will require that all emergency equipment and communication procedures be in place.

During the confined space entry, the process technician will work closely with the people performing the work. Contract technicians on the entry team will not be aware of all the hazards in the confined space unless the process technician documents them and tells others. The permit must be displayed near the job at all times and removed after the job is completed. After the work on the confined space is initiated, the process technician works with the confined space entry team.

Confined Space Entry Team and Process Technician

The confined space entry team is composed of an entry supervisor, a standby or attendant, and an entrant. Each member of this team will have specific training prior to performing confined space work. The entry supervisor will ensure that conditions are safe prior to admitting his or her team. During the operation the entry supervisor will frequently monitor work conditions. The standby is required to observe conditions, know hazards and recognize exposure to hazards, keep unauthorized people out, stay in constant contact with entrant, order workers out of the area if a hazard is present, coordinate rescues, and remain outside the confined space until relieved. The entrant should recognize the early signs of exposure to a hazard. For example, **oxygen deficiency** can cause breathing difficulty, loss of muscle control, ringing in the ears, feelings of well being, mental confusion, and death. Entrants should be familiar with PPE and its limitations. The entrant should leave the confined space as soon as one of these situations is detected.

Control of Hazardous Energy (Lockout/ Tagout) CFR 29 1910.147

The purpose of the hazardous energy standard is to protect employees from the hazards associated with the accidental release of uncontrolled energy. The lockout/tagout procedure is a standard designed to isolate a piece of equipment from its energy source. OSHA requires employers to have a written energy isolation program and provide training to new employees upon initial assignment and every two years thereafter. Equipment modifications and new unit startups require the existing workforce to have additional lockout/tagout training.

The chemical processing industry harnesses energy from seven basic forms: electrical, pneumatic, hydraulic, compressed gases and liquids, gravity, and spring tension. The two most common classifications of energy are kinetic and potential. Kinetic energy is closely associated with movement whereas potential energy is related to stored energy. Every year the chemical processing industry records severe injuries because process technicians fail to disconnect electrical equipment, dissipate residual energy, stop the equipment, restart the equipment accidentally, and clear the area before equipment startup.

OSHA has established a six-step procedure for locking out a piece of equipment. The first step is the preparation for shutdown phase. During this phase the type of energy being isolated must be identified and the specific hazards controlled. Phase two involves shutting down the equipment. Phase three is the isolation step, which involves closing valves, shutting down main disconnects and circuit breakers, and disconnecting pneumatic, electric, hydraulic and compressed gas and liquid lines. Phase four is the application of lockout/tagout devices (see Figure 9-4) to breakers and the disconnection of switches, valves, and energy isolating devices. Phase five is directed at the control of stored energy. Pressure must be relieved, grounding cables connected, elevated equipment supported, and moving parts stopped. Phase six is the verification step. The term *lock-tag-try* is applied when the electrically disconnected equipment is checked by attempting to start the equipment at the local start-stop switch. If the procedure has been performed correctly, the equipment will not start. The six-step procedure for locking out a piece of equipment includes the following procedures:

1. Preparation for shut down
2. Shut down of the equipment
3. Isolation of equipment—closing valves, shut down main disconnects and circuit breakers, disconnect pneumatic, electric, hydraulic, and compressed gas and liquid lines
4. Application of lockout/tagout devices to breakers and disconnect switches, valves, and energy-isolating devices

Figure 9-4 *Lockout/ Tagout Permit*

5. Control of stored energy—relieve pressure
6. Verification step—lock-tag-try

All of this information should be recorded in an **electrical lockout logbook**.

One of the more common procedures performed by a process technician is the isolation and lockout/tagout of a pump or compressor motor. Although each plant or facility is different, a number of common steps can be found. The pump system is designed to provide a constant liquid flow rate to other process systems. In most cases a back-up or redundant pump is provided in the event of a pump failure. During most operations the alternate pump can be lined up if a motor fails or needs to be worked on. The pump system includes a feed tank, piping and valves, the pump, and control valve. One of the most important parts in the system is the motor. The motor is classified as a driver that has a fixed coupling attached to the rotating shaft of the pump. The motor is designed to operate at a fixed speed measured in revolutions per minute (RPM). Figure 9-5 is an illustration of a simple pump-around system with two possible places to attach a lock during an energy

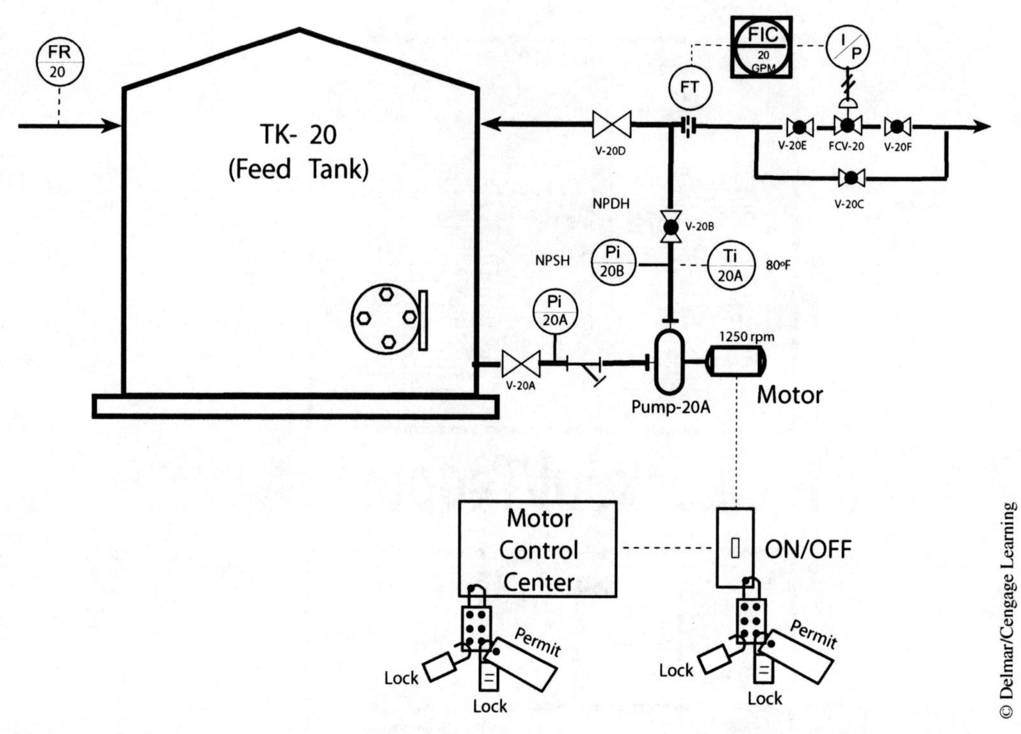

Figure 9-5 *Pump-around System*

isolation or lockout/tagout procedure. The typical steps in the isolation of hazardous energy include the following:

1. Line up the alternate system.
2. Ensure work area is clean.
3. Ensure that electrical power is off to the motor being worked on.
4. Double-block and bleed the isolated equipment.
5. Follow purging requirements if needed.
6. Ensure the breaker closest to the driver is in the off position.
7. Fill out the permit.
8. Place permit and lock on nearest breaker.
9. Place soft copy of permit on display board in control room.
10. Go to local motor control center (MCC) and lockout breaker.
11. Notify supervisor of permit status, and coordinate work with maintenance department.
12. Ensure maintenance technician places lock on breaker.
13. Have each maintenance person working on the equipment place a lock on the breaker and remove it when the work is finished.
14. Coordinate all work activities on the equipment and report to supervisor.
15. Check-out and ensure the equipment is operational.
16. Complete documentation, and remove soft permit from control room and at local breakers.

Opening or Blinding Permits

The purpose of the opening or blinding permit is to limit accidents through improved communications and structured guidelines for removing blinds, installing blinds, or opening vessels, lines, and equipment. Figure 9-6 is an example of an opening or blinding permit.

Examples of work covered under opening and blinding permits include the installation and removal of blinds, removal of hatch covers, and opening equipment. To initiate an opening or blinding permit, the process technician performs a safety inspection, secures the area for work, isolates the equipment, and fills out the opening and blinding permit. After the permit is filled out, the technician speaks with the craftsperson about the work to be performed, reviews the hazards, and hangs a hard copy of the permit near the unit with the craft-person's signature. The soft copy of the permit is displayed in the control room. Figure 9-7 is an example of a vessel with three blinds installed and a hatch cover.

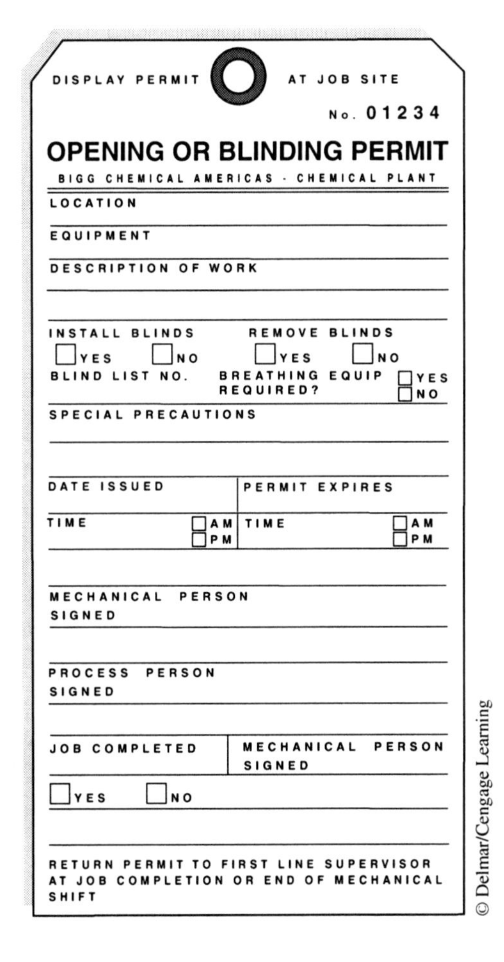

DISPLAY PERMIT AT JOB SITE

No. 01234

OPENING OR BLINDING PERMIT

BIGG CHEMICAL AMERICAS - CHEMICAL PLANT

LOCATION

EQUIPMENT

DESCRIPTION OF WORK

INSTALL BLINDS ☐ YES ☐ NO

REMOVE BLINDS ☐ YES ☐ NO

BLIND LIST NO.

BREATHING EQUIP REQUIRED? ☐ YES ☐ NO

SPECIAL PRECAUTIONS

DATE ISSUED

PERMIT EXPIRES

TIME ☐ AM ☐ PM

TIME ☐ AM ☐ PM

MECHANICAL PERSON SIGNED

PROCESS PERSON SIGNED

JOB COMPLETED ☐ YES ☐ NO

MECHANICAL PERSON SIGNED

RETURN PERMIT TO FIRST LINE SUPERVISOR AT JOB COMPLETION OR END OF MECHANICAL SHIFT

Figure 9-6 *Opening or Blinding Permit*

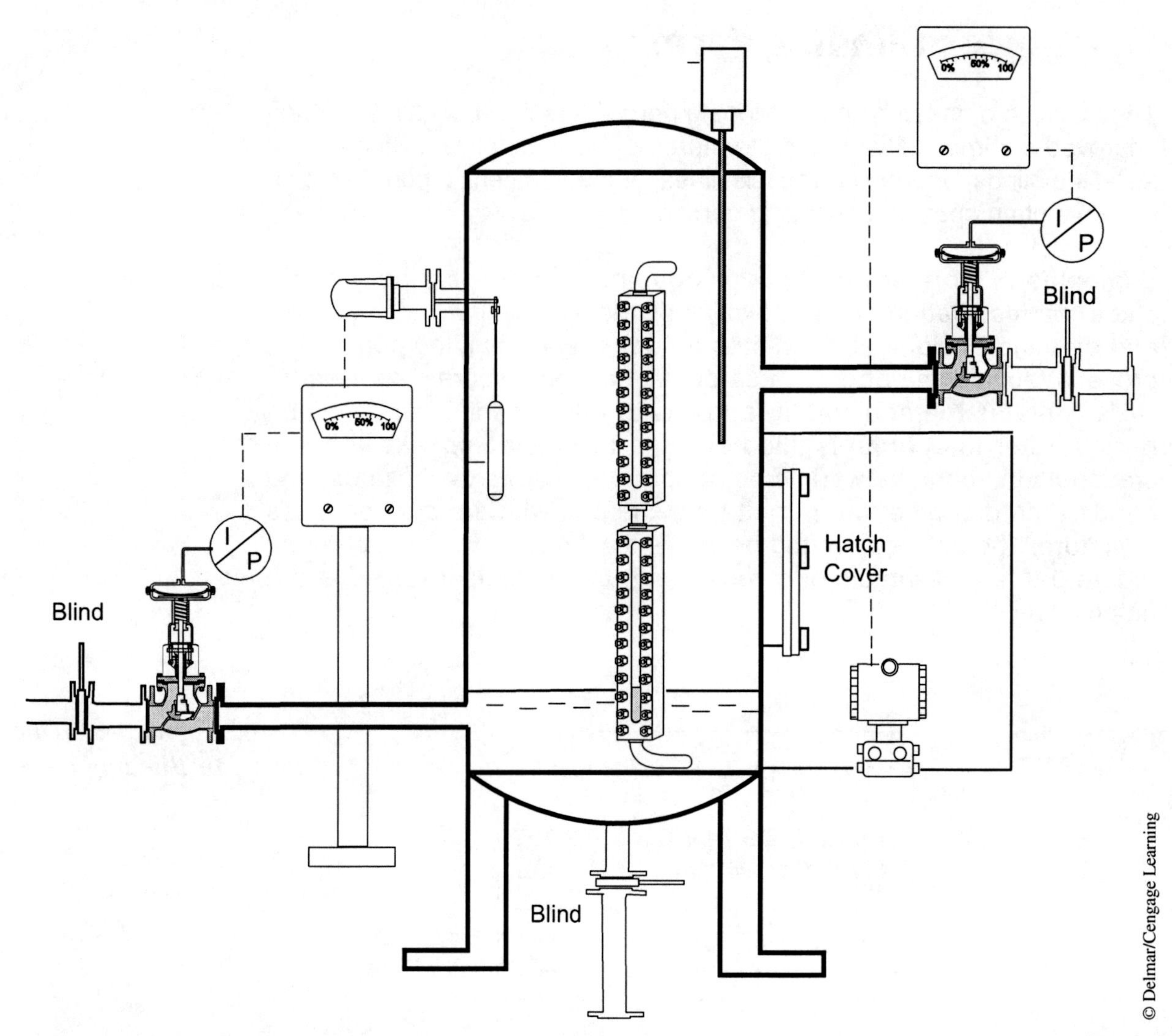

Figure 9-7 *Opening or Blinding a Vessel*

Routine Maintenance Permits

The routine maintenance permit system is designed to prevent accidents by improving communications between process technicians and the mechanical craftsperson doing the mechanical work. In order to issue the routine maintenance permit, the technician is required to check out the area where the work will be performed. Safety hazards must be identified and the permit displayed near the work site. The soft copy of the permit is displayed in the control room while the hard copy is hung near the job site. Both the process technician and the craftsperson sign the permit. Figure 9-8 is an example of a routine maintenance permit.

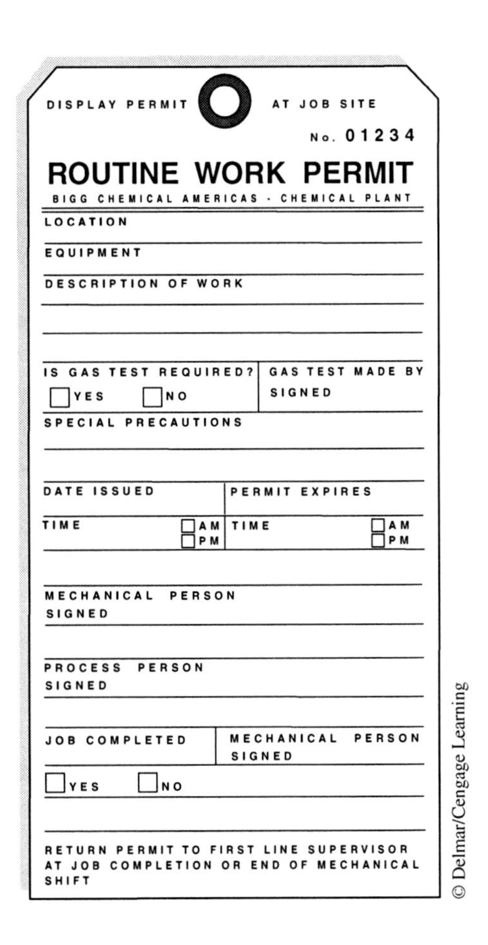

DISPLAY PERMIT AT JOB SITE

No. 01234

ROUTINE WORK PERMIT

BIGG CHEMICAL AMERICAS · CHEMICAL PLANT

LOCATION

EQUIPMENT

DESCRIPTION OF WORK

IS GAS TEST REQUIRED? ☐ YES ☐ NO

GAS TEST MADE BY SIGNED

SPECIAL PRECAUTIONS

DATE ISSUED | PERMIT EXPIRES

TIME ☐ AM ☐ PM | TIME ☐ AM ☐ PM

MECHANICAL PERSON SIGNED

PROCESS PERSON SIGNED

JOB COMPLETED ☐ YES ☐ NO

MECHANICAL PERSON SIGNED

RETURN PERMIT TO FIRST LINE SUPERVISOR AT JOB COMPLETION OR END OF MECHANICAL SHIFT

Figure 9-8 *Example of a Routine Maintenance Permit*

The process technician's responsibilities include the following:

1. Perform safety inspection prior to issuing permit.
2. Fill out routine maintenance permit.
3. Speak with craftsperson about the work to be performed.
4. Review hazards.
5. Hang hard copy of permit near unit with craftsperson's signature.
6. Post soft copy of permit in control room.
7. Extend permit if necessary.

The mechanical craftsperson's responsibilities include the following:

1. Initiate the cold work permit.
2. Perform safety inspection.
3. Sign cold work permit.
4. Perform authorized work.
5. Secure permit extension.
6. Perform housekeeping after job.
7. Complete work, sign card, and return it to process representative.
8. Secure process approval and update of permit board.

Summary

Each plant has its own permitting systems that address routine work and maintenance, hot work, confined space entry, isolation of hazardous energy or lockout/tagout, opening piping, blinding lines, removing chemical waste, and unplugging lines.

The purpose of a hot work permit is to protect personnel and equipment from explosions and fires that might occur from hot work performed in an operational area. Hot work is defined as any maintenance procedure that produces a spark or excessive heat, or requires welding or burning. Examples of hot work include portable grinders, open fires, welding equipment, energized electrical circuits, internal combustion engines, electric motors, hot plates, Coriolis meters, turbine meters, concrete busters, soldering irons, dry sandblasting, and so on. The hot work permit has multiple layers of protection.

A confined space is defined as a space large enough so a person can enter, has restricted entry, and is not designed for continuous occupancy. This includes excavations, sewers, pits, reactors, boilers, furnaces, distillation columns, strippers, absorbers, vessels, tanks, silos, blenders, drums, piping, pumps, compressors, heat exchangers, extruders, or any space that restricts entry or allows the head of the worker to go below the top of the confined space. A confined space may contain a hazardous atmosphere, chemical, or asphyxiate. Inside a refinery or chemical plant, a variety of confined spaces, vessels, piping, excavations, power-driven equipment, and so on exist. The permit to enter is designed to protect employees from oxygen-deficient atmospheres, hazardous conditions, power-driven equipment, and toxic and flammable materials. Hazards from a confined space include asphyxiation, engulfment, heat stress, electrocution, moving equipment, getting wedged in a narrow space, falling, and so on. These hazards can be removed when the steps for confined space entry are followed.

The purpose of the hazardous energy standard is to protect employees from the hazards associated with the accidental release of uncontrolled energy. The lockout/tagout procedure is a standard designed to isolate a piece of equipment from its energy source. OSHA requires employers to have a written energy isolation program and provide training to new employees upon initial assignment and every two years thereafter.

Review Questions

1. Describe the key points associated with preparing a hot work permit.
2. List the steps a technician takes when preparing a motor for energy isolation or lockout/tagout.
3. Explain the hazards associated with confined space entry.
4. Describe the procedure associated with preparing a system for completing the opening or blinding permit.
5. Describe the steps associated with an unplugging permit.
6. Define the term “cold work or routine maintenance permits.”
7. Describe the hazards associated with confined space entry.
8. A permit to enter is designed to protect a technician from what hazards?
9. Explain the purpose of the hot work permit.
10. List examples of work that is covered by opening or blinding permits.
11. Explain the relationship and responsibilities of a process technician and the mechanical technician performing the work.
12. List examples that would require a hot work permit.
13. List examples that would require a confined space permit.
14. List examples that would require the use of a lockout/tagout procedure.
15. Safe oxygen concentrations must be between what two ranges?
16. List the symptoms associated with oxygen deficiency.
17. Define the term *inerting*.
18. List the people associated with the hot work procedure.
19. Identify which permits could be associated with the term *vapor pressure.*
20. Describe the concerns and responsibilities of a process technician when preparing a permit.

Personal Protective Equipment

OBJECTIVES

After studying this chapter, the student will be able to:

- Identify personal protective equipment found in a plant.
- Describe when and how to use personal protective equipment.
- Identify typical workplace hazards.
- Discuss the methods of hazard exposure prevention.
- Contrast emergency response and personal protective equipment.
- Describe the four levels of personal protective equipment.
- Describe the principles of hearing protection.
- Identify personal protective equipment outerwear worn by technicians.
- Explain the general limitations of personal protective equipment.
- Contrast engineering and environmental controls with personal protective equipment.
- Explain the purpose of respiratory protection.
- Describe air-purifying respirators.
- Describe air-supplying respirators.
- Describe air-purifying, half-face respirators.
- Describe air-purifying, full-face respirators.
- Describe an air-supplying self-contained breathing apparatus (SCBA).
- Describe an air-supplying hose line respirator.
- Explain the steps required to take care of and use a respirator.
- Analyze and contrast the limitations of each type of respirator.
- Review the procedures for donning and doffing air-purifying respirators.

Key Terms

- **Air-purifying respirator**—mechanically filters or absorbs airborne contaminants.
- **Air-supplying respirator**—provides the user with a contaminant-free air source.
- **Don a respirator**—put on a respirator.
- **Engineering and environmental controls**—hard technology improvements designed to make the work environment safer. Examples of this include soundproofing, installing guards to rotating equipment, automating product transfers, and so on.
- **Filter life**—the estimated amount of service time of an air-purifying filter.
- **Fit testing**—a critical part of a respiratory protection program that provides respirator training, matches respirators to technician face structure, identifies specific chemical hazards in assigned areas, and provides information on how to locate and dispose of filters, cartridges, and respirators.
- **Four levels of PPE**—OSHA and EPA have identified four levels (A, B, C, D) of personal protective equipment that could be required during an emergency situation. Level A provides the most protection; level D requires the least.
- **IDLH**—immediately dangerous to life and health.
- **Improper filter**—a term applied to the selectivity characteristics of a respirator. Process technicians must select the correct respirator to remove a specific contaminant.
- **Negative fit test**—To perform a negative fit test, you must pull the straps on the respirator snug to ensure a tight seal. By placing the open palm over the respirator's inlet(s), a vacuum should be created when you inhale. Hold the vacuum for ten seconds to ensure a good seal.
- **NIOSH**—National Institute for Occupational Safety and Health. Responsibilities include testing and certifying protective devices and research in safety and health issues in the workplace.
- **Overpowering**—a term applied to atmospheric concentrations that exceed the limitations of the respirator.
- **Positive fit test**—is performed by exhaling and covering the exhaust ports of a respirator. If the respirator has a good seal, a positive pressure will fill the face piece.
- **PPE**—personal protective equipment is used to protect a technician from hazards found in a plant.
- **Respiratory protection**—a standard designed to protect employees from airborne contaminants.
- **Selectivity**—this term applies to the specific compound or contaminant that a respirator is designed to remove.
- **STS**—Standard Threshold Shift is a term used to describe an audiometric test that is compared to prior tests to determine a measurable hearing loss.
- **TECP**— Totally Encapsulating Chemical Protective suit.

Personal Protective Equipment

The human body has over 19 square feet of surface area and inhales over 3,000 gallons of air per day. Because chemical exposure comes through inhalation, ingestion, and skin contact, protective measures need to be in place. Personal protective equipment (PPE) provides an effective means for protecting technicians from hazardous situations. **Engineering and environmental controls** provide another layer of protection. **PPE** is designed to be used in environments that place technicians in contact with a hazardous situation. The primary purpose of PPE is to prevent exposure to hazards when engineering or environmental controls cannot be used.

Most of the hazards found in the chemical processing industry have been assessed and PPE that will provide some degree of protection have been developed. (See Figure 10-1.) Modern hazard management controls workplace hazards with engineering and environmental systems. This limits or reduces the amount of exposure a technician may have to a hazardous condition.

Figure 10-1 *Personal Protective Equipment*

Outerwear

Typical outerwear worn by process technicians include:

- *Safety hats* provide protection to the head. They may become brittle over time. They must be replaced when the time stamp is exceeded. Hard hats should not be altered: holes drilled, webbing removed, or painted.
- *Safety glasses* provide protection to the eye from front impact. Safety glasses with side shields provide minimal protection from windblown objects. It should be noted that most of the injuries reported in a plant are related to the eye. Safety glasses do not provide an airtight fit around the eyes. Airborne contaminants can still find their way into the eye. Caution should be exercised when working with chemicals or in dusty environments. Another limitation that clear safety glasses have is the lack of protection they provide from ultraviolet or infrared light sources, such as during welding or cutting.
- *Fire-retardant clothing* is designed to protect a technician for a limited amount of time from a flash fire or heat source. The clothing is specially treated and will resist bursting into flames upon contact but will not protect a technician during sustained periods of exposure. The material does not breathe like cotton and is uncomfortable.
- *Safety shoes* provide protection from falling or rolling objects. A typical safety shoe policy requires all technicians to wear safety shoes that have all leather uppers and steel toes. Canvas type shoes, thongs, sandals, open-toed shoes, or high heels are not allowed in the plant. Some of the limitations of a safety shoe include the following: chemical contamination through absorption and chemicals that enter through the top of the shoe.
- *Hearing protection* protects technicians from noise that can permanently damage the inner ear. Earplugs and earmuffs are commonly used to keep exposures below 82 decibels for 12 hours or 85 decibels for 8 hours. Earplugs and muffs are only as effective as the technician's knowledge about correctly using them. If they are used incorrectly they will not provide enough protection.
- *Gloves* are designed to provide protection from chemicals or abrasive or cutting surfaces. Gloves are made of a variety of compositions and are designed to work in specific situations. Some chemicals will penetrate gloves with certain compositions. Glove protection can be split into two categories: chemical and physical. Chemically resistant gloves are designed to provide protection from specific chemicals. For example, nitrile gloves are approved to work with benzene, Methyl Tertiary Butyl Ether (MTBE), toluene, and xylene, and black plastic polyvinyl chloride (PVC) gloves can be used when working with oils and other liquids. Physical hazards to the

hand can be avoided by wearing leather or cotton gloves, which provide protection from sharp or abrasive surfaces.
- *Face shields* are designed to prevent chemical splashes and contaminants from getting into the eyes. The face shield provides greater coverage area than monogoggles. They will not protect the technician from airborne chemicals, fumes, or ultraviolet or infrared light sources. Face shields do not provide an airtight fit around the eyes. Airborne contaminants can still find their way into the eye. Caution should be exercised when working with chemicals or in dusty environments.
- *Chemical monogoggles* are designed to prevent chemical splashes and contaminants from getting into the eyes. Monogoggles will not protect the technician from airborne chemicals, fumes, or ultraviolet or infrared light sources.
- *Slicker suits* provide protection from rain and inclement weather. Slicker suit material is typically not fire resistant. They can be worn when there is a potential for exposure to an acid, caustic, or chemical.
- *Radios* are used to provide emergency and operational information to other technicians and supervisors.
- *Respirators* provide protection from chemicals and airborne contaminants. They come in two basic designs: air purifying and air supplying. Respirators are chemically and contaminant-selective, and the right one must be selected for a specific job.
- *Chemical suits* provide protection from specific chemicals.
- **Totally encapsulating chemical protective suits (TECP)** come in a variety of shapes and designs. They are individually unique and must be used when working with specific substances and in specific situations. TECP suits provide the maximum amount of protection when engineering controls are not sufficient. This type of suit has a self-contained air system and can be used for a limited time.

Hand Protection

Many of the gloves used by chemical technicians have a wide variety of operational limitations; one type of glove cannot be used for every job. The primary difference in gloves can be determined by: chemical resistance, type of material, thickness of material, and compatibility with toxic and hazardous substances. Standard materials include: butyl rubber, natural rubber, neoprene, nitrile synthetic latex, polyvinylalcohol (PVA), polyvinyl-chloride (PVC), and viton. Table 10-1 illustrates the type and chemical resistance chart for selecting a pair of gloves.

Hand protection is a concept that starts with good training and personal awareness of the many hazards found in the workplace. Rings should be left at home because they can easily catch on ladders or in moving machinery. Glove selection includes the right type and the right size for the job. Worn gloves should be replaced with new ones, and gloves that have been

Table 10-1
Selecting Gloves

Glove Type	Chemical Resistance Guide
Butyl rubber (High-density polymer)	Aldehydes, Gas, Ketones, Esters, Organic Acids
Natural Rubber	Acids, Alcohols, Caustics, Epoxy Resins
Neoprene	Acids, Alcohols, Caustics, Solvents, Oils, Metal Salts, Greases, Oxidizers, Peroxides
Nitrile	Aromatic, Petroleum and Chlorinated Solvent Benzene, Toluene, MTBE, Xylene, and so on.
PVA (PVA is water-soluble)	Organic Solvents (Aromatics, Ketones, Chlorinated)
PVC	Acids, Caustics, Oils, Gas, General Solvents and Chemicals
Viton	Aromatic and Chlorinated Solvents, Acids, Water-Based Chemical Solutions, PCBs, Oxidizers, Peroxides

Table 10-2
Hand Protection Guide

Hazard	Glove Selection
Cold	Cotton, Leather, Insulated Plastic and Rubber
Heat	Neoprene-Coated Asbestos, Nomex, Kevlar
Chemicals	Natural Rubber, Neoprene, Butyl Rubber, Viton, PVC, PVA
Electricity	Rubber-insulated gloves CSA Standard Z259.4-M1979
Radiation	Lead-Lined Plastic, Rubber, and Leather
Sharp Edges	Metal Mesh, Staple-Reinforced Leather

contaminated with chemicals should be disposed of correctly. Glove charts will provide all of the information you need to be able to intelligently select the right glove. A hand injury can have short-term and long-term effects on your work team and your career. Table 10-2 shows a helpful guide to selecting the proper hand protection.

Eye Protection

Wearing standard eye protection is a requirement in every plant; however, this cannot prevent windblown objects or chemical spray from finding entry into the eyes. Eye and face protection are serious responsibilities shared between managers and employees. A variety of hazards pose a threat to the face and to the eyes and require special procedures. The two main aspects of these procedures are the following: wearing approved eye protection where minimum eye protection is insufficient; and wearing approved face protection where face and eye injury is a concern. Management should identify the specific jobs and areas where hazardous conditions exist. This information should be clearly communicated with all affected personnel through signs, training, and procedures. The company is responsible for providing approved eye and face protection equipment and for specifying

Figure 10-2 *Approved Eyewear*

where it should be used. Figure 10-2 shows examples of approved eyewear worn by process technicians.

Foot Protection

Over 70,000 toe injuries occur each year in the United States from accidental falls, slips, trips, and heavy object impact. Many of these injuries could have been prevented by wearing proper foot protection. This section describes the required use of approved safety shoes and foot protection at the job site. Safety shoes are designed to protect the foot from impact and compression forces. Approved footwear must match the following guidelines:

- The design and composite of the soles must be chemical resistant, puncture proof, slip resistant, abrasion resistant, and provide cushioning support.
- Footwear must pass performance tests for impact and compression according to ANSI Z41.1-1967 (R1972) and meet Class 75 requirements for men's and women's shoes.

The management and communication of this standard is similar to the eye protection standard. Management, safety, and operations personnel determine areas where protection is required. A communication program is established to inform all affected personnel. The company should ensure available access to required company footwear.

Head Protection

Hard hat requirements have developed in response to hazards found in the workplace. The construction, design, and materials have continued to improve as advances in plastics technology evolve. Hard hats were originally made from aluminum or steel. This material had several design problems that the use of plastic resolved. Hard hats or caps use a webbed suspension to cushion the effects of an impact. The structural support of the hat is determined by each of the elements of the design being in place. For example, holes should never be drilled in the cap or paint applied that could weaken the exterior. The adhesive on the back of some stickers will adversely affect the hat. Hard hats and suspensions need to be inspected frequently to ensure the connections between the suspension and hard hat are in place, the adjustment is correctly aligned, and the interior and exterior are clean. The hard hat should be replaced if it is cracked, chipped, worn, or exceeds reasonable age limits.

Ear Protection

Process technicians exposed to high noise levels can permanently lose their hearing. Noise is measured in decibels (dB) that are used as a relative measure of sound. Process equipment and systems can generate high noise levels. The chemical processing industry (CPI) regulates noise levels over an eight-hour, time-weighted average (TWA) work shift to be around 85 dBA. Exposure should never exceed 115 dBA. The CPI has serious responsibilities for protecting their employees' hearing. Earplugs and earmuffs are required operating equipment that must be used during exposure. The following definitions are vital in understanding the importance of hearing protection:

- Administrative control—procedures that are in place to limit technician exposure to noise.
- Audiogram—a sound test used to graph or chart a technician's hearing level. (Process technicians are tested periodically to check hearing loss.) **STS**—Standard Threshold Shift is a term used to describe an audiometric test that is compared to prior tests to determine a measurable hearing loss.
- Engineering control—reduction of noise through noise absorbing equipment, enclosures, and equipment system redesign.
- Hearing protection—use of ear plugs and earmuffs.
- Hearing conservation program—includes training, hearing protection, exposure monitoring, audiometric testing, and noise abatement.

Skin Protection

Protective clothing is designed to prevent exposure to hazardous environments. Examples of protective clothing include the following: acid hoods and suits, rubber aprons, gloves, and chemical-resistant suits. Chemical plants and refineries process and manufacture a variety of chemicals. Because close contact with this material occurs on a frequent basis, flame-resistant clothing (FRC) and chemical-resistant clothing are considered to be part of a typical work uniform. In areas where exposure levels increase, the level of protection increases.

Hazards in the Workplace

Hazards in the chemical processing industry can be classified as unplanned situations that can injure or kill a process technician. As shown in Figure 10-3, examples of hazards include burning, exploding, cutting, electrocution, impaling, tripping, falling, slipping, crushing, loud noises, or exposure to a chemical. Exposure to chemical hazards may expose a technician to cancer-causing chemicals, toxic chemicals, asphixiants, chemical burns, or irritants.

Hazard Prevention

Some of the methods used by industrial manufacturers to prevent exposure include engineering or environmental controls, personal protective equipment, safe work practices, good housekeeping, and training. Engineering controls include reduction of noise through the use of noise or sound absorbing material, new technology automating the processes, and so on. Safe work practices include written procedures and effective employee training. Engineering controls and safe work practices are the preferred methods of hazard prevention by OSHA. When these two methods cannot be used, personal protective equipment should be used to prevent exposure. PPE is designed to protect process technicians from chemical and physical hazards.

Emergency Response—Four Levels of PPE

As shown in Figure 10-4, emergency response has four levels of personal protective equipment according to the Environmental Protection Agency and the Occupational Safety and Health Administration. Level A requires the highest level of PPE protection by requiring a technician to don a Totally Encapsulating Chemical Protective (TECP) suit. The TECP suit is used in situations where inhalation and skin absorption of hazardous chemicals are possible. TECP suits provide a limited, positive air supply to the technician. The TECP suit comes in a variety of designs and material compositions. The chemical suit is designed for working with specific chemicals and in specific situations. Level B deals with chemical exposures that are not considered to be extremely toxic unless they are absorbed through the skin. In this case a non-airtight chemical protective suit may be worn. Typically the openings on a non-airtight chemical suit are taped to limit exposure. Level C

Figure 10-3
Hazards in the Workplace

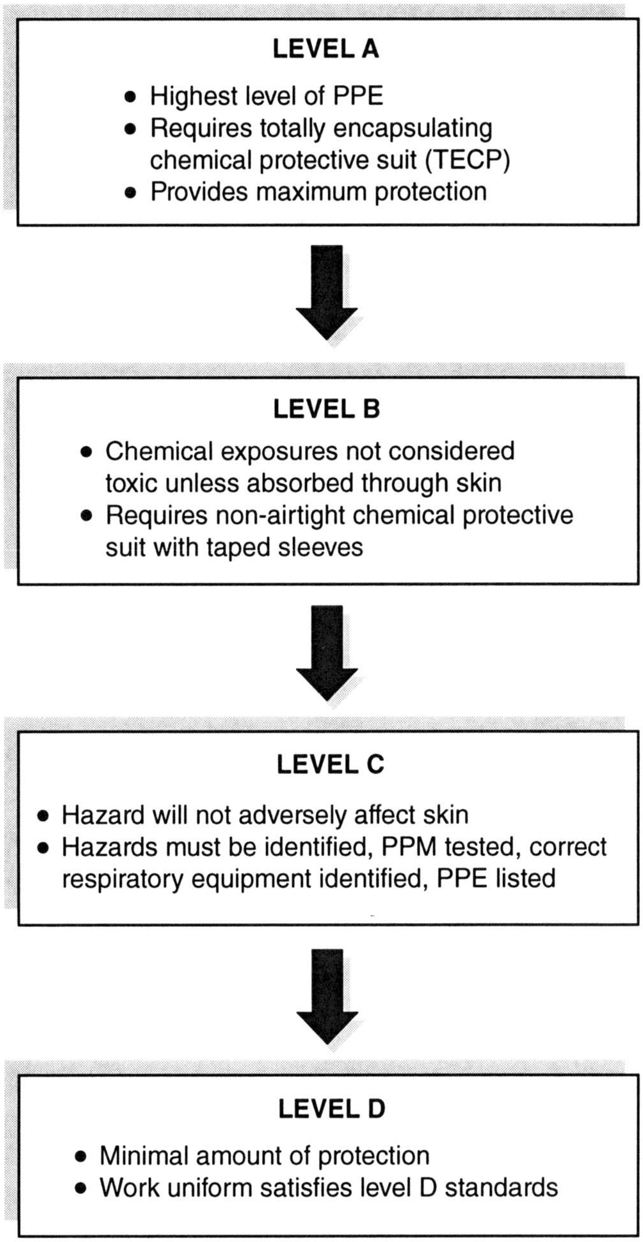

LEVEL A

- Highest level of PPE
- Requires totally encapsulating chemical protective suit (TECP)
- Provides maximum protection

LEVEL B

- Chemical exposures not considered toxic unless absorbed through skin
- Requires non-airtight chemical protective suit with taped sleeves

LEVEL C

- Hazard will not adversely affect skin
- Hazards must be identified, PPM tested, correct respiratory equipment identified, PPE listed

LEVEL D

- Minimal amount of protection
- Work uniform satisfies level D standards

Figure 10-4
Four Levels of PPE

is used when the hazard is determined to not adversely affect exposed skin. Before level C protection is chosen, the following steps must be completed:

1. The chemical hazard or contaminant must be identified.
2. The concentration (parts per million) must be identified.
3. Correct respiratory protection must be identified.
4. A list of suggested PPE must be provided, for example, chemical-resistant gloves.

Level D provides the minimal amount of protection to a process technician. This level takes into account that engineering controls are in place and that the technician is safe working in this environment. Level D protection is determined by individual companies because the standard personal protective equipment is the work uniform. This may include fire-retardant clothing, safety glasses, hard hat, steel-toed shoes, hearing protection, radio, and so on.

Written Respiratory Protection Programs

The Occupational Safety and Health Administration requires employers who use and issue respirators to develop a written respiratory protection program. The purpose of respiratory protection is to protect employees from environments that could be hazardous. Employees must receive proper training in respiratory protection. Typically the industrial hygiene department administers the written respiratory protection program and coordinates employee training. Technicians are fit tested for each respirator they will use on their units. **Fit testing** is a procedure that ensures the respirator forms a good seal on the technician's face. Because the facial structure of each technician is different, it is impossible to create a *one design fits all* respirator. During fit test training the technician learns the following:

- Which respirators will be used in their assigned area
- Which respirators fit their faces
- How to properly **don a respirator**
- How to select a respirator
- Specific hazards in their assigned units

Process technicians use two basic types of respirators: (1) air purifying and (2) air supplying.

Air-Purifying Respirators

Air-purifying respirators come in two designs, half-face and full-face. Half-face air-purifying respirators are designed to cover the mouth and nose, and full-face respirators form a positive seal around the eyes, nose, and mouth. These respirators are designed to remove specific contaminants or organic vapors from the air. These concentrations may range from 5 to 50 times the normal exposure limit of dust, mists, fumes, organic vapors, ammonia, and acid gases allowed by law. Filter life is a term used to estimate the amount of service time of an air purifying filter.

Air-Supplying Respirators

Air-supplying respirators come in two designs: self-contained breathing apparati (SCBA) and hose line respirators. These respirators are designed to be used in oxygen-deficient atmospheres, and to also provide protection from dust, mists, fumes, organic vapors, ammonia, and acid gases.

Selecting a Respirator

Process technicians select respirators for use with specific substances. Because respirators are selective in what they will remove and the protection they will provide, a number of issues must be addressed before a respirator can be selected for a specific job:

- Type of contaminant or hazard
- Airborne chemicals or contaminant present
- Hazards associated with the chemical(s)
- Physical characteristics of the chemical(s)
- Air-purifying or air-supplying respirators
- Oxygen-deficient environment
- Specific procedure that will be used
- Personal protective equipment required
- Permits required
- First aid
- Length of time to perform task(s)
- Exposure time limit
- Concentration of chemicals present

Ultimately, you will need to ask yourself, "Will the respirator I choose protect me from the environment I will be working in?" Figure 10-5 describes the key elements of a respirator program CFR 29 1910.134.

After the known contaminants are identified, the process technician is required to reference the material safety data sheet (MSDS) for specific information. For more information, refer to HAZCOM.

Process technicians should always be concerned about the environment they are working in and the chemicals they are handling. For this reason the various processes inside a plant are carefully monitored. Industrial hygienists periodically sample and monitor the conditions that exist at various units over extended periods of time. Historical data is combined with current trends to determine the most effective procedure for limiting hazards.

NIOSH-approved air-purifying respirators are color coded so filters, cartridges, and respirators can easily be matched up. In most plants, the contaminant they are designed to remove can be found written on the band.

Caring for and Using Respirators

Respirators should always be stored in plastic bags to keep dust and dirt from accumulating on them. OSHA inspectors will cite the company for the improper storage of a respirator. For this reason respirators should never be carelessly discarded or left lying around. Site safety inspectors and

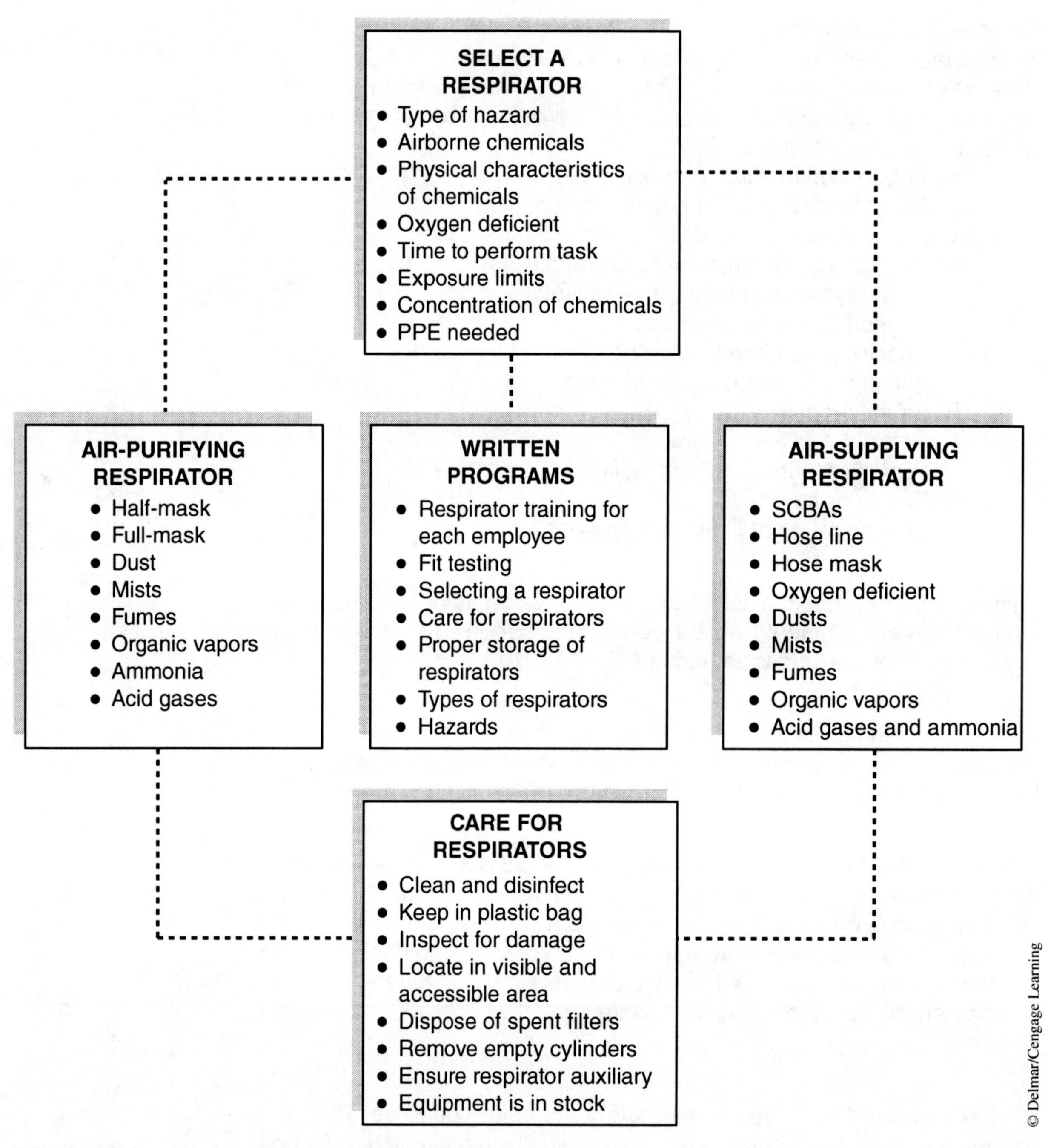

Figure 10-5 *Respirator Program CFR 29 1910.134*

first and second-line supervisors will notice this infraction quickly. Respirators should be located in areas where they are visible and accessible.

When a respirator is used, it is the responsibility of the technician to perform the following functions:

- Carefully clean and disinfect the device.
- Return the respirator to a plastic bag.

- Ensure the elastomer around the face piece seal is not folded or bent.
- Visually inspect for damage.
- Properly dispose of spent filters and cartridges.

The company will typically supply the procedure and the cleaning materials required for this assignment. Used filters and cartridges should be properly disposed of. Check your local procedure to ensure that you are complying with your plant's requirements. If the filter is contaminated with chemicals, it may fall under a special disposal plan.

Periodic training is required for those using respirators in routine job tasks. Self-contained breathing apparatus (SCBA) users typically train monthly on donning and using the air-supplying respirator. Most companies require a technician to be able to don a SCBA in 60 seconds or less.

When a respirator is used, the technician should perform a positive or negative fit test. To perform a **negative fit test** the straps on the respirator should be pulled snug to ensure a tight seal. By placing the open palm over the respirator's inlet(s), you should be able to create a vacuum when you inhale. Hold the vacuum for 10 seconds to ensure a positive seal. The **positive fit test** is performed by exhaling and covering the exhaust ports. If the respirator has a good seal, a positive pressure will fill the face piece. If the respirator is damaged, it should be immediately replaced.

Air-purifying respirators remove contaminants from the atmosphere by placing a filtering medium between the contaminant and the respiratory tract of a process technician. Air is pulled through a canister, cartridge, or filter and into the respirator mask during normal respiration. The filtering media will purify the air using one of two methods: mechanically filtered or chemically adsorbed.

- Air-purifying respirators are designed to be used on specific contaminants and compounds. Extreme care should be used when selecting a respirator. The **filter life** of a respirator is determined by two factors: service time and contaminant concentration. The term **overpowering** is used to describe atmospheric conditions that exceed the design limitations of a respirator. A respirator that makes respiration difficult or allows a small percentage of the contaminant or compound to enter into the mask should be replaced immediately. Improper filter is a term applied to the selectivity characteristics of a respirator. Process technicians must select the correct respirator to remove a specific contaminant.

The oxygen content of an atmosphere is another important factor that a process technician needs to consider. Air-purifying respirators are not designed to work in atmospheres with less than 19.5% oxygen content. The term **IDLH** is applied to oxygen-deficient atmospheres that are immediately

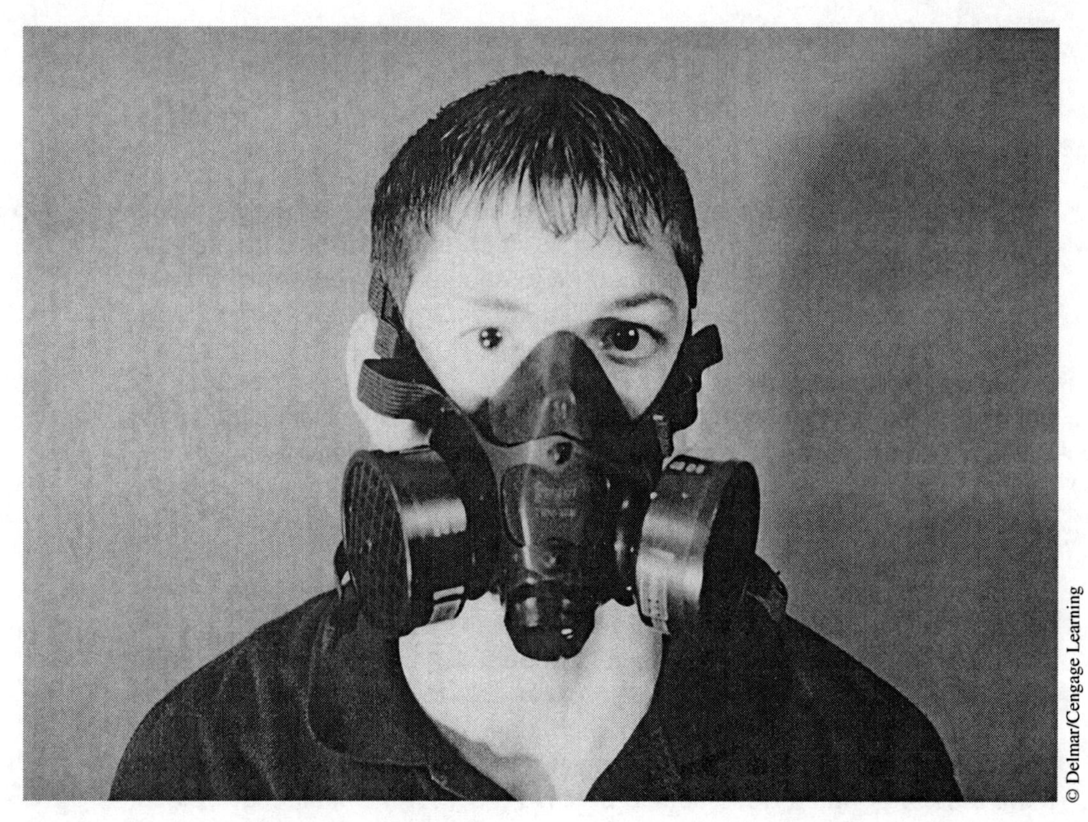

Figure 10-6 *Air-Purifying Respirator*

dangerous to life and health. Air-purifying respirators should not be confused with air-supplying respirators. Respirators have limitations that must be fully understood prior to performing a task that requires the use of a respirator.

Air-purifying respirators come in two basic designs: half-face and full-face. The half-face respirator fits securely over the mouth and nose whereas the full-face respirator forms a seal over the face of a process technician. Respirators come in a variety of shapes and designs. Figure 10-6 shows a half-mask, air-purifying respirator.

The chemical processing industry purchases respirators from a wide array of suppliers. Some common examples of different respirators include the following:

- North 7700 Half-Face Respirator—removes particulates from air; used in concentrations up to 10 times exposure limit of dust, mist, fumes, organic vapors, ammonia, and acid gases.
- Scott 66 Half-Face Respirator—removes particulates from air; to be used in concentrations up to 10 times exposure limit of dust, mist, fumes, organic vapors, ammonia, and acid gases.

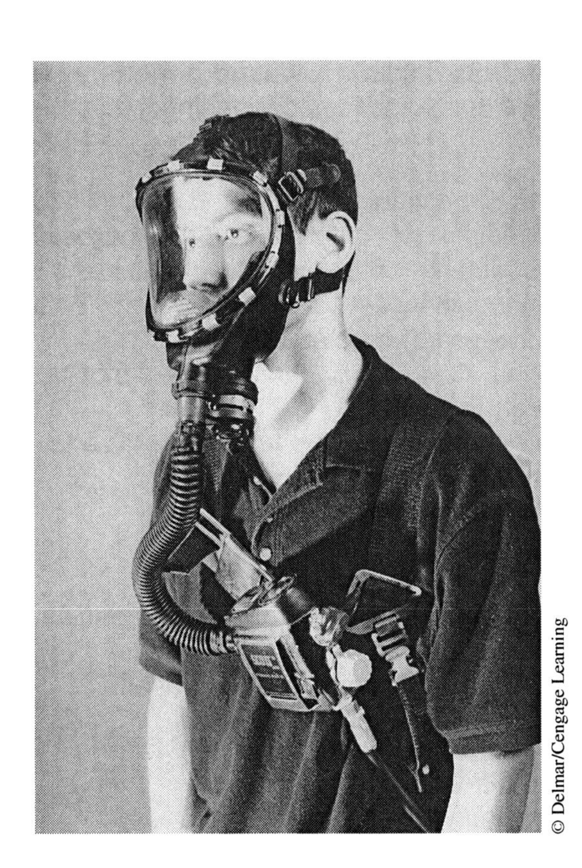

Figure 10-7
Full-Mask Air-Supplying

- North 7600 Full-Face Respirator—removes particulates from air, to be used in concentrations up to 50 times exposure limit of dust, mist, fumes, organic vapors, ammonia, and acid gases.
- MSA Ultravue Full-Face Piece—used for escape.
- 3M 9920 Dust-Fume Mist Respirator—disposable dust mask, removes particulates from air, to be used in concentrations up to 5 times exposure limit of dry materials, dusts, mists, aerosols, and welding fumes. Figure 10-7 is an example of a full mask respirator, air-supplying (SCBA).

Air-supplying respirators are used in environments that are suspected of having toxic concentrations of chemicals or are oxygen-deficient.

Self-Contained Breathing Apparatus

Most self-contained breathing apparati (SCBA) have full-face pieces, connecting hose, cylinder, and harness and weigh between 35 and 45 pounds. One of the first things new technicians notice after donning the air pack is the way it redistributes their center of gravity. This takes a little time to get used to and could cause an inexperienced technician to trip. Another disadvantage experienced with the SCBA is found in the face piece. On hot or humid days the mask fogs up and collects perspiration.

Three other limitations include reduced peripheral vision, increased difficulty in getting into tight spaces or openings, and muffled and unclear communication.

SCBAs are located throughout the plant and are typically located near areas where they could possibly be needed. SCBA boxes are clearly marked and should be inspected monthly. Process technicians should check the following things during an inspection:

- Check five-year hydrostatic test date.
- Inspect physical condition of SCBA—dents, rust, worn parts, cracks, and so on.
- Ensure that the pressure gauge indicates a full charge.
- Make sure the cylinder valve is off.
- Close the red bypass valve.
- Close the yellow mainline regulator valve.
- Ensure that the regulator cover is secure.
- Unthread breathing tube and ensure threading.
- Check diaphragm.
- Don SCBA and ensure proper operation.

SCBAs provide a self-contained air supply that will last 30 to 60 minutes. Figure 10-8 is an illustration of an SCBA. The air supply in a SCBA will vary depending upon a number of factors:

- Physical condition of the technician
- Physical size of the technician
- Training and experience of the technician
- Total charge in the cylinder
- Atmospheric pressure
- Environmental conditions—fear, excitement, emotional factors, and physical exertion
- Carbon dioxide concentration greater than 0.04% in the compressed air cylinder

Hose Line Respirators

The basic components of a typical hose line respirator include a face piece, hose line, harness assembly, and air supply. These are illustrated in Figure 10-9. Some of the more common limitations associated with hose line respirators include the following:

- The mask fogs up and collects perspiration.
- Peripheral vision is reduced.
- The mask increases difficulty in getting into tight spaces or openings.
- Communication becomes muffled and unclear.
- The weight of the hose line can be uncomfortable.

Hose line operators are required to wear emergency escape units. These units are designed to provide five minutes or more of emergency air to

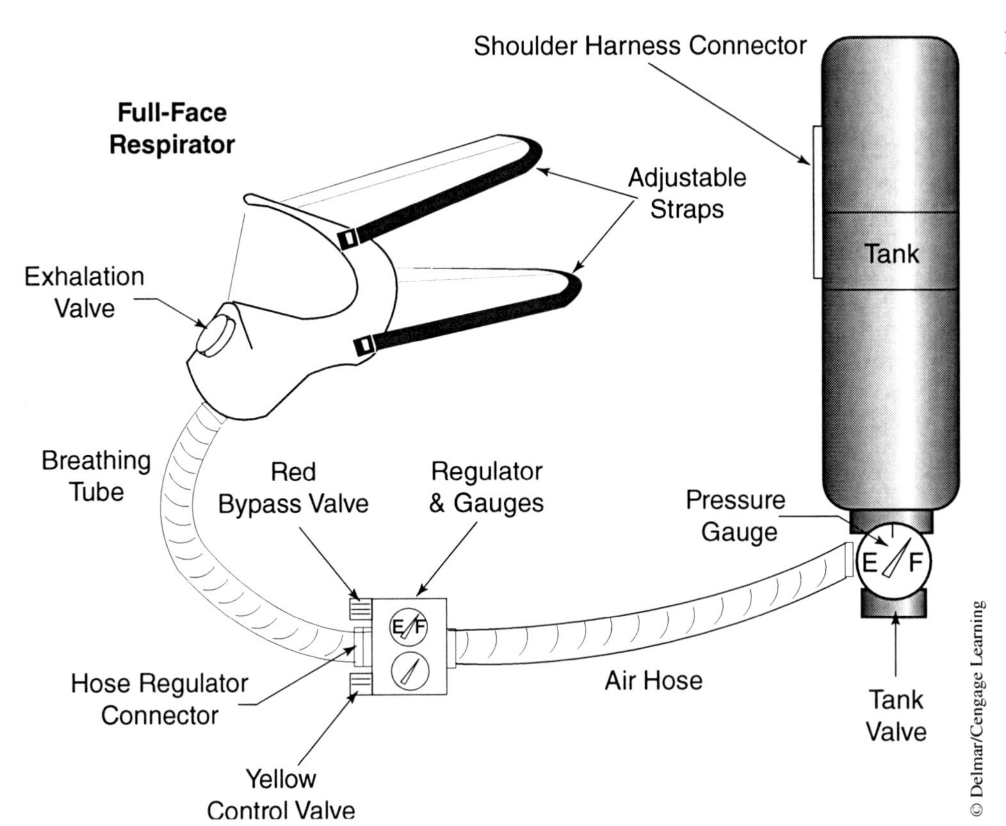

Figure 10-8 *SCBA*

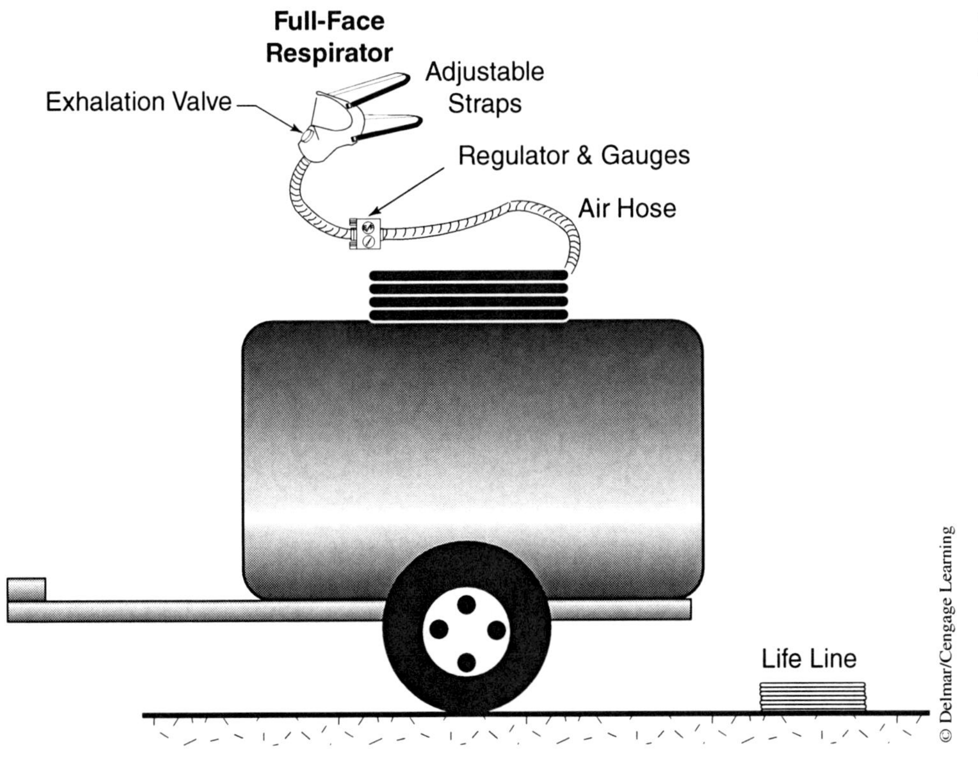

Figure 10-9 *Hose Line Respirator*

allow the technician to escape quickly. Typically hose line respirators are used in environments that are immediately dangerous to life and health. If air is interrupted to the face piece, the technician can use the emergency escape unit to leave the area.

Types of hose line respirators and face pieces include the following:

- Scott Ska-Pak—trailer mounted air supply
- Scott-o-Vista—face piece: cone shaped lens and net-type webbing with two straps
- Scottoramic—face piece: flat lens with five rubber straps

Summary

Some of the methods used by industrial manufacturers to prevent exposure include: engineering or environmental controls, personal protective equipment (PPE), safe work practices, good housekeeping, and training. Engineering controls and safe work practices are the preferred method of hazard prevention by OSHA. When these two methods cannot be used, PPE should be used to prevent exposure. Chemical exposure occurs through inhalation, ingestion, injection, and absorption or skin contact. Protective measures need to be in place. PPE provides an effective means for protecting technicians from hazardous situations. Engineering and environmental controls provide another layer of protection. PPE is designed to be used in environments that place technicians in contact with hazardous situations. The primary purpose of PPE is to prevent exposure to hazards when engineering or environmental controls cannot be used. Typical outerwear worn by process technicians includes the following:

- Safety hats
- Safety glasses
- Fire-retardant clothing
- Safety shoes
- Hearing protection
- Gloves
- Face shield
- Chemical monogoggles
- Slicker suits
- Radio
- Respirators
- Chemical suits

Emergency response has four levels of personal protective equipment: Level A requires the highest level of PPE protection. Level B deals with chemical exposures that are not considered to be extremely toxic unless they are absorbed through the skin. Level C is used when the hazard is determined to not adversely affect exposed skin. Level D provides the minimal amount of protection, and this is provided by the work uniform.

The Occupational Safety and Health Administration requires employers who use and issue respirators to develop a written respiratory protection program. Employees must receive proper training in respiratory protection. Process technicians use two basic types of respirators: (1) air purifying and (2) air supplying. Air-purifying respirators come in two designs: half-face and full-face. Air-purifying respirators remove contaminants from the atmosphere by placing a filtering medium between the contaminant and the respiratory tract of a process technician. The filtering media will purify the air using one of two methods: mechanically filtered or chemically adsorbed. Air-supplying respirators come in two designs: self-contained breathing apparati (SCBA) and hose line respirators. These respirators are designed to be used in oxygen-deficient atmospheres. Respirators are selective in what they will remove and the protection they will provide. The following should be considered prior to selecting a respirator:

- Type of contaminant or hazard
- Health hazards associated with the chemical(s)
- Physical hazards of the chemical(s)
- Air-purifying or air-supplying respirator
- Oxygen-deficient environment
- Length of time to perform task(s)
- Exposure time limit
- Concentration of chemicals present

After the known contaminants are identified, the process technician is required to reference the material safety data sheet (MSDS) for specific information.

Respirators should always be stored in plastic bags to keep dust and dirt from accumulating on them.

When a respirator is used, it is the responsibility of the technician to perform the following functions:

1. Carefully clean and disinfect the device.
2. Return the respirator to a plastic bag.
3. Ensure the elastomer around face piece seal is not folded or bent.
4. Visually inspect for damage.
5. Properly dispose of spent filters and cartridges.

Review Questions

1. Identify the PPE found in a plant.
2. Describe when and how to use PPE.
3. Identify typical workplace hazards.
4. Discuss the methods of hazard exposure prevention.
5. Contrast emergency response and PPE.
6. Describe the four levels of PPE.
7. Describe the principles of hearing protection.
8. Identify PPE outerwear worn by technicians.
9. Explain the general limitations of PPE.
10. Contrast engineering and environmental controls with PPE.
11. Air-purifying respirators come in two basic types. List them.
12. Air-supplying respirators come in several different types. List them.
13. Describe the limitations of air-purifying respirators.
14. Describe the advantages of using an air-supplying respirator.
15. Define *IDLH*.
16. Define the term *filter life*.
17. Define *overpowering*.
18. In what atmospheres will air-purifying respirators work?
19. Contrast the terms *selectivity* and *respirators*.
20. Contrast the terms *don* and *doff* in relation to *respirators*.
21. Describe fit testing.
22. Contrast negative and positive fit testing.

Engineering Controls

Objectives

After studying this chapter, the student will be able to:

- Describe the key elements of risk evaluation.
- Describe the different types of alarms and indicators used in the CPI.
- Explain how floating roof tanks work.
- Discuss how closed-loop sampling systems work.
- Describe the term *noise abatement*.
- Describe interlocks and automatic shutdown devices.
- Explain the principles of process containment.
- Describe the principles of effluent and waste control.
- Illustrate and explain how a flare system works.
- Describe deluge and explosion suppression systems.
- Identify basic pressure relief devices.

Key Terms

- **Chemical process evaluation**—analyzes the hazardous properties of reactants, which are products that might be formed under certain conditions, and the environmental effects of those products
- **Equipment design evaluation**—is designed so that the failure of one or more devices will not result in a disaster. Process equipment must comply with safety codes, government regulation, standards, and current industry practices.
- **Hazardous materials evaluation**—includes a detailed analysis of all the properties of the materials handled, stored, and processed in the plant. This process looks at: (1) quantities, (2) physical properties, (3) toxicity, (4) stability hazards, (5) corrosiveness, and (6) impurities.
- **Interlock**—a device that will prevent an operational action unless a specific condition has been satisfied.
- **Operator practices and training evaluation**—operational procedures, training for operating technicians and supervisors, startup and shutdown procedures, permit system, housekeeping and inspection, chemical hazard recognition, emergency response, use of PPE, and auditing.
- **Physical operations evaluation**—includes chemical processes that change state. Examples of this include: distillation, absorption, agitating, centrifuging, crushing and grinding, crystallization, evaporation, extraction, filtering, granulation, leaching, spraying, mixing, and milling.
- **Plant location and layout evaluation**—key elements include: drainage and runoff control, climatic conditions, effects of uncontrolled releases, community capability and emergency response, plant accessibility, available utilities, gate security, hazardous unit placement, and spacing of equipment. Additional elements to this evaluation include: NEC-regulated electrical installations, clearly marked exits, building ventilation, fire walls, fire spread considerations, foundation and subsoil loadings, and administrative building location.
- **Redundancy**—a process that uses two or more devices to shut down a system.
- **Risk evaluation**—a process that is used to consider all of the potential risk factors found in a chemical process. Primary areas of concentration include: hazardous materials, chemical process, physical operations, equipment design, layout and location, and training.

Risk Evaluation

Some of the primary steps in the design and operation of a new chemical facility include a process called **risk evaluation**. There are many risk factors to consider in a chemical process—some that are not readily apparent—that could possibly lead to a major disaster. The risk evaluation system includes the evaluation of six areas:

- Hazardous materials
- Chemical process
- Physical operations
- Equipment design

- Plant location and layout
- Operator practices and training

Hazardous materials evaluation includes a detailed analysis of all the properties of the materials handled, stored, and processed in the plant. This process looks at: (1) quantities, (2) physical properties, (3) toxicity, (4) stability hazards, (5) corrosiveness, and (6) impurities. Examples of this process include evaluation of the following:

- Quantities—production, storage, handling, determine if loss of containment could result in a fire, explosion, toxicity, corrosion, and environmental.
- Physical properties—flashpoint, boiling point, melting point, explosive limits, and vapor pressure.
- Toxicity—threshold limit values.
- Stability hazards—reactivity, self-polymerization, and spontaneous combustion of amounts found in the plant.
- Corrosiveness—assign level of personal protective equipment (PPE).
- Impurities—considers the effects on materials. Determine if impurities could enhance fire, explosion, stability, toxicity, or corrosion of plant materials.

The **chemical process evaluation** analyzes the hazardous properties of reactants, which are products that might be formed under certain conditions, and the environmental effects of those products. There is a wide variety of chemical processes, gas processes, and refinery processes. Processes that have certain hazardous characteristics must be evaluated. Some of these characteristics include explosive reaction or detonation, spontaneous polymerization, reaction with water or air, operation at high pressures and temperatures, production of exothermic reactions, possibility of releasing toxic or flammable vapors, operation near explosive range, and presentation of a dust or mist hazard. Laboratory and pilot plant testing is always conducted prior to these operations to ensure that all primary hazards have been identified. This includes the use of process flow diagrams and descriptions of chemical reactions and the development of standard operating procedures. This last process, the development of operating procedures, should provide for emergency response to special conditions and the identification of all possible hazards.

The **physical operations evaluation** includes chemical processes that change state. Examples of this include distillation, absorption, agitating, centrifuging, crushing and grinding, crystallization, evaporation, extraction, filtering, granulation, leaching, spraying, mixing, and milling. These processes have the ability to introduce hazards that are not present before the material changes state. Areas of specific concern include the following:

- Dusting—dispersion of toxic and combustible solids
- Heat transfer—heating up of unstable chemicals
- Pressure—generated by mixing of unstable chemicals

- vaporization—diffusion of toxic and flammable liquids and gases
- Spraying—atomizing flammable, toxic, or combustible liquids
- Mixing—oxidizing agents with combustible materials
- Separation—of protective dilutants and inertants from hazardous materials
- Generation—static charge

Modern equipment design has developed as a result of the vast knowledge accumulated in chemical and refinery processing. Knowledge gained from past experience has been applied to present-day manufacturing. The **equipment design evaluation** is an integral part of hazard prevention. Equipment systems are designed so that the failure of one or more devices will not result in a disaster. Process equipment must comply with safety codes, government regulation, standards, and current industry practices. New construction and installation must be completed using the strictest guidelines. Emergency shutdowns and safeguards should be installed to provide adequate protection. Equipment and system modifications should not violate original design requirements. New unit construction should have equipment to safely handle any temperatures or pressures generated by the process. A failsafe analysis of electrical and control loop systems ensures that all systems will fail in the safe mode. An inspection and maintenance program is required to be in place prior to starting up the system.

The **plant location and layout evaluation** is an important part of community protection. The key elements of this evaluation include: drainage and runoff control, climatic conditions, effects of uncontrolled releases, community capability and emergency response, plant accessibility, available utilities, gate security, hazardous unit placement, and spacing of equipment. Additional elements to this evaluation include NEC-regulated electrical installations, clearly marked exits, building ventilation, fire walls, fire spread considerations, foundation and subsoil loadings, and administrative building location.

Operator practices and training evaluation is the final part of the risk evaluation. Operational failures have been identified as the most frequent cause of industrial disasters. The essential parts of this section include the following: operational procedures, training for operating technicians and supervisors, startup and shutdown procedures, permit system, housekeeping and inspection, chemical hazard recognition, emergency response, use of PPE, and auditing.

Design and Operation of Plants for Safety

Emergency preparedness is a complex process that takes place before construction starts. The safety design team can minimize the chance for catastrophic incidents during plant operations. Several factors make this system work. It is important that clear responsibility and accountability for this process be defined early. The risk evaluation team should be composed

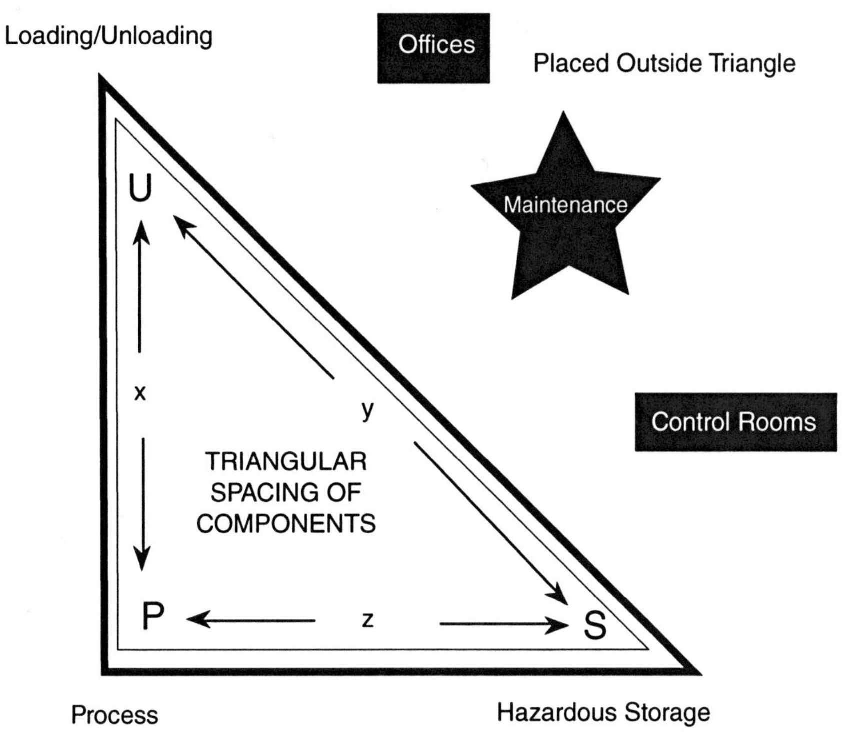

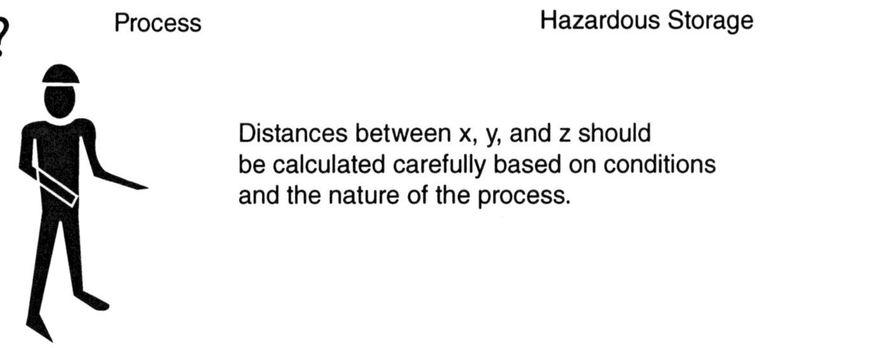

Figure 11-1
Triangular Equipment Placement

of experienced people from operations and design. The risk evaluation team will apply the principles of emergency planning to the original plant design. This will include adequate spacing between process equipment, specifically storage and loading facilities. The team will evaluate and limit the storage of hazardous materials. Where it is possible, liquefied gases will be stored under low temperatures and pressures. The team will also evaluate the use of open structures and will recommend new designs for total storage containment and other safety features like the use of water curtains. Figure 11-1 illustrates several elements this team would be looking for as it sequentially works through each step of the risk evaluation.

Alarms and Indicators

The most effective tools used by a process technician in the operation of a large chemical facility are alarms and indicators. Control rooms are the central hub for the location of most alarms and indicators; however,

Figure 11-2 *Alarms and Indicators*

in smaller operations, these devices can be located near the equipment. Critical alarms are attached to panels that will flash and emit an audible alarm. This will always generate an immediate response from operating personnel. The primary variables controlled by a process technician include the following: pressure, temperature, flow, level, and analytical variables. In addition to the panel alarms, advanced computer systems are used to monitor the process. These control systems provide a picture of the process that is not clearly visible from the field. Figure 11-2 is a photo of a typical control room.

Fire Alarms and Detection Systems

Early warning fire detection systems provide the first critical minutes needed to respond to a fire. Most fires start out small and can be controlled and extinguished easily. A fire in the advanced stages is difficult to put out and can result in a catastrophic event. Training is one of the most important means for deterring the development of a fire. Each process technician should be aware of exactly what to do in the event of a fire. Although each plant has its own system for fighting fires, the basic procedure includes the following: notifying your supervisor or shift members about the fire; giving your name, location, and the extent of the fire; and then responding early to fight the fire or assist the fire team. Most technicians are given basic fire extinguisher training and are well aware of how to put out a small fire.

Automatic fire alarm systems come in three types: fixed-temperature type, rate-of-rise type, and nuclear detector device. A fixed-temperature type alarm is activated when the temperature reaches a predetermined value. This type of alarm may sound when the temperature reaches 150°F, 175°F, or 200°F. Rate-of-rise fire alarms activate when there is a rapid temperature increase in the area. Nuclear detector alarms are sensitive to the by-products of combustion and do not require heat to activate. These automatic systems eliminate the risk that a fire will go undetected. This type of system should emit a reliable signal, reach all personnel, and be easily recognized.

In addition to these systems, water sprinklers, carbon dioxide, and Halon gas systems may be activated as well. Common examples of other detectors include smoke detectors, temperature switches, and carbon monoxide detectors.

Toxic Gas Alarms and Detection Systems

The purpose of toxic gas detection and alarm systems is to detect the presence of a toxic or hazardous gas in sufficient concentrations (parts per million—ppm) to be potentially dangerous to personnel. These alarms may be located on the outside perimeter of a plant or in areas where certain hazardous chemicals exist. These devices are frequently checked and periodically replaced. Toxic gas alarm systems may include an audible signal, a water sprinkler system, or a deluge system to suppress the toxic release. Examples of hazardous gases include the following: hydrogen sulfide, hydrogen fluoride, phosgene, chlorine, and so on.

Redundant Alarm and Shutdown Devices

Operating a large chemical complex requires the use of redundant systems that provide emergency shutdown and warning protection. **Redundancy** is a process that uses two or more devices to shut down a system. Redundant alarm and shutdown devices are required in certain hazardous situations where it has been determined that a specific condition poses a threat to the safety of the operation. This may include multiple (2) electric signals sent to a computer system that is programmed to activate an alarm or shutdown. It may include two separate switches or detection devices connected to multiple alarms. It may include the use of pressure-relief devices and shutdown equipment. The purpose of automatic shutdown devices is to protect personnel and equipment. These devices typically respond to the primary variables controlled by a process technician and measurements that are considered high or low.

Interlocks and Automatic Shutdown Devices

An **interlock** is a device that will prevent an operational action unless a specific condition has been satisfied. An interlock will prevent damage to equipment and personnel. This is accomplished by stopping or preventing the start of certain equipment functions unless a preset condition has been met.

There are two types of interlocks: softwire and hardwire. Softwire interlocks are contained within the logic of the Distributed Control System DCS software. Hardwire interlocks are physical arrangements. The hardwire interlock usually involves electrical relays that operate independent of the control computer. In many cases they run side by side with the computer interlocks. However, hardwire interlocks cannot be bypassed. They must be satisfied before the process they are part of can take place.

A permissive is a special type of interlock that controls a set of conditions that must be satisfied before a piece of equipment can be started. Permissives deal with startup items whereas hardwire interlocks deal with shutdown items.

Process Containment and Upset Controls

There are a number of elements in controlling process containment and in the design of upset controls. These three elements include the following:

- Reduction in hazardous materials storage inventory
- Storage of liquefied gases and the conditions under which these should be stored
- Plan for a design that provides total containment

The first element is the reduction in hazardous materials storage inventory. Control in this case is associated with the philosophy that the best way to reduce the release of hazardous materials is to reduce the quantity stored inside the facility. Reducing inventory will result in a safer manufacturing facility; however, it will also require more frequent shipments of raw materials and better coordination between suppliers and customers. It is also essential that the number of transfer nozzles and ports be limited to improve safety. In this case, one large reinforced tank is preferred over several smaller ones. Hose connections should not be increased in size to accommodate higher flows and pressures, because this would increase the hazard.

The second element pertains to the storage of liquefied gases and the conditions under which these should be stored. For safety considerations it has been determined that keeping liquefied compressed gases (LPGs) at low temperatures and pressures is a much better operating practice. At higher temperatures greater pressures are exerted, which could lead to hazardous conditions: explosions or fires. In the liquid state, materials can easily be transferred using centrifugal pump systems. Liquids that are moving between vapor and liquid state present a series of hazardous processes for process technicians and engineers. Table 11-1 illustrates the hazards associated with liquefied gases stored at various conditions. It is also much safer to store hazardous materials under refrigerated conditions.

Table 11-1 *Liquefied Gas Loss of Containment*

Release	Temperature	Pressure	Result
LPG	Atmospheric	Atmospheric	Slow Evaporation
LPG	Refrigerated	Atmospheric	Small Flash/Slow Evaporation
LPG	Atmospheric	Pressure	Large Flash/Slow Evaporation

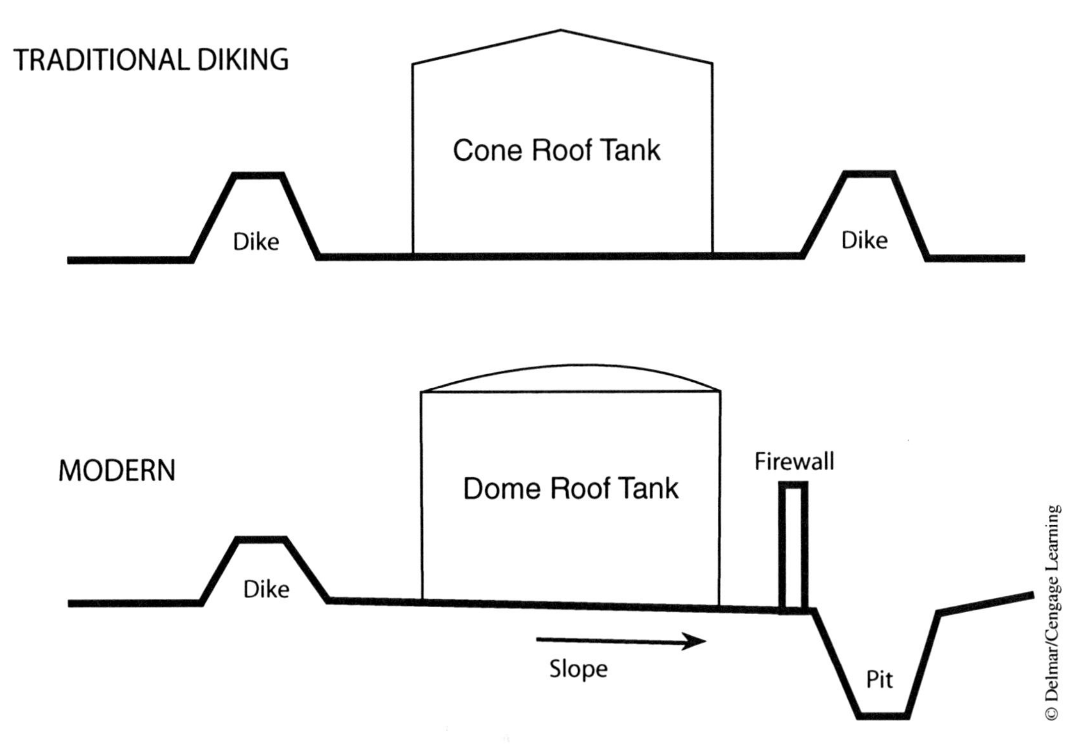

Figure 11-3 *Traditional and Nontraditional Diking Systems (Totally Self-Contained)*

The third element is for chemical facilities to plan for a design that provides total containment. This may include earthen dikes, concrete pits or sumps, firewalls, and sloped surfaces designed to move the flammable material away from the tank. Figure 11-3 shows modern and traditional diking designs for containment. The design should ensure that spills or leaks do not build up or accumulate under the equipment. Sumps are used to safely contain and remove fluids during a loss of containment.

Closed Systems/Closed-Loop Sampling

Closed systems/closed-loop sampling is a procedure where a bypass loop is installed that allows small part of the stream to be diverted from the main flow. In this type of system, an inline analyzer, container, or small sample port is installed. Figure 11-4 shows an illustration of a closed-loop sampling system. Closed-loop sampling offers a number of advantages that include waste reduction and environmental release control.

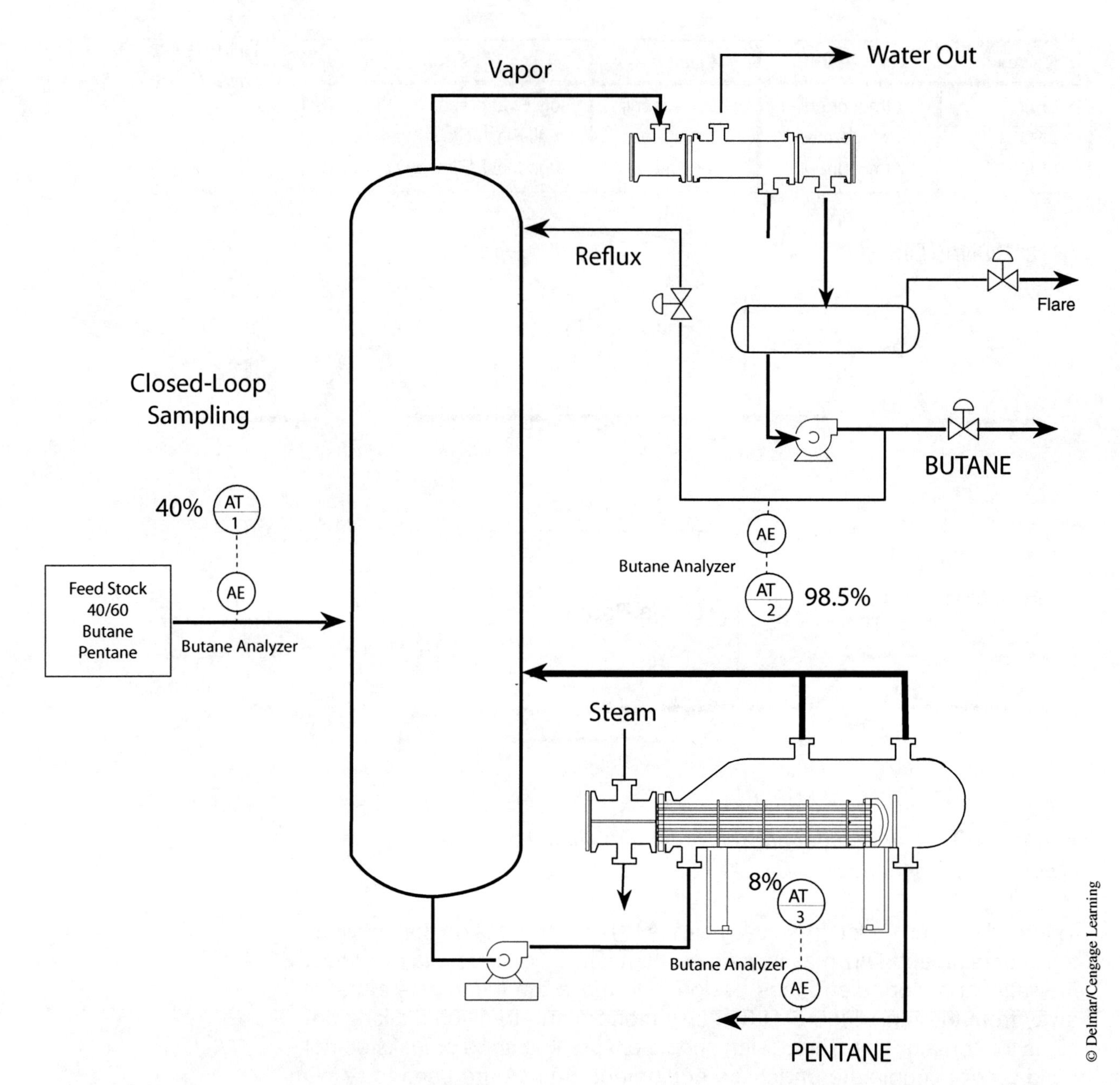

Figure 11-4 *Closed-Loop Sampling System*

Floating Roof Tank and Ventilation Systems

Internal or external floating roof storage tanks are used for storing materials at atmospheric pressure. A floating roof tank is designed to literally float on the surface of the liquid. The seal between the floating roof and liquid prevents vapors from leaving and entering the atmosphere. The primary purpose of a floating roof is to reduce vapor losses. Some tanks are designed with a

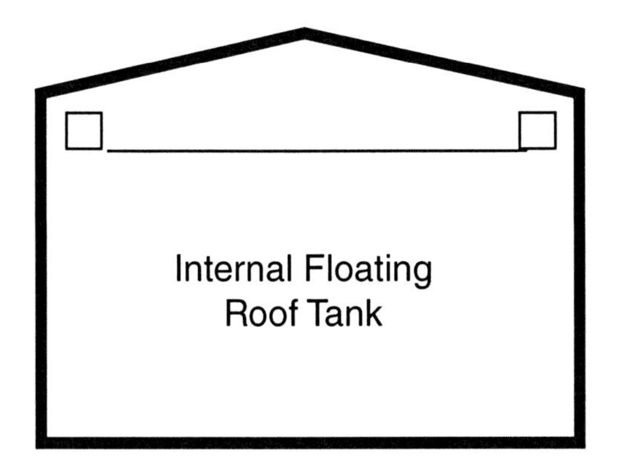

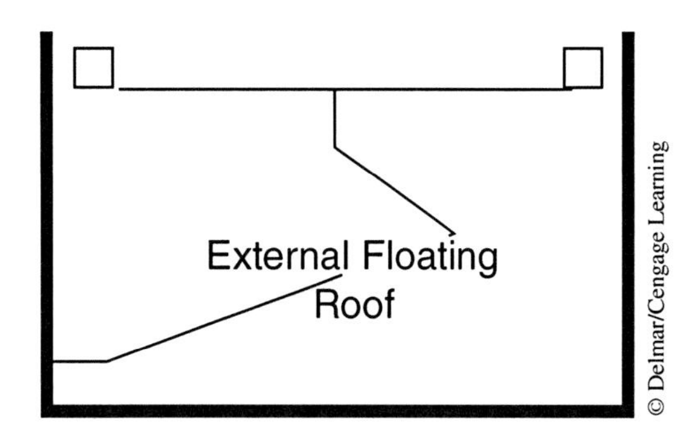

Figure 11-5 *Floating Tank Designs*

nitrogen purge that fills the void between the seal and the liquid. These types of systems are employed when hazardous materials are involved and hazardous vapors are formed. Internal floating roofs are used in areas of heavy snowfall because the weight of the snow would affect the seal on an external floating roof. Figure 11-5 shows external and internal roof designs. Floating roof tanks include a code stamp that identifies high pressure and temperature ratings, manufacturer, date, and type of metal, storage capacity, and special precautions. Because this type of system is considered to be above-ground storage, the American Society of Mechanical Engineers (ASME) Code, Section V111, governs vessels that have pressures greater than 15 psig.

Ventilation systems are usually designed to protect process technicians from organic vapors, harmful additive dusts, and fumes. Vent hoods operate by pulling a slight vacuum above the area producing the fumes. This procedure contains the airborne impurities in a closed system and allows for the safe disposal of particulates. Plastics technicians mix various additives into large ribbon blenders. This procedure produces excess particulates that easily flow into the open atmosphere without a ventilation system. Process technicians should use a redundant system of proper respiratory protection and ventilation system to remove impurities out of the air and keep these chemicals off the skin and out of the lungs. A vent hood uses a fan, piping, hood, hopper, and cyclone.

Ventilation systems are also used to provide fresh air into a control room or building. Positive pressure is typically maintained on a control room by recirculating the air inside the room. This procedure has been found useful in keeping airborne toxic substances outside the control area.

Effluent Control and Waste Treatment

sWastewater treatment is an important function in the plant. New technicians are typically assigned to this area in order to train on a variety of complex systems. Surface water is brought into the chemical or refinery complex in large quantities and stored in large settling basins. Suspended solids are allowed to drop out, and the water is sent into the plant for further treatment. Raw water is sent to filtering systems and then to a variety of areas including cooling towers, boilers, firewater, industrial, and so on. Additional water treatment is taken care of at or near the units that will be using the water. After this water has been used in industrial processes, it needs to be cleaned up for discharge. This may include removing heat and impurities from the process. Figure 11-6 shows an example of a clarifier, a device used to clean up water. Aeration basins are used with microorganisms to eat the hydrocarbons contained in wastewater. Retention time, a good supply of bugs, aerators to agitate the water, and a little phosphoric acid keep the system operating effectively. Sewer systems in the plant are connected to the wastewater treatment center. This area is often called the environmental control unit. Contaminated water is sent to the aeration basin for treatment. Excess water is held up in retention ponds or lagoons. Plant design should include proper control of all water run-off systems. Figure 11-7 shows how a settling basin works as raw water enters the plant.

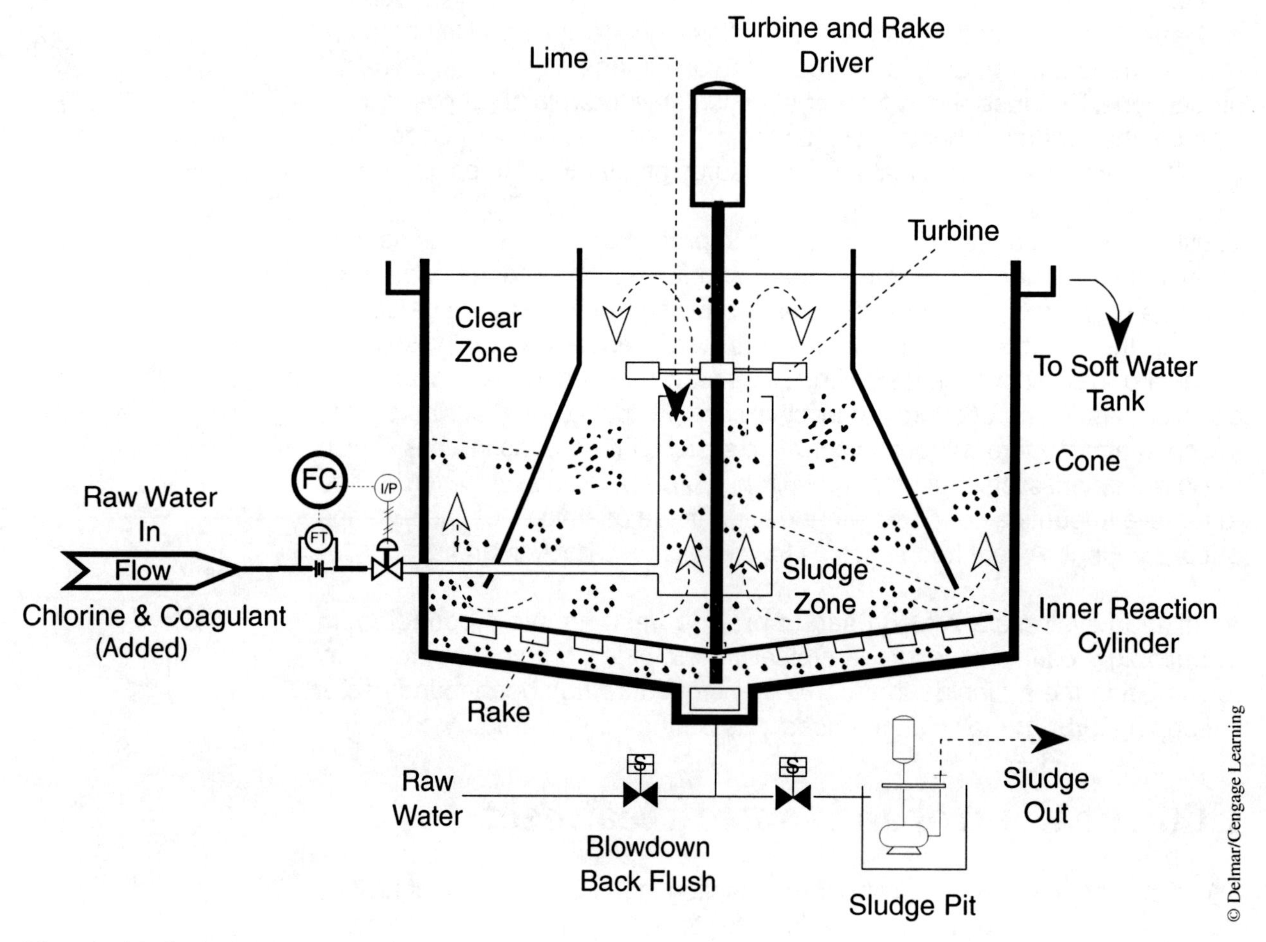

Figure 11-6 *Clarifier*

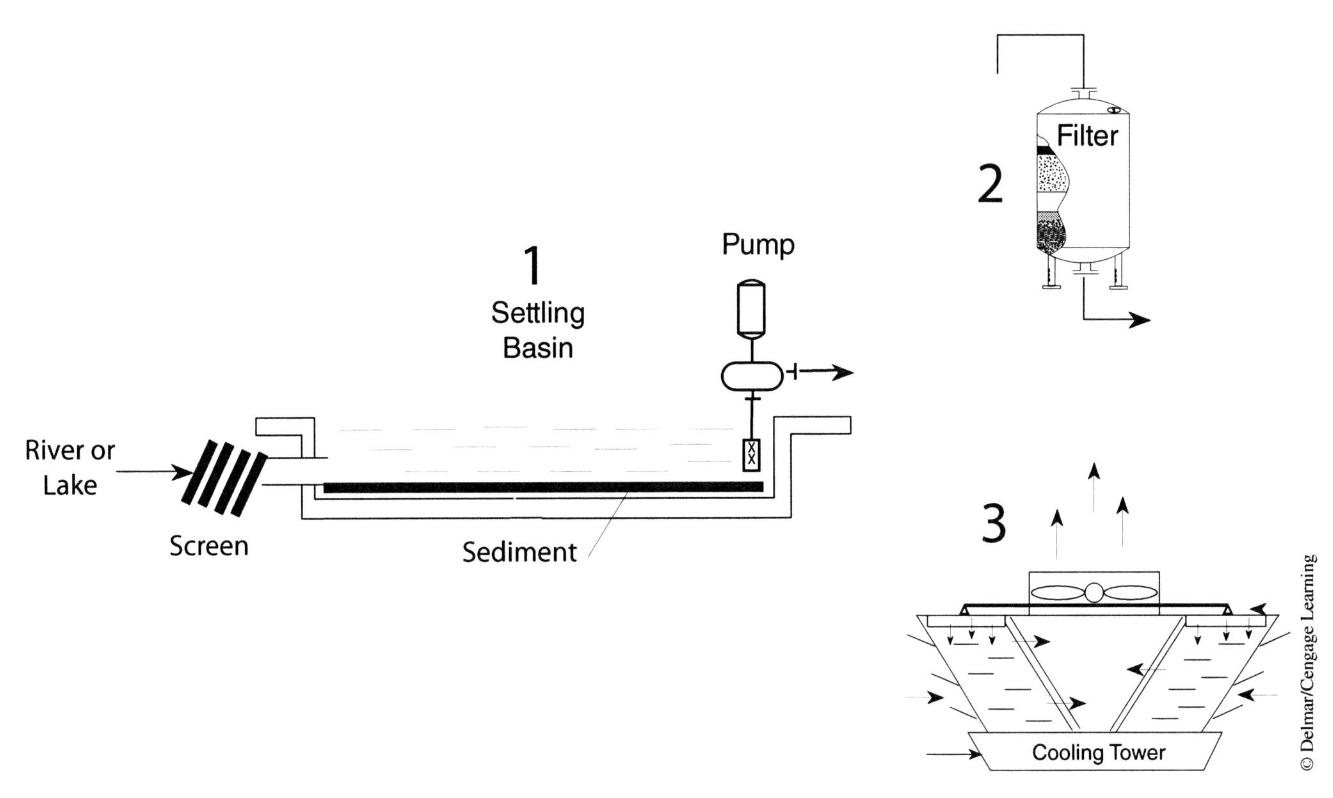

Figure 11-7 *Settling Basin*

Noise Abatement

Many devices in the process industry produce noise. Noise abatement is an engineering procedure designed to reduce or limit this noise through the use of modern technology. Examples of this technology include mufflers or silencers, noise reducing insulation, equipment designed to produce less noise, and heavily insulated doors.

Compressors generate high noise levels and typically use silencers on the inlet and suction side. Large piston compressors may be mounted inside an insulated sound-proof housing. Process technicians working in high noise areas may have sound-proof work stations located near the equipment. As a technician leaves an area that has sound-reducing paneling, such as a control room, kitchen, or sound island, hearing protection is quickly put in place to reduce the hazardous effects of noise.

Flares

One of the most important systems in a chemical processing system is a flare system that is designed to burn waste gases and control pressure generated by process upsets. A flare system is connected to every system

Figure 11-8 *Flare System*

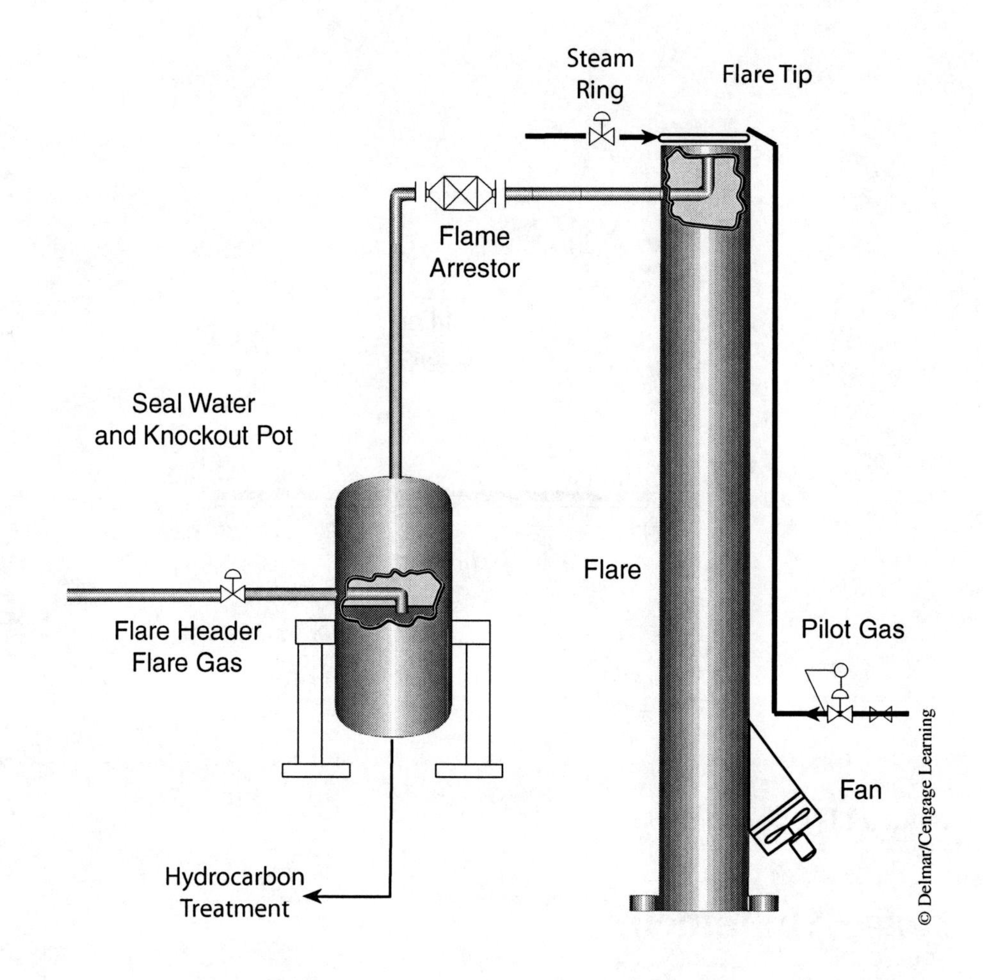

in the plant. This elaborate system is composed of numberless safety and relief valves, rupture discs, small and large piping, and one central header that leads to the flare. In some flare systems a knockout drum is provided with a water seal to separate liquids and vapors. The flare is a hollow pipe that allows the vapors to rise and burn cleanly at the flare tip. Figure 11-8 illustrates the basic components of a flare system. A fan may be installed at the base of the flare to increase airflow through the flare. Steam is used to disperse the hydrocarbons as they burn. More steam is used during periods of high discharge. A small pilot light burns continually at the tip of the flare and is used primarily to ignite the vapors.

Pressure Relief Devices

Pressure relief devices come in three different forms: safety valves for gases, relief valves for liquids, and rupture discs for both. All of these devices are considered to be automatic because they will work whether a technician is present or not. Figure 11-9 shows two of these devices. They are typically mounted on the equipment being protected and are sized to address maximum flow rates. Because vapors respond differently than liquids,

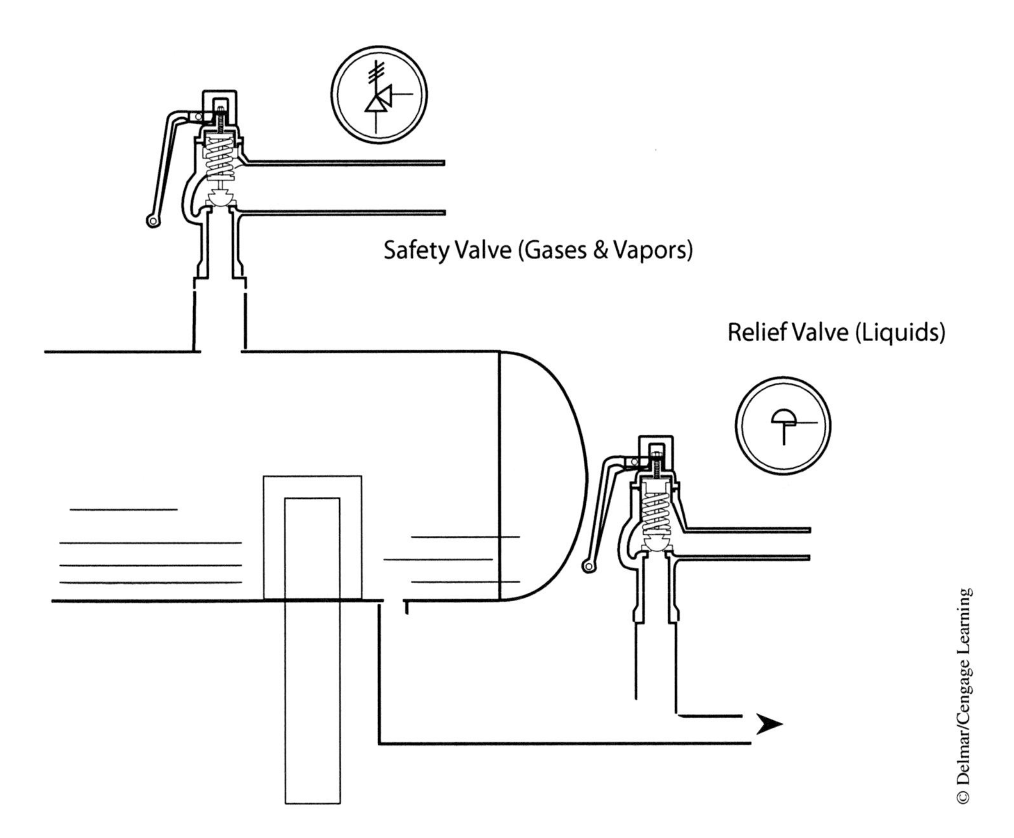

Figure 11-9 *Pressure Relief Devices*

the discharge port is much larger to accommodate increased flow rates. Pressure inside liquids is transferred equally in all directions and the discharge port can be the same size as the entry port.

Deluge Systems and Explosion Suppression Systems

A deluge system is designed to dump or spray large quantities of water for the purpose of extinguishing a fire. Similar systems are used to suppress toxic releases or hydrocarbon spills. Explosion suppression systems are composed of walls erected to contain an accidental explosion. These types of systems have been tested and used in process plant design. This includes control rooms and process areas.

Summary

Emergency preparedness is a complex process that takes place before construction starts on a new chemical facility, and can minimize the chance for catastrophic incidents during plant operations. The risk evaluation team is composed of experienced operations and design personnel who have clear responsibility and accountability for the process. This team will infuse emergency planning into the design of the plant, use adequate spacing for process plants, storage, and loading, limiting hazardous materials storage, storing liquefied gases at low pressures and temperatures, the use of open structures and water curtains, and design storage for total containment. The risk evaluation system includes six evaluation areas:

- Hazardous materials
- Chemical process
- Physical operations
- Equipment design
- Plant location and layout
- Operator practices and training

Operating a large chemical complex requires the use of redundant systems that provide emergency shutdown and warning protection. Redundancy is a process that uses two or more devices to shut down a system. Redundant alarm and shutdown devices are required in certain hazardous situations where it has been determined that a specific condition poses a threat to the safety of the operation. The purpose of automatic shutdown devices is to protect personnel and equipment. The three elements in controlling process containment and in the design of upset controls include the following:

- Reduction in hazardous materials storage inventory
- Storage of liquefied gases and the conditions under which these should be stored
- Plan for a design that provides total containment

An interlock is a device that will prevent an operational action unless a specific condition has been satisfied. An interlock will prevent damage to equipment and personnel. Pressure relief devices come in three different forms: safety valves for gases, relief valves for liquids, and rupture discs for both. All of these devices are considered to be automatic. Noise abatement is an engineering procedure designed to reduce or limit noise through the use of modern technology.

One of the most important systems in a chemical processing system is a flare system that is designed to burn waste gases and control pressure generated by process upsets.

Wastewater treatment is an important function in the plant. After water has been used in industrial processes, it needs to be cleaned up for discharge. This may include removing heat and impurities from the process. Aeration basins use microorganisms to eat the hydrocarbons contained in wastewater. Sewer systems in the plant are connected to the wastewater treatment center. Contaminated water is sent to the aeration basin for treatment. Excess water is held up in retention ponds or lagoons. Plant design should include proper control of all water run-off systems.

Review Questions

1. Describe how interlocks are used in automatic shutdown.
2. Identify the equipment used in noise abatement.
3. Identify the standard pressure relief devices.
4. Describe the three elements in controlling process containment.
5. Explain the term *process redundancy* and how it is used to protect equipment and personnel.
6. Describe the basic components of risk evaluation.
7. Explain how alarms and indicators are used in the chemical industry.
8. Identify the basic equipment used in a flare system, and explain how the system works.
9. Explain the reason for using a floating roof design.
10. Describe the stages of effluent control and wastewater treatment.
11. Explain the term *engineering controls*.
12. List the key elements of plant location and layout evaluation.
13. List the essential parts of *operator practices and training evaluation*.
14. List the integral parts of *equipment design evaluation*.
15. List examples of the *physical operations evaluation*.
16. Provide a detailed analysis of the *hazardous materials evaluation*.
17. Explain the primary purpose of toxic gas alarms and detection systems.
18. Describe the importance of fire alarms and detection systems.
19. List the three types of automatic fire alarm systems.
20. Describe closed-loop sampling.

Administrative Controls

Objectives

After studying this chapter, the student will be able to:

- Describe the key elements of administrative control.
- Explain the steps used in performing a job safety analysis.
- Describe Responsible Care® and Community Awareness and Emergency Response.
- Identify the essential parts of the Hazards and Operability Study (HAZOP).
- Compare and contrast a comprehensive process hazards analysis (PHA) system and HAZOP.
- Describe government-mandated training for process technicians.
- List the three things required to become a qualified first aid provider.
- Describe the importance of good housekeeping.

Key Terms

- **Administrative controls**—can be described as the programs and activities used to control industrial hazards.
- **Chemical Manufacturing Association (CMA)**—organized Responsible Care® in 1988 in order to address the public's concern about the use and distribution of chemicals.
- **Community Awareness and Emergency Response (CAER) programs**—designed to respond to communities' concerns about the manufacture and use of chemicals.
- **First aid**—the immediate, temporary care given to an accident victim.
- **Hazards and Operability Study (HAZOP)**—a system designed to identify hazards to technicians, equipment, operations, and environment.
- **Industrial hygienists**—collect samples from the work environment to determine hazardous conditions.
- **Mutual aid agreement**—provides a formal accord between industry and outside emergency response organizations in the event of a catastrophic release or situation.
- **Process hazards analysis (PHA)**—a structured brainstorming system used to identify hazards. HAZOP is an example of a PHA.
- **Qualified first aid provider**—an employee qualified by a certified medical group to administer first aid.

Introduction to Administrative Controls

Administrative controls can be described as the programs and activities used to control industrial hazards. This includes the policies, procedures, plans, principles, rules, agreements, and systems used in administrative control. Policies are guiding principles. A procedure is a sequential list of steps included to carry out an action. Plans and agreements are constructive methods used to carry out an action between different groups. **Mutual aid agreements** are written agreements between industry and outside emergency response organizations in the event of a catastrophic release or situation. Principles are used to establish a set of rules or guidelines. These rules are statements of how something is to be completed. Systems are associated with principles and rules and are described as organized sets of related principles.

Written programs are influenced by government and regulatory guidelines, company-specific guidelines, and unit-specific requirements. Examples of written programs include the following:

- Hazard Communication Program (HAZCOM) CFR 29 1910.1200
- Community Awareness and Emergency Response (CAER)
- Process Hazards Analysis (PHA)
- Hazards and Operability Study (HAZOP)
- Incident Command Systems (ICS)
- Plant permit systems

- Operator training
- Housekeeping
- Audits and inspections
- Mutual aid agreements
- Accident investigations
- Industrial hygiene monitoring
- Fugitive emissions monitoring

Hazard control utilizes evaluation, recognition, and removal of workplace hazards. The basic techniques used to remove industrial hazards can include the application of engineering controls, administrative controls, or the use of personal protective equipment (PPE).

Community Awareness and Emergency Response

The **Chemical Manufacturing Association** organized Responsible Care® in 1988 in order to address the public's concern about the use and distribution of chemicals. This program is frequently referred to as **Community Awareness and Emergency Response (CAER).** The guiding principles of Responsible Care® include the following:

- Respond to community concerns about chemicals and operations.
- Produce chemicals that can be disposed of safely.
- Report health, chemical, and environmental hazards and required PPE.
- Incorporate safety, health, and environmental considerations into new products and processes.
- Operate an environmentally safe plant.
- Conduct safety, health, and environmental research on products.
- Work with customers on the transportation, storage, and disposal of chemicals.
- Resolve handling and disposal problems.
- Create responsible laws, regulations, safeguards, and standards.
- Promote the principles of Responsible Care®.

Job Safety Analysis

Many techniques are used to develop safe work practices and procedures for jobs that have been identified as hazardous. A job safety analysis, or JSA, is the most common approach used by operations personnel. The primary steps used in this procedure are: (1) observe the people doing the job, (2) document the steps in the procedure, and (3) validate the procedure with subject matter experts. During this process, the reviewer looks for the following potential hazards:

- Being struck by or injured by equipment
- Rotating equipment
- Falling, slipping, or tripping
- Physical stress
- Poor visibility

Hazards and Operability Study

The Hazards and Operability Study (HAZOP) is a comprehensive **process hazards analysis (PHA)** system designed to identify hazards to technicians, equipment, operations, and environment. There are several benefits to using this type of study, including product optimization, productivity, profitability, and the identification of operational hazards. These benefits indirectly influence company morale, project improvements, and schedules.

Training and Mandated Training

Government-mandated training for process technicians covers a variety of topics and includes formal classroom and hands-on training that takes place upon initial assignment to a process unit, and training that takes place annually, every two years, or every third year. The following list includes most of the essential, mandated training topics and reference numbers. This is not the entire list; however, it does include most of the common topics taught at industrial facilities.

- Asbestos awareness and gasket removal—required upon initial assignment and annually each year thereafter (CFR 1910.1200/1001 and CFR 1926/1101)
- Benzene—required upon initial assignment (CFR 1910.1028).
- Blood-borne pathogens—required upon initial assignment (CFR 1910.1030)
- Bunker gear—required upon initial assignment and annually each year thereafter (CFR 29 1910.120)
- Control of hazardous energy (lockout/tagout)—required upon initial assignment and every two years thereafter (CFR 1910.147)
- Department of Transportation (DOT)—CFR 49 171-177
- Electrical training for unqualified persons—required upon initial assignment and every two years thereafter (CFR 1910.332)
- Fire extinguisher—required upon initial assignment and annually each year thereafter (CFR 29 1910.157)
- Hazard Communication (HAZCOM)—required upon initial assignment (CFR 29 1910.1200)

- Hazardous Waste Operations and Emergency Response (HAZWOPER) Awareness, Operations—required upon initial assignment; annually required annually thereafter (CFR 29 1910.120)
- Occupational noise exposure—required upon initial assignment; audiometric testing required annually thereafter (CFR 29 1910.95)
- Permit required for confined spaces—required upon initial assignment (CFR 29 1910.146)
- Personal protective equipment (PPE)—required upon initial assignment (CFR 29 1910.133 and 135)
- Powered industrial trucks—required upon initial assignment (1910.178)
- Process safety management (hot work)—required upon initial assignment (CFR 29 1910.119)
- Respiratory protection—required upon initial assignment and annually each year thereafter (CFR 29 1910.134)
- Scaffold user safety inspections—required upon initial assignment (CFR 29 1926.450–454)
- Specifications for accident prevention signs and tags—required upon initial assignment (CFR 29 1910.145)

Housekeeping

New process technicians are taught early in their work careers about the importance of housekeeping. Housekeeping prevents accidents and increases productivity. Process technicians can prevent accidents by removing items that clutter up the work site. Trash, debris, tools, and other items should be moved to the correct place. Water hoses are used to wash down the concrete mat and prevent dirt and debris from accumulating. A janitorial staff works with operating personnel to complete some required housekeeping tasks. Excess oils and greases that accumulate on equipment and floors should be cleaned up quickly. Chemical spills should be cleaned up and reported according to procedure. Good housekeeping improves the image of the company and the morale of plant employees.

Safety Inspections and Audits

The primary objective of a safety inspection is to identify unsafe work conditions and to ensure compliance with plant and government regulations. Audits and safety inspections are proactive in nature and design. Engineering and management should establish a system for inspecting operating units and facilities. Inspectors are specifically looking at emergency equipment and critical systems that pose a serious hazard should they fail. The frequency of the inspections should be determined by past experiences with the systems. Local inspections should cover the following: a record-keeping system, inspection training, a removal system for defective equipment, and a follow-up system. Process technicians assigned to the inspection team should perform inspections on schedule, record results, ensure the problems cited are corrected, and document any defects. Inspections are frequently conducted by local personnel who are familiar with the operation and maintenance of the facility. Audits are typically conducted by outside groups from corporate or regulatory agencies. Both systems use inspection and audit checklists during the process. (See Figure 12-1.)

Monitoring Equipment

Industrial monitoring comes in two monitoring systems: industrial hygiene and fugitive emissions. **Industrial hygienists** collect samples from the work environment to determine hazardous conditions. Examples of these activities include the following: ergonomic studies, noise monitoring, and toxic substance sampling. Fugitive emissions monitoring compares samples taken in the field to company, EPA, and government regulations. Testing equipment used for process monitoring includes the following: gas detection equipment, lower explosive limit (LEL) monitors, O_2 meters, and personal monitoring devices (dosimeters). This equipment is used to ensure a safe work environment, detect leaks, and measure a process technician's exposure to hazardous substances.

LEL monitors are used to monitor and detect leaks. These devices work by drawing in a sample and heating it up. LEL monitors measure process samples in three unique ways: metal oxide semiconductor (MOS), oxidized, and thermal conductivity. MOS detectors are designed to absorb combustible gases that generate a change in electrical conductivity that can be measured. Oxidized detectors measure the heat released by burning combustible gases. Thermal conductivity detectors vary from the other devices by measuring variations in thermal conductivity in combustible atmospheres.

O_2 detectors are classified as coulometric and polargraphic detectors. Coulometric detectors take a sample of the atmosphere and pass it over a coulometric cell that reacts with any contaminants producing an electrical current. Polargraphic oxygen detectors measure oxygen and carbon monoxide in the atmosphere.

There is a large array of monitoring equipment in addition to those devices previously mentioned, as shown in Figure 12-2. A short list of these detectors includes the following:

- Mercury vapor monitors—ultraviolet analyzers designed to measure mercury vapor concentrations
- Direct–reading colorimetric tubes and badges
- Flame ionization detectors

Sample Inspection Checklist

Date:____________ Shift No.:________ Technician:____________________________

EQUIPMENT	ITEM CHECKED (√) OR RECORDED (R)		
AS-520	Proper oil level (gearbox and blower)	√	
	Pressure	R	
P-520 OFFSPEC	Proper oil level (gearbox)	√	
	Suction pressure	R	
	Discharge pressure	R	
	No unusual noise on pump	√	
TK-520		√	
	Level	R	
	Area clean	√	
Ex-520	Steam pressure	R	
	Bypass closed	√	
	Shell inlet/out lined up	√	
	TIC-520 set @ 150°F	√	
	Air system on	√	
	CV-520 operating	√	
Ex-202	Steam lined-up to shell @ valve and TIC 202 (200°F)	√	
	Visual check on rotameter	√	
C-202	Upper steam tracing on	√	
	Lower steam tracing on	√	
	Direct inject steam on	√	
	Level on column	R	
	Bottom temperature	R	
	Differential pressure	R	
TK-202	Level	√	
	Flow line-up from TK-520	√	
	Temperature	R	
P-202	Proper line-up to Ex-202-C-202	√	
	Suction/discharge pressure	R	
TK-530	Proper dye level in reservoir sight glass (local)	√	
	Lined-up to P-530 and P-202	√	
Over-Head System	Ex-204 lined-up, Ex-205 lined-up	√	
	P-209 lined-up to 3-way and reflux	√	
Steam Traps	No unusual noises	√	
	Pulsing	√	

Comments: Describe what the problem is. Highlight (*) major operational or quality concerns.

Figure 12-1 *Inspection Checklist*

COMBUSTIBLE GAS INDICATOR

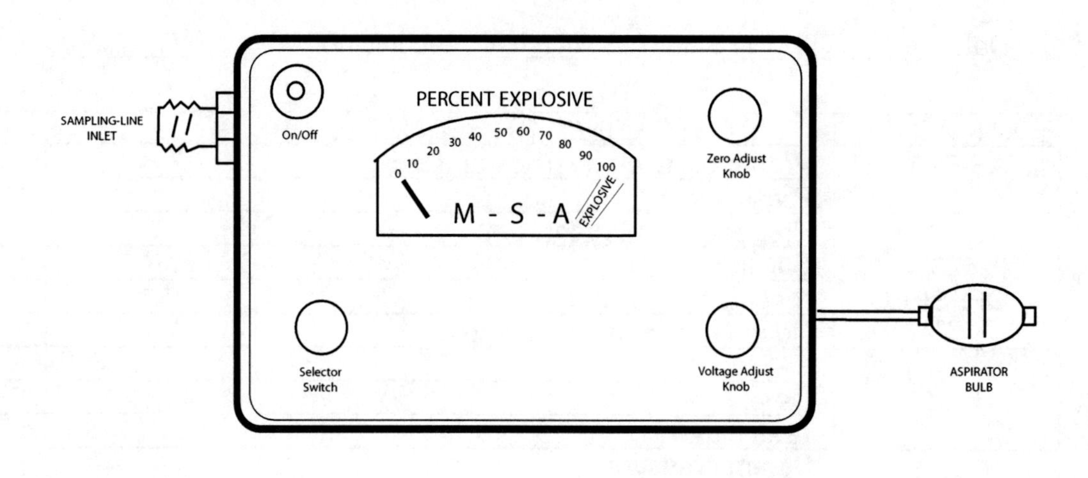

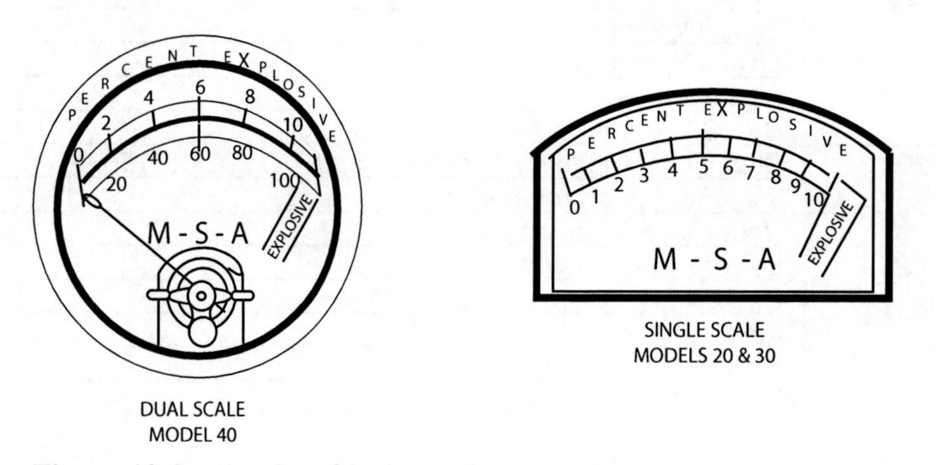

Figure 12-2 *Combustible Gas Indicator*

- Photonionization detectors
- Infrared analyzers
- Ultraviolet analyzers
- Gas chromatographs
- Ion mobility spectrometers
- Particulate monitors

First Aid

OSHA's General Industry Standard, CFR 29 1910.151 Medical Services and First Aid, is used by the chemical processing industry to establish a **first aid** program. First aid can be defined as the immediate, temporary care given to an accident victim. Despite the combined efforts of plant

Figure 12-3 *First Aid Station*

personnel, accidents and injuries occasionally occur. OSHA has acknowledged the need for trained workers prepared to administer first aid. A **qualified first aid provider** is described as an employee qualified by a certified medical group to administer first aid. Plant management is responsible for coordinating certified training programs, documenting recordable injuries, and supplying medical supplies. In order to qualify as a first aid provider, a technician must perform the following functions:

- Complete the Red Cross standard first aid, multimedia, eight-hour course
- Complete the cardiopulmonary resuscitation (CPR) eight-hour course
- Attend his or her company's first aid administrator lecture or seminar

Recertification is needed in CPR and Red Cross in one-year and three-year intervals. The voluntary duties of a qualified first aid responder include the following: administer first aid, keep medical records, maintain housekeeping and inventory on medical kits and stations, and apply the plant's hospital and emergency coordination plan when needed. Figure 12-3 shows a typical first aid station.

Summary

Administrative controls can be described as the programs and activities used to control industrial hazards. These include the policies, procedures, plans, principles, rules, agreements, and systems used in administrative control. Mutual aid agreements are written agreements between industry and outside emergency response organizations in the event of a catastrophic release or situation. Examples of written programs include the following:

- Hazard Communication Program (HAZCOM) CFR 29 1910.1200
- Community Awareness and Emergency Response (CAER)
- Hazards and Operability Study (HAZOP)
- Operator training
- Audits and inspections

The Chemical Manufacturing Association organized Responsible Care® in 1988 in order to address the public's concern about the use and distribution of chemicals. A number of techniques are used to develop safe work practices and procedures for jobs that have been identified as hazardous.

A job safety analysis (JSA) is the most common approach used by operations personnel. The primary steps used in this procedure are as follows: (1) observe the people doing the job, (2) document the steps in the procedure, and (3) validate the procedure with subject matter experts.

The Hazards and Operability Study (HAZOP) is a comprehensive system designed to identify hazards to technicians, equipment, operations, and environment. Benefits of performing a HAZOP include the following: product optimization, productivity, profitability, and the identification of operational hazards. These benefits indirectly influence company morale, project improvements, and schedules.

Government-mandated training for process technicians covers a variety of topics and includes formal classroom and hands-on training that takes place upon initial assignment to a process unit and training that takes place annually, every two years, or every three years. The following list includes most of the essential, mandated training topics and reference numbers: Hazard Communication HAZCOM—CFR 29 1910.1200, Hazardous Waste Operations and Emergency Response—HAZWOPER, CFR 29 1910.120, Respiratory Protection—CFR 29 1910.134, Personal Protective Equipment—CFR 29 1910.133 and 135, the Control of Hazardous Energy (lockout/tagout)—CFR 1910.147, and so on.

Housekeeping prevents accidents and increases productivity. Process technicians can prevent accidents by removing items that clutter the worksite. Trash, debris, tools, and other items should be moved to the correct place. Excess oils and greases that accumulate on equipment and floors should be cleaned up quickly. Chemical spills should be cleaned up and reported according to procedure. Good housekeeping improves the image of the company and the morale of plant employees.

The primary objective of an audit and safety inspection is to identify unsafe work conditions and to comply with plant and government regulations.

Inspections are frequently conducted by local personnel who are familiar with the operation and maintenance of the facility. Audits are typically conducted by outside groups from corporate or regulatory agencies. Engineering and management should establish a system for inspecting operating units and facilities. Inspectors are specifically looking at emergency equipment and critical systems that pose a serious hazard should they fail. Local inspections should include the following: a record-keeping system, inspection training, removal system for defective equipment, and a follow-up system.

Industrial monitoring comes in two systems: industrial hygiene and fugitive emissions. Industrial hygienists collect samples from the work environment to determine hazardous conditions. Examples of these activities include the following: ergonomic studies, noise monitoring, and toxic substance sampling. Fugitive emissions monitoring compares samples taken in the field to company, EPA, and government regulations. Testing equipment used for process monitoring includes gas detection equipment, LEL monitors, O_2 meters, and personal monitoring devices (dosimeters).

A qualified first aid provider is an employee qualified by a certified medical group to administer first aid. Plant management is responsible for coordinating certified training programs, documenting recordable injuries, and providing medical supplies. In order to qualify as a first aid provider, a technician must do the following: complete the Red Cross standard first aid, multimedia, eight-hour course; complete the Cardiopulmonary Resuscitation (CPR) eight-hour course; and attend his or her company's first aid administrator lecture or seminar. The voluntary duties of a qualified first aid responder include the following: administer first aid, keep medical records, maintain housekeeping and inventory on medical kits and stations, and apply the plant's hospital and emergency coordination plan when needed.

Review Questions

1. Explain the steps to becoming a qualified first aid provider.
2. Describe the effect good housekeeping has on safety.
3. List the various types of monitoring equipment.
4. Compare and contrast Responsible Care® with Community Awareness and Emergency Response.
5. Describe how a job safety analysis is conducted.
6. List the two major monitoring systems used in industrial monitoring.
7. Identify the primary objective of a safety inspection.
8. List the most essential mandated training topics covered in mandated training.
9. Compare and contrast industrial audits and inspections.
10. Explain the term *administrative controls*.
11. Identify the essential parts of the Hazards and Operability Study (HAZOP).
12. Define the term, *first aid*.
13. Describe mutual aid agreements.
14. List 13 written programs influenced by the government and regulatory guidelines, company-specific guidelines, and unit-specific requirements.
15. Explain the terms, training and mandated training.
16. List each of the mandated training requirements.
17. Explain the purpose of Fugitive emission monitoring.
18. Describe the testing equipment used for process emissions monitoring.
19. Explain the purpose of LEL monitors and MOS detectors.
20. O_2 detectors are classified as coulometric and polargraphic detectors. How do these devices work?

Regulatory Overview: OSHA, PSM, and EPA

OBJECTIVES

After studying this chapter, the student will be able to:

- Describe the process safety management (PSM) standard.
- Explain the written procedures requirement of the PSM standard.
- Identify the critical components of PSM action plans.
- Review the employee training issues contained in the PSM standard.
- Describe the process requirement of PSM.
- Review the three management issues covered under the PSM standard.
- Describe the audit section of the PSM.
- Explain the Occupational Safety and Health Act.

Key Terms

- **Code of Federal Regulations (CFR)**—contains all of the permanent rules and regulations of OSHA and is produced in paperback format once a year.
- **Compliance audits**—conducted by OSHA auditors to ensure compliance with governmental rules and regulations.
- **Emergency response plan**—a written plan that documents how specific individuals should respond during an emergency situation.
- **Environmental Protection Agency (EPA)**—established in 1970 to develop environmentally sound policies and national standards, support research and development, and enforce environmental regulations.
- **Federal Register**—a publication that (1) produces information on current OSHA standards, and (2) shows all adopted amendments, deletions, insertions, and corrections to government standards.
- **Flow diagram**—a simplified process drawing that uses standard symbols and diagrams to identify equipment and flows.
- **Hot work**—defined as welding, cutting, or using a spark-producing device.
- **Mechanical integrity**—a term that applies to the soundness of a plant process.
- **National Institute for Occupational Safety and Health (NIOSH)**—one of the three primary agencies created under the Occupational Safety and Health Act.
- **Occupational Safety and Health Act of 1970**—the purpose of this act is to (1) remove known hazards from the workplace that could lead to serious injury or death and (2) ensure safe and healthful working conditions for American workers. The Occupational Safety and Health Act applies to four broad categories: agriculture, construction, general industry, and maritime. There are three primary agencies responsible for the administration of the Occupational Safety and Health Act: National Institute for Occupational Safety and Health (NIOSH), Occupational Safety and Health Administration (OSHA), and Occupational Safety and Health Review Commission (OSHRC).
- **Occupational Safety and Health Administration (OSHA)**—one of the three primary agencies created under the Occupational Safety and Health Act.
- **Occupational Safety and Health Review Commission (OSHRC)**—one of the three primary agencies created under the Occupational Safety and Health Act.
- **Process hazard analysis (PHA)**—designed to identify the causes and consequences of fires, vapor releases, and explosions.
- **Process safety management (PSM) standard**—designed to prevent the catastrophic release of toxic, hazardous, or flammable materials that could lead to a fire, explosion, or asphyxiation.

Occupational Safety and Health Act

In 1970, a landmark piece of legislation was passed that required the chemical processing industry to make safety and health on the job a matter of federal law. The **Occupational Safety and Health Act** brought in sweeping changes that affected 4 million American businesses and, more importantly, 57 million employees and their families. In 1969 there were 2.5 million disabling injuries and 14,000 deaths that were directly linked to safety and health violations. The purpose of the Occupational Safety and Health Act is to (1) remove known hazards from the workplace that could lead to serious injury or death and (2) ensure safe and healthful working conditions for American workers. The coverage of the legislation is extensive in scope. The Occupational Safety and Health Act applies to four broad categories: agriculture, construction, general industry, and the maritime industry.

Figure 13-1 shows the three primary agencies responsible for the administration of the Occupational Safety and Health Act.

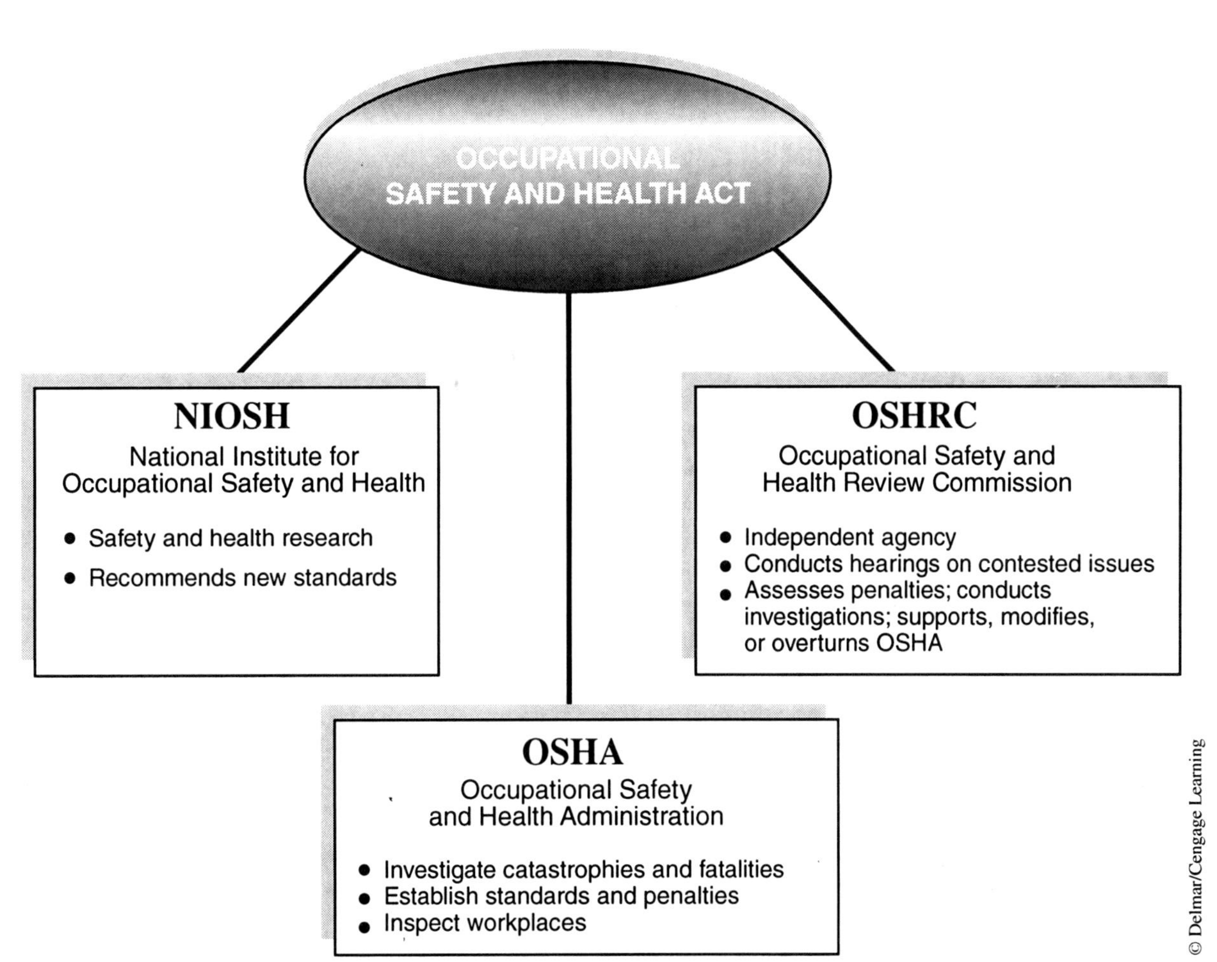

Figure 13-1 *OSHA*

1. **NIOSH—National Institute for Occupational Safety and Health.** This agency is responsible for safety and health research and performs the following functions:
   - Develops criteria for handling toxic substances
   - Researches safety and health issues
   - Recommends new safety and health standards
2. **OSHA—Occupational Safety and Health Administration.** This agency is responsible to do the following:
   - Investigates catastrophes and fatalities
   - Establishes new safety and health standards and penalties
   - Inspects workplaces
3. **OSHRC—Occupational Safety and Health Review Commission.** This independent agency performs the following functions:
   - Conducts hearings for situations on noncompliance that are contested
   - Assesses penalties, conducts investigations, and supports, modifies, or overturns OSHA findings

The best source for information on OSHA standards is the **Code of Federal Regulations (CFR).** The CFR contains all of the permanent rules and regulations of OSHA and is produced in paperback format once a year. Another publication that produces information on current OSHA standards is the **Federal Register.** The Federal Register is a publication that shows all adopted amendments, deletions, insertions, and corrections to government standards. This document is published daily and provides information to the public on federal activities.

When an OSHA inspection is conducted, prior warning is not given to the plant. The following steps are typically followed:

1. Opening conference—The inspector meets with employer and explains the purpose of visit. The inspector is accompanied by a company representative on the plant walk-through tour. The inspector is required to meet with a reasonable number of employees.
2. Inspection tour—The inspector will review plant injury and illness records, walk through, and review operations. It is a good idea to fix the problem on the spot. This shows the inspector that the company is serious about conforming to safety standards.
3. Closing conference—The inspector notes violations.

Process Safety Management (CFR 29 1910.119)

After the Bhopal incident in 1984, and the Phillips plant and ARCO explosions in 1989 and 1990, OSHA and the **Environmental Protection Agency (EPA)** went to work on a new standard that would limit the

EMPLOYEE PARTICIPATION
- Written program
- How employees will access hazard identification system
  - identify hazards
  - gather information
  - communication system

OPERATIONS PROCEDURES
- Operations and maintenance
- Reflect current work practices
- Process properties
- Hazards
- Startup, shutdown
- Change of chemicals

PROCESS SAFETY
- Process flow diagram
- Equipment, process description, limitations
- Consequences of deviation
- Safety and relief devices
- Electrical classifications
- Characteristics of chemicals
- Process chemistry
- Mixing chemicals

EMPLOYEE TRAINING
- Process overview
- Training records and method
- Attendance and competency
- Training materials reflect current work practices
- Control access to unit
- Refresher training
- Contractors must inform and train their employees and document that training

Figure 13-2 *The PSM Standard*

possibility of these events happening again. After a tedious investigation into the causes of the Bhopal 1984, Institute West Virginia 1985, Phillips 1989, ARCO 1990, BASF 1990, IMC 1991, Pepcon, and Shell Oil industrial explosions, fires, and vapor releases, the government issued the **process safety management (PSM) standard** (as shown in Figure 13-2). This document was modeled after the API 750 industry consensus standard. Key elements of the PSM standard include the following:

- Employee participation
- Process safety information
- Process hazard analysis
- Operation procedures
- Employee training
- Emergency planning and response
- Pre-startup safety review
- Mechanical integrity
- Hot work permit
- Management of change
- Incident investigation
- Contractors
- Compliance audits
- Trade secrets

The PSM standard has had a widespread, systematic impact on the chemical processing industry. From the moment the PSM *Final Rule* was implemented in 1992, a change occurred that restructured the occupation of the process technician. Job descriptions have been rewritten, and the process technician of the future looks much different from the one of the 1970s and 1980s.

Since 1992, industrial manufacturers have initiated partnerships with their local community colleges and universities. The traditional apprentice training function has gravitated from industrial training managers toward the regionally accredited halls of academia. This transition has saved the chemical processing industry millions in training budget dollars. Company training managers and coordinators were then able to focus on site-specific on-the-job training programs and government-mandated safety and environmental training. The site-specific training function will never move into the community college or university environment because there is such a variety of industrial processes. Only a handful of colleges have pilot plant facilities and bench-top units.

This transition can be documented all across the United States as certain elements of the PSM standard require a higher level of certification. This higher educational credential and need for some training redundancy has changed the occupation of the process technician from one who has little, if any, formal training to one who has the preferred status of a two-year process technology A.A.S. degree. A standardized college curriculum exists in at least one state, with several other states positioning themselves for educational standardization.

The five sections of the PSM standard that have had the greatest impact on process technician apprentice training are the following: 03.03 Employee Participation, 03.04 Process Safety Information, 03.05 Process Hazard Analysis, 03.06 Operation Procedures, and 03.07 Employee Training. It is virtually impossible for a new apprentice technician to successfully address these sections of the PSM standard without formal training.

Employee Participation

Employees need to be involved in planning how company employees and contractors will participate in the **process hazard analysis.** This team effort will include employees from all levels. Access to information needs to be readily available to all employees and contractors.

The employee participation section has four important parts:

- How to identify hazards
- How to *gather* crucial process and hazard information
- How to *communicate* crucial process and hazard information
- The written procedure and how employees will access it

Process Safety Information

The process safety information section includes information on the accidental mixing of chemicals and the products they produce, potential hazards, and safe handling procedures. This section provides an accurate assessment of the physical data associated with chemicals: toxic, reactive, corrosive, and explosive limits of specific substances. A simplified **flow diagram** of the process should be included in this section of the standard so a new person can easily see the various steps of the process. The complete list of the items that should be included in the process safety information include the following:

- Process flow diagram
- Equipment and process description
- Operational limits of equipment
- Consequences of deviations
- Safety and relief system design
- Electrical classifications
- Physical characteristics of chemicals: toxic, reactive, corrosive, explosive, and limits
- Process chemistry
- Inadvertent mixing of chemicals under a variety of conditions: temperature, pressure, and flow rates

Process Hazard Analysis

The process hazard analysis section is designed to identify the causes and consequences of fires, vapor releases, and explosions. All of the hazards on a unit need to be identified. During the startup, check for new equipment or a major process change. The equipment and construction must conform to established standards. Potential hazards and risks should be identified and discussed prior to implementation. All of the operational procedures, safety procedures, and training must be in place prior to startup.

A job hazard analysis is often referred to as a process used to assess risk. The basic steps include the identification of unacceptable risks and the process used to eliminate or control these risks. A hazard is defined as a condition or practice that could contribute to an undesirable or unplanned event or as the potential for harm. A hazard analysis reviews and analyzes an operating process from start to finish. The primary focus is on the identification of hazards associated with each job task. Other areas include the work environment, required tools, and the relationship between the technician and the task.

A job hazard analysis starts with the selection of a reasonable cross section of employees with experience in the job. An employee mixture of the new and innovative and mature and experienced can produce good results. The next step requires the use of a modified brainstorming session where tasks and associated hazards are identified. The session runs on until it naturally runs out of energy. At this point the tasks and hazards are placed in major categories. Tasks that pose an immediate danger to the technician's health or life are identified and require immediate action. Hazards need to be ranked in order of severity.

In order to conduct a job hazard analysis, the following detective work is required:

- List what can go wrong.
- Identify the consequences.
- Explain how it could arise or occur.
- Select the contributing factors.
- Determine the frequency that the hazard occurs.

Here is an example of a job hazard analysis form.

An example of a JHA form

Job Location:________________ Safety Analyst: Date:

Task Description:

Hazard Description:

Hazard Controls:
1.
2.
3.

Operation Procedures

Written procedures fall into two specific categories: operating and maintenance. Standard operating procedures (SOPs) apply to process technicians and must accurately reflect current work practices, process properties, hazards, process change of chemicals, and startup and shutdown procedures.

Maintenance procedures apply to maintenance employees and must identify the departments' collective consensus for how a specific task should be done. The chemicals that are used in the process should be well documented in the maintenance procedure. The **mechanical integrity** of the equipment must be ensured and any deficiencies identified and repaired. Any change in the process or equipment requires written procedures and training before work can be done. An evergreen feature for keeping all operational and maintenance documentation up to date should be built into the procedure system.

Employee Training

The Occupational Safety and Health Administration (OSHA) and the **Environmental Protection Agency (EPA)** believe that the key to preventing catastrophic emergencies inside the chemical processing industry is to provide adequate employee training. This was the conclusion of the governmental groups that investigated the Phillips Chemical Company and ARCO vapor release and explosions. The employee training aspect of the PSM standard includes seven sections:

1. Process overview.
2. Training records and method used to administer training. (You must document attendance and competency achieved.)
3. Chemicals used in the process.
4. Description of how access to and from the process unit is controlled.
5. Training materials that reflect current work practices.
6. Refresher training provided.
7. Contractors must inform and train their employees and document the training.

Contractors

Under the law, contractors fall under the same requirements of the PSM standard as company employees. This text applies all of the requirements of the standard to contract employees.

Pre-startup Safety Review

The pre-startup safety review and process hazard analysis work together during the evaluation of the startup process. During this phase the equipment and construction are carefully scrutinized. A risk assessment is made as each area is studied. It is during this phase that operational and safety procedures are written and tested.

Mechanical Integrity

Mechanical integrity is a term that applies to the soundness of a plant process. Mechanical integrity is monitored during the hazard analysis and prestartup section of this standard and during operational rounds and routine maintenance.

Hot Work Permit

The primary purpose of the **hot work** permit is to protect personnel from fires and explosions that could result from hot work that is performed in their area. Hot work is defined as welding, cutting, or using a spark-producing device. Specific written procedures must be in place to comply with this section of the standard.

Management of Change

The PSM standard requires that a formal system be in place that makes deviation from standard operational procedures structured with layers of approval required. This system must be documented and the proper training set in place for each employee.

Incident Investigation

Incident investigation is a procedure that takes place when a near-catastrophic or catastrophic events occur. This part of the PSM standard has three parts:

1. Assemble incident investigation team within 48 hours.
2. Address all of the findings.
3. Correct all of the action items identified during the investigation.

Emergency Planning and Response

Emergency planning and response is a complex plan that affects the company and community. Within the company are several levels of response. The first category deals with emergency plans that impact plant employees: what they do, their designated meeting points, and their key contacts. The next level deals with people who are responding to the emergency: their roles and responsibilities. The final level deals with the community, the plant, and the emergency. The PSM standard requires that these sections be documented and appropriate training completed. Written action plans are required.

Compliance Audits

OSHA inspectors will verify the compliance of each section of the PSM standard. A typical compliance audit has an opening meeting and a walkthrough to observe on-site conditions. The auditor will require documentation for a records review and will select one or more areas to conduct employee interviews. A closing meeting will be held to discuss the findings of the compliance audit.

Trade Secrets

Many companies have specific trade secrets that need to be in place to protect them from their competitors. This section of the standard allows companies to protect themselves by entering into confidentiality agreements with their employees.

Environmental Protection Agency

The Environmental Protection Agency (EPA) was established in 1970 to protect the environment from pollution. The creation of the EPA brought 15 federal programs under one umbrella. The EPA is most noted for its management of the Comprehensive Environmental Response, Compensation and Liability Act of 1980 (Superfund). In 1983 the EPA came under severe criticism from Congress for their handling of the Superfund. The synergy created by Congress and special interest groups made it clear that the public opinion about a clean environment vetoed the interests of the chemical processing industry. The enforcement of these new, tough environmental regulations put many industrial manufacturers out of business.

Initially the EPA focused on recycling and cleaning up open dump sites. Today the government has passed over 12 environmental laws that impact air, water, and land. These laws span cradle-to-grave, or in other words, they hold chemical manufacturers responsible from the time the chemical is created until it is disposed of. The EPA regulates water quality, pesticides used on farms, oil spills, everything that goes into the ground, and many other things.

Summary

The PSM standard has had a significant impact on the process technician and the chemical processing industry. From the moment the PSM *Final Rule* was implemented in 1992, a change occurred that restructured the occupation of the process technician. The key elements of the PSM standard that have had the greatest impact on process technician training are the following:

- Employee Participation (03.03)—Employees need to be involved in planning how company employees and contractors will participate in the process hazard analysis. This team effort will include employees from all levels. Access to information needs to be readily available to all employees and contractors. The employee participation section has four important parts:
  - How to identify hazards
  - How to *gather* crucial process and hazard information
  - How to *communicate* crucial process and hazard information
  - The written procedure and how employees will access it
- Process Safety Information (03.04)—process flow diagram, equipment and process description, operational limits of equipment, consequences of deviations, safety and relief system design, electrical classifications, physical characteristics of chemicals:
  - toxic
  - reactive
  - corrosive
  - explosive
  - limits
  - process chemistry
  - inadvertent mixing of chemicals under a variety of conditions like temperature, pressure, and flow rates.
- Process Hazard Analysis (03.05)—designed to identify all of the hazards on a unit; the causes and consequences of fires, vapor releases, and explosions; operational procedures; safety procedures; and training.
- Operation Procedures (03.06)—must accurately reflect current work practices, training procedures, process properties,

hazards, process change of chemicals, and startup and shutdown procedures.

- Employee Training (03.07)—The government believes that the key to preventing catastrophic emergencies inside the chemical processing industry is adequate employee training. Upon initial assignment, employees cover through the following topics:
  - A process overview (This will require an understanding of symbols and diagrams, equipment, and typical processes.)
  - Training records and method used to administer the training.
  - Attendance must be documented and competency achieved.
  - Chemicals used in the process.
  - Methods to control access to and from the process unit reviewed.
  - Training materials that reflect current work practices and operational procedures.
  - Refresher training that occurs based upon government mandates.
  - Contractors must inform and train their employees and document that training.

Review Questions

1. Explain the purpose of the process safety management (PSM) standard.
2. What is emergency response?
3. Define the term *hot work.*
4. In your opinion, what is the most important section of the PSM standard?
5. Describe the PSM standard, and list in order of importance each section.
6. Describe the impact the PSM standard has had on process technicians as well as on the chemical processing industry.
7. Define the term *Code of Federal Regulations (CFR).*
8. Explain the importance of compliance audits conducted by OSHA auditors.
9. Describe an emergency response plan.
10. Describe the Environmental Protection Agency (EPA). Identify when it was established and its primary purpose.
11. Explain the purpose of the Federal Register.
12. Explain the importance of community college safety, process, and engineering programs in preparation for reading and understanding *flow diagrams*.
13. Explain the term *mechanical integrity*.
14. Describe the Occupational Safety and Health Act.
15. Describe the Occupational Safety and Health Administration.
16. Describe the primary purpose of NIOSH.
17. Explain how the Occupational Safety and Health Review Commission works.
18. List the key steps in conducting a process hazard analysis.
19. Describe the employee training section of the PSM standard.
20. Describe the Comprehensive Environmental Response, Compensation and Liability Act of 1980.

HAZWOPER

Objectives

After studying this chapter, the student will be able to:

- Describe the operating hazards found in the chemical industry.
- Describe the HAZWOPER first responder, awareness level.
- Describe the HAZWOPER first responder, operations level.
- Describe emergency response.
- Identify potential hazards encountered during an emergency situation and the impact each could have.
- Describe the Incident Command System.
- Describe a fall protection system
- Describe the equipment used in fall protection.
- Identify safe work practices used with fall protection.
- Complete a fall protection inspection checklist.
- Describe the proper cleaning and storage procedures associated with fall protection equipment.

Key Terms

- **Anchor point**—a tie-off connection device used to secure the free end of a full body harness lanyard.
- **Auto-ignition**—the temperature at which a liquid will spontaneously ignite without a spark or flame.
- **Blind tracking**—keeps a record of all blind installations. The unit blind book prevents unit startup upsets and improves efficiency.
- **Blinding**—a term applied to the installation of slip blinds between pipe flanges. Blinding isolates a process stream and allows a craftsperson to work on a piece of equipment safely.
- **Buckles, D-rings, and snap hooks**—auxiliary equipment found on the full body harness, lanyard, and anchor point.
- **Designated equipment owner**—the process technician who operates a piece of equipment or process.
- **Donning and doffing**—terms used to describe putting on personal protective equipment and taking off personal protective equipment.
- **Emergency response**—a procedure initiated by the loss of containment for a chemical or the potential for loss of containment that results in an emergency situation requiring an immediate response. *Emergency response* drills are carefully planned and include preparations for worst-case scenarios. Examples include the following: vapor releases, chemical spills, explosions, fires, equipment failures, hurricanes, high winds, loss of power, and bomb threats.
- **First responder—awareness level**—individuals who are trained to respond to a hazardous substance release, initiate an emergency response, evacuate the area, and notify proper authorities.
- **First responder—operations level**—an individual who has been trained to respond with an aggressive posture during a chemical release by going to the point of the release and attempting to contain or stop it.
- **Hazards that will initiate an emergency response**—have been determined by the chemical processing industry to be any of the following: (1) explosion, (2) fire, (3) vapor release, (4) toxic chemical release, (5) large product or chemical spill, and (6) loss of containment of radioactive material.
- **HAZMAT**—hazardous materials.
- **HAZMAT response team**—the hazardous materials response team falls under the operations level of emergency response. This team receives specialized training so they can perform work to control and handle a hazardous chemical spill or release.
- **HAZWOPER**—Hazardous Waste Operations and Emergency Response.
- **Incident command system**—a military-type system designed to respond to an emergency.

- **Incident commander**—one who is responsible for organizing and coordinating response activities and who is surrounded by a formal organization with defined lines of authority and responsibility that provide information and carry out orders.
- **Levels of response**—the chemical processing industry has two levels of emergency response: (1) first responder—awareness level and (2) first responder—operations level.
- **Loss of containment**—the chemical processing industry has defined the following situations as the primary causes for loss of containment: (1) pipe or flange failure, (2) pump seal failure, (3) explosions, (4) fires, (5) overfilled tanks, (6) overpressured tanks, and (7) overturned drums or containers.
- **Polyester full body harness**—a safety device designed to evenly distribute the forces of an accidental fall.
- **Polyester lanyard**—a tie-off rope that is attached to a full body harness.
- **Process representative**—the first-line supervisor or designated representative who owns the equipment.
- **Safe haven**—a designated area that is safe from vapor releases.

Fall Protection

Fall protection is a standard designed to reduce or eliminate injuries from accidental falls. The standard requires fall protection to be used when a technician needs access, movement, or works at an elevation of 6 feet or higher. The only exception to this is if proper facilities for working at 6 feet or higher already exist. Examples of situations that would require a process technician to comply with the standard are working in suspended baskets, working in motorized lifts, standing on portable ladders, and working on a pipe rack.

The essential elements of a fall protection system include the following:

- Written fall protection program. (See Figure 14-1.)
- Basic fall protection equipment:
    - Full body harness—a safety device designed to evenly distribute the forces of an accidental fall to the strongest muscles in the body. This protects vital organs and limits the possibility of a serious injury.
    - **Lanyard**—a tie-off rope that is attached to a full body harness. It is designed to support a minimum of 5,000 pounds. The fall protection standard requires that the lanyard be a maximum of 6 feet long, have double locking **snap hooks** on each end, and include a shock-absorbing device that is attached to the harness **D-ring**.

Figure 14-1 *Fall Protection*

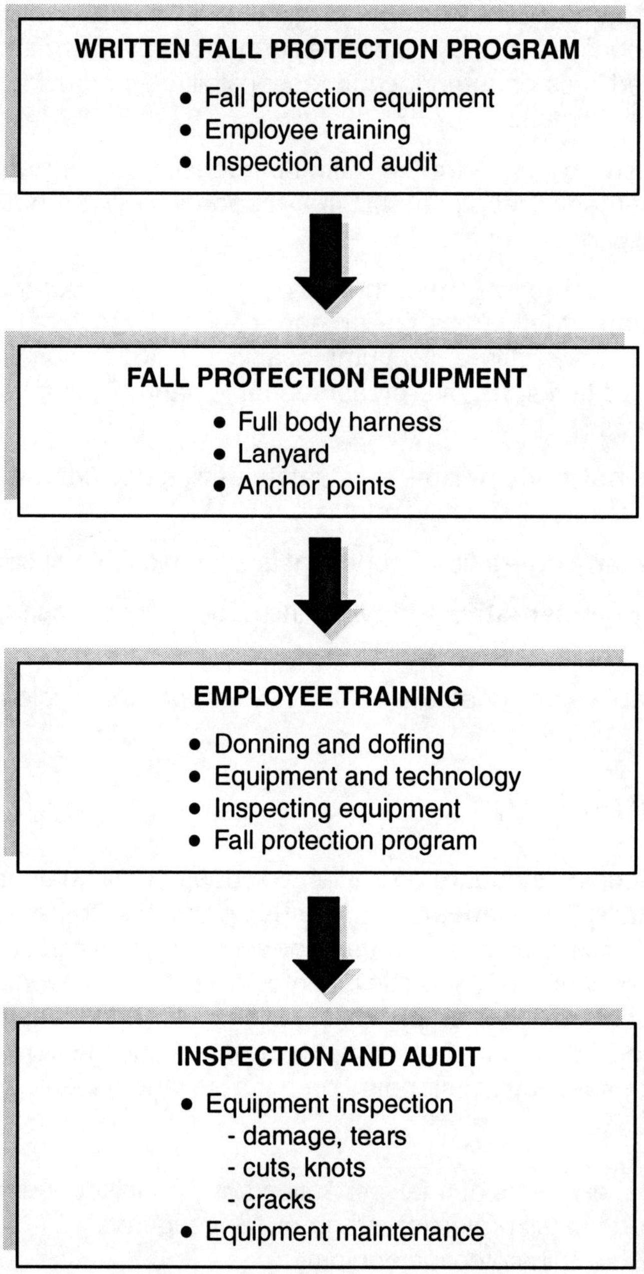

WRITTEN FALL PROTECTION PROGRAM
- Fall protection equipment
- Employee training
- Inspection and audit

FALL PROTECTION EQUIPMENT
- Full body harness
- Lanyard
- Anchor points

EMPLOYEE TRAINING
- Donning and doffing
- Equipment and technology
- Inspecting equipment
- Fall protection program

INSPECTION AND AUDIT
- Equipment inspection
  - damage, tears
  - cuts, knots
  - cracks
- Equipment maintenance

- **Anchor point**—a tie-off connection device used to secure the free end of a full body harness lanyard. Examples of anchor points include the following: pre-engineered eye bolts, cable, slide rail, structural steel, or pipe. The anchor point should be strong enough to support 5,000 pounds.

- Employee training—all employees should be given training on the fall protection program, **donning and doffing** fall protection equipment, and inspecting the equipment.

- Inspection and audit program—visually inspect equipment for damage. Knotted lanyards cannot be used because their tensile strength is reduced by 50%. Harnesses should be inspected for wear at the **buckles** and D-ring. If fiber wear is detected or a cut or tear is found in the harness material, it should be replaced immediately. Check D-rings and buckles for sharp edges, cracks, dents, burrs, or corrosion. Check snap hooks and ensure they operate correctly.

After a full body harness and lanyard has been used, it should be returned to its appropriate location. Harnesses can be cleaned with a wet soapy sponge.

A typical procedure used to don a full body harness includes the following:

1. Remove harness from container and visually inspect.
2. Hold harness by D-ring and shake so straps will fall into place.
3. Put the harness on as you would a jacket.
4. Secure leg straps.
5. Adjust straps so weight is transferred to the designated locations.
6. Visually inspect and secure the lanyard to the D-ring double locking hook and shock-absorbing device.
7. Secure lanyard to anchor point with double locking hook as soon as you reach your work location.

Hoisting Equipment

Hoisting systems are used to safely raise, lower, and move loads that are too heavy to be moved manually. Process technicians should receive the correct training prior to operating the hoist. Improperly installed hoists can bind up and fall from the ceiling mount. If a load is not centered correctly, the entire mechanism can collapse and fall to the mat. Load capacity should be clearly specified on the hoisting equipment. The most difficult aspect of the job is estimating the load. Prior to use, the hoist system should be carefully inspected. The load should be carefully balanced and all personnel cleared from the area. Hoisting lifts should be smooth and made in small increments. Loads should be suspended no longer than necessary. Tension should be removed from the suspension cables as soon as the load is down. Plant maintenance and the engineering staff should periodically inspect the system and ensure structural integrity. A preventive maintenance program should keep the hoisting system in good working order for many years.

HAZWOPER

Government-mandated **HAZWOPER** CFR 29 1910.120 training is required for technicians upon initial assignment and annually each year after.

Figure 14-2
HAZWOPER

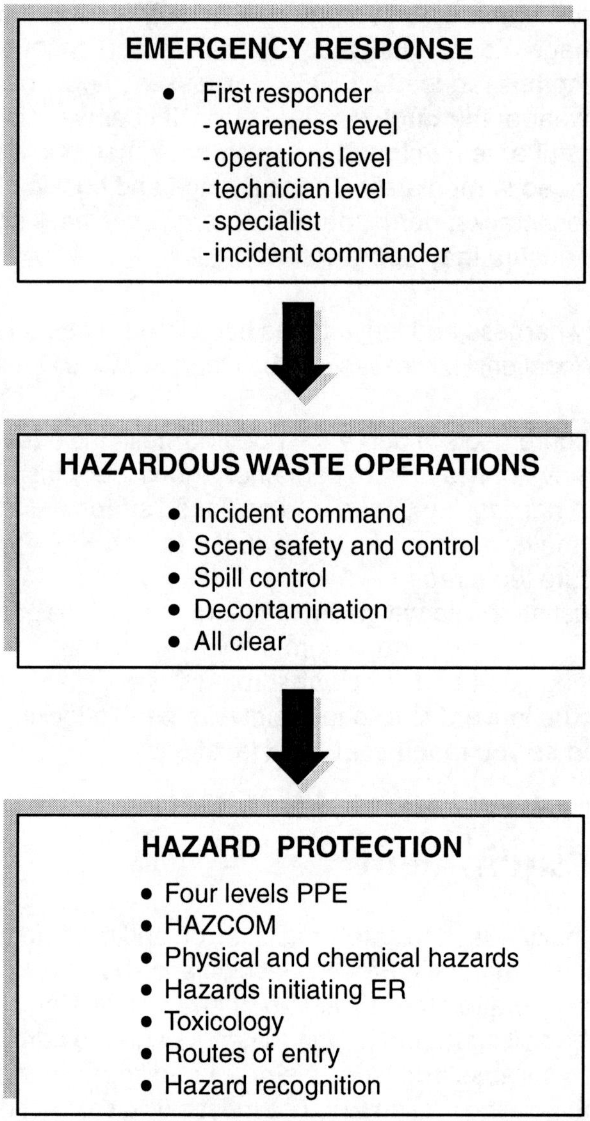

EMERGENCY RESPONSE

- First responder
  - awareness level
  - operations level
  - technician level
  - specialist
  - incident commander

HAZARDOUS WASTE OPERATIONS

- Incident command
- Scene safety and control
- Spill control
- Decontamination
- All clear

HAZARD PROTECTION

- Four levels PPE
- HAZCOM
- Physical and chemical hazards
- Hazards initiating ER
- Toxicology
- Routes of entry
- Hazard recognition

The term HAZWOPER is used to describe OSHA's Hazardous Waste Operations and Emergency Response standard. (See Figure 14-2.) The standard covers two important parts of a plant's operation: **emergency response** and hazardous waste operations. The CFR 29 1910.120 requires that all individuals who respond to an emergency situation have at least 24 hours of training. Refresher training is also covered under the standard, whereas plant-specific requirements may include additional training. HAZWOPER is categorized as follows:

- Emergency Response—**first responder awareness level**, **first responder operations level**, and hazardous materials technicians and specialist level

- Hazardous Waste Operations—**incident command system**, scene safety and control, spill control and containment, decontamination procedures, emergency termination, or all clear
- Hazard Protection, Prevention and Control—terms and definitions, personal protective equipment (PPE) levels, identifying hazardous materials and **hazards initiating an emergency response**, avoiding hazards, entry of hazardous materials into the body, and use of unit monitors and field survey instruments

The HAZWOPER standard defines five levels for emergency responders. These five levels include the following:

1. First responder—awareness level
2. First responder—operations level
3. Hazardous materials technician level
4. Hazardous materials specialist
5. On-scene **incident commander**

The information in this chapter will reflect objectives specific to the first responder—awareness level and first responder—operations level.

First Responder—Awareness Level

Process technicians play a key role in the operation and maintenance of chemical plants and refineries. Due to their close proximity to the operation and the chemicals found in the process, they are likely to be the first to witness a release. The first responder—awareness level is directed at individuals who witness or discover a hazardous chemical release and who have received emergency response training. Properly trained technicians know how to recognize a hazardous chemical release, the hazards associated with this release, how to initiate the emergency response procedure, and how to notify appropriate personnel.

First Responder—Operations Level

HAZWOPER CFR 29 1910.120 training is required for technicians upon initial assignment and annually after that. When a hazardous chemical release occurs, the process technicians working on the unit specific to the release will attempt to respond. Such efforts are directed at preventing the spread of the release and saving lives, equipment, community, and environment. It should be noted that the process technicians operating a unit will know more about that specific process than anyone else. It is virtually impossible to pull in a highly trained emergency response team who would know the unit a fraction as well as the technician. As long as the situation is under minimal control, the process technician is the best first line of defense. First responders for the operations level learn the following:

- How to respond to a release and how to control, contain, and confine the release
- Communication procedures between plant personnel

- Structure of emergency response system
- How to use personal protective equipment
- Standard emergency response and termination procedures

Events that would trigger an emergency response include the following:

- Loss of containment
- Punctured 55-gallon drum
- Pump or compressor seal failure
- Overflowing tank
- Pipe or vessel leak
- Explosion or fire
- Gas release or vapor release
- Toxic chemical spill or release

Emergency Response

The chemical manufacturing industry defines emergency response as a **loss of containment** of a chemical or the potential for loss of containment that results in an emergency situation requiring an immediate response. Examples of emergency response situations include fires, explosions, vapor releases, and reportable quantity chemical spills. The **levels of response** have been determined by the chemical processing industry to be as follows:

- First responder—awareness level: individuals who are trained to respond to a hazardous substance release, initiate an emergency response, evacuate the area, and notify proper authorities
- First responder—operations level: an individual who has been trained to respond with an aggressive posture during a chemical release by going to the point of the release and attempting to contain or stop it
- Hazardous materials—technician level
- Hazardous materials specialist
- On-scene incident commander

Refinery and petrochemical plant employees who are likely to discover or witness a chemical release fall under the scope of the awareness level whereas those employees who take preventive measures to control and secure the release fall under the guidelines of the operations level. Emergency response procedures are applied to every individual working for a company. The chemical processing industry typically has the following groups working in and around facilities:

- Process and lab technicians
- Maintenance technicians—instrument, electricians, mechanics, and so on
- Construction and janitorial technicians
- Engineers and chemists
- Management and administrative staff
- Safety and security

Each of these groups is trained to respond to an emergency situation. During an emergency situation, process technicians (PTs) who fall into the first responder—awareness level are required to complete specific procedures. By definition PTs are trained to respond to a hazardous substance release, initiate an emergency response, evacuate the area, and notify proper authorities.

The following is an example of an emergency response. On July 4, 2010, a process technician was completing a routine checklist when she noticed a large vapor cloud escaping from the top of a nearby unit's loop reactor. The chemicals escaping from the reactor were composed of an extremely flammable and explosive material. Five minutes earlier a small fire had been reported over the radio in the warehouse. Auto-ignition of some oily rags was suspected. Auto-ignition is a term used to describe the temperature at which a liquid will spontaneously ignite without a spark or flame. The warehouse was downwind of the release. The technician recognized the potential hazard and the need to initiate her plant's emergency response plan.

The following steps are typically followed during an emergency situation:

1. The technician immediately radios her supervisor about the release.
2. The first-line supervisor (FLS) notifies the supervisors at the loop reactor and the warehouse about the problems. They are already aware of the situation because of process instrumentation and are responding. They confirm that the situation is serious. The fire in the warehouse has been extinguished; however, hot surfaces are still exposed. The loop reactor FLS initiates the emergency response situation. The plant alarm system for a vapor release is sounded. The emergency response teams mobilize.
3. Process technicians from each unit begin the evacuation of their units to the designated **safe havens**.
4. Roll is taken and reported to a central incident commander. All plant personnel are accounted for.
5. Technicians trained to respond to the fire and release are mobilized.
6. When the fire and vapor release is contained the all-clear is sounded.

Hazardous Waste Operations

Hazardous waste operations in the chemical processing industry involve the use of a complex **incident command system** (ICS). The ICS coordinates all emergency response activities. Under the Industry Cooperation on Standards and Conformity Assessment (ICSCA), a scene safety and control system is coordinated. Decontamination, spill control and containment, and an emergency termination program also fall under the incident command system.

The ICS is led by an incident commander. The incident commander is responsible for organizing and coordinating response activities and is surrounded by a formal organization with defined lines of authority and responsibility that provide information and carry out orders.

Hazardous waste operations are typically classified into three categories:

1. Small hazard that does not affect the whole unit.
2. Medium-sized hazard that impacts one or more operating units.
3. Large hazard that impacts the plant and community.

The organizational structure for a typical incident command system is as follows:

- Incident commander—direct on-scene activities
- On-scene commander—establish command post, first response activities
- Scene specialist—conducts risk assessment
- Planning chief—coordinates unit operations
- Operations chief—directs **HAZMAT,** firefighting
- Medical chief—provides industrial hygiene, medical first aid
- Employee welfare chief—provides financial and material assistance to community
- Logistics chief—refuels vehicles, delivers supplies
- Security chief—secures scene and maintains order
- Communications chief—initiates radio and telephone communications
- Public affairs chief—provides information to media
- Spill containment, cleanup chief—coordinates containment and cleanup
- Safety chief—assesses hazard and prevents unsafe acts

Scene Safety and Control

In order to reduce the possibility of accidental spread of hazardous chemicals into areas outside of the affected unit, a three-zone system has been established. (See Figure 14-3.)

- Hot zone—the area around the incident where contamination has occurred. (Emergency response activities that occur in the hot zone require the appropriate PPE and the buddy system.)
- Warm zone—used to decontaminate technicians leaving the hot zone.
- Cold zone—a staging area where the incident command post is established.

Spill Control and Containment

- Chemical spills and vapor releases require different containment procedures. Most operating units have been designed with spill control or containment in mind. Absorbents,

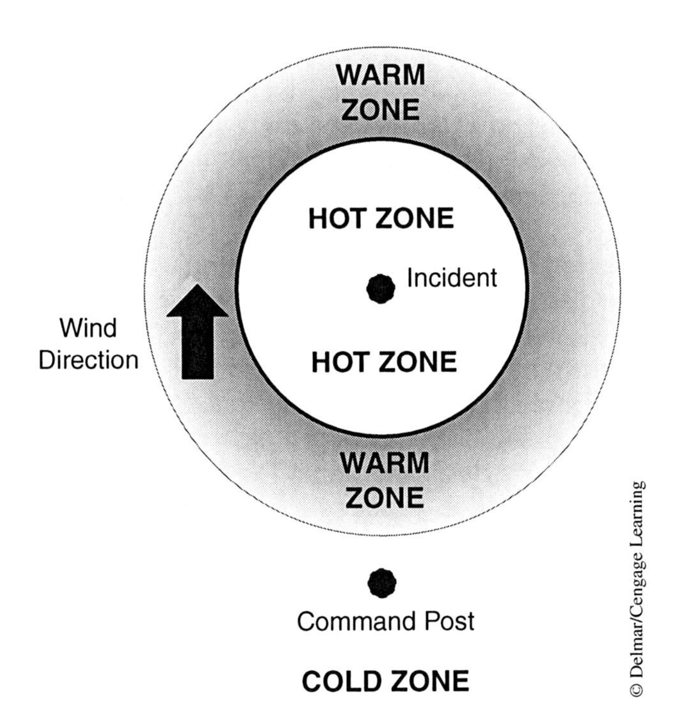

Figure 14-3
Scene Safety and Control

adsorbents, curbing and diking, segregated sewers, earthen dams, and fire monitors are devices used to control chemical spills and releases.

Decontamination Procedures

In a decontamination procedure, anything that comes into contact with a hazardous chemical is neutralized in the warm zone. The typical way of doing this is through disposal of the contaminated PPE. Contaminated workers are first rinsed off with fresh water before doffing their protective equipment. Dirty clothing is placed in a pile whereas dirty tools are placed in plastic bags for solid waste collection and cleanup.

Hazard Identification

During a chemical spill or release, the identification of the material is important. Because time is critical during an emergency response situation, the following procedure must be completed quickly and from a safe distance:

1. Determine the location of the release. A number of chemicals can be identified through this procedure.
2. Analyze the composition of the material being released. Determine whether it is a gas, vapor, liquid, solid, or a combination.
3. Identify other characteristics, such as odor, color, and physical characteristics.
4. Report and discuss with area technicians and supervision.

Hazards Initiating an Emergency Response

Hazards that will initiate an emergency response have been determined by the chemical processing industry to be the following: explosion, fire, vapor release, toxic chemical release, large product or chemical spill, and loss of containment of radioactive material.

Loss of Containment The chemical processing industry has defined the following situations as the primary causes for loss of containment: pipe or flange failure, pump seal failure, explosions, fires, overfilled tanks, overpressured tanks, and overturned drums or containers. Blinds are used to block off lines that could possibly leak. Blinding is a term applied to the installation of slip blinds between pipe flanges. Blinding isolates a process stream and allows a craftsperson to work on a piece of equipment safely. Blind tracking is a process that keeps a record of all blind installations. The unit blind book prevents unit startup upsets and improves efficiency.

Personal Protective Equipment (PPE)

For more information on this subject, refer to HAZCOM: PPE Levels A, B, C, and D.

Unit Monitors and Field Survey Instruments—Identifying Hazardous Materials

Draeger pumps are portable vacuum-type devices used to collect a representative gas or vapor sample to determine if a known contaminant is present. The Draeger pump has a bellows to draw air into a glass tube filled with a chemically treated sorbent. The activated media in the glass tube will change color if the suspected chemical is present in the air. Each glass tube is scaled or graduated so it can be read in parts per million. Process technicians need specialized training in order to operate a Draeger vacuum pump properly.

Explosimeters, commonly called combustible gas meters, are used to determine whether there are sufficient concentrations of a combustible gas mixture to produce a fire or explosion. Explosimeters operate by drawing an air sample into a combustion chamber and igniting it. If the air sample ignites, combustion level registers.

Fixed unit monitors are designed to test for combustible or toxic substances such as hydrogen sulfide or chlorine. Organic vapor analyzers and oxygen analyzers are used to detect hydrocarbons and oxygen levels. Fixed detectors are strategically located around the unit and will send an audible alarm to the control room in the event that a target substance is detected. Geiger counters are used to measure the level of radioactivity in an atmosphere by counting the number of ions given off by a source and the rate at which they are lost.

Bunker Gear

Several years ago a fire broke out in a plant that required the process technicians to call in a firefighting team to assist. After a short time an explosion occurred which resulted in a huge fire ball that engulfed several technicians and firefighters. The heat from the fire immediately killed the technicians but inflicted only minor injuries on the firefighters. Upon closer investigation it was determined that the bunker gear the fire fighters were wearing provided superior protection over the flame-retardant clothing (FRC) that the technicians were wearing.

Requirements of the bunker gear standard fall under CFR 29 1910.120 and are provided for the purpose of training individuals who will respond to a major fire or vapor release with a fire team. Most chemical companies and refineries are giving new employees this training upon initial assignment and annually thereafter. Bunker gear personal protective equipment consists of the following:

- Gloves—special heat-resistant gloves.
- Boots—special heat-resistant material.
- Coats—special heat-resistant material, comes in small, medium, large, extra large, all latches are engaged, none can be left open, the collar should be pulled up straight and the strap fastened.
- Helmet—special heat-resistant material with a face shield, inner liner should be pulled down over the ears, and the chin strap secured. (A rotary adjustment knob is located on the back of the helmet to ensure the helmet sits snugly on the head.)

Cutting, Welding, and Brazing

Cutting, welding, and brazing are processes that take place frequently in the chemical processing industry. Process technicians are responsible for controlling and minimizing any potential hazards. Cutting, welding, and brazing produce toxic fumes, generate ultraviolet radiation and electricity, and can easily start a fire. Cutting is described as a process that severs or removes metal using heat from the combustion of hot fuel gases to generate a chemical reaction between the metal, oxygen, and flame. Welding is the process of connecting metals using heat and allowing the metals to flow together. Brazing is defined as a process of soldering using an infusible alloy. Adequate ventilation and preparation are needed before any brazing, cutting, or welding occurs.

Summary

The term HAZWOPER is used to describe OSHA's *Hazardous Waste Operations and Emergency Response* standard. HAZWOPER is broken down into the following areas:

- Emergency Response
- Hazardous Waste Operations
- Hazard Protection, Prevention, and Control

Emergency response is the loss of containment of a chemical or the potential for loss of containment that results in an emergency situation requiring an immediate response. The levels of response have been determined by OSHA to be the following:

- First responder—awareness level
- First responder—operations level
- Hazardous materials technician level
- Hazardous materials specialist
- On-scene incident commander

Hazardous waste operations in the chemical processing industry involve the use of a complex incident command system (ICS). The ICS coordinates all emergency response activities. In order to reduce the possibility of accidental spread of hazardous chemicals into areas outside of the affected unit, a three-zone system has been established: hot zone, warm zone, and cold zone.

Fall protection is a standard designed to reduce or eliminate injuries from accidental falls. The standard requires fall protection to be used when a technician needs access, movement, or works at an elevation of 6 feet or higher. The only exception to this is if proper facilities for working at 6 feet or higher already exist.

Hoisting systems are used to safely raise, lower, and move loads that are too heavy to be moved manually. Process technicians should receive the correct training prior to operating the hoist.

Cutting, welding, and brazing produce toxic fumes, generate ultraviolet radiation and electricity, and can easily start a fire. Cutting is described as a process that severs or removes metal using heat from the combustion of hot fuel gases to generate a chemical reaction between the metal, oxygen, and flame. Welding is the process of connecting metals using heat and allowing the metals to flow together. Brazing is defined as a process of soldering using an infusible alloy.

Review Questions

1. Explain the terms cutting, welding, and brazing.
2. Explain the hazards of working with hoisting systems.
3. What is the purpose of HAZWOPER?
4. Describe why a process technician needs to have emergency response training.
5. The chemical processing industry has five basic levels of emergency response; name them.
6. Contrast the first responder awareness level and first responder operations level.
7. How does the incident command system work for minor, medium, and major vapor releases and fires?
8. Define emergency response.
9. Describe hazardous waste operations.
10. Describe the difference between bunker gear and the typical uniform and safety equipment worn by a process technician.
11. Describe Draeger pumps and explosimeters.
12. Describe the principles of spill control and containment.
13. List the equipment used in fall protection.
14. Explain the oxygen requirements for human life and the hazards associated with the low and high ends of the scale.
15. List the steps of hazard recognition.
16. List the command structure for incident command (ICS).
17. Describe scene safety and control.
18. Define the term *safe haven*.
19. Describe the *fall protection* program.
20. List the steps you would take if you noticed a severe vapor release in your plant.

Process System Hazards

OBJECTIVES

After studying this chapter, the student will be able to:

- Describe the operating hazards found in the chemical industry.
- Explain key terms and definitions used in basic process principles.
- Describe and apply the basic principles of pressure.
- Define fundamental chemistry terms.
- Describe the fundamental principles of chemistry.
- Describe and use a chemical equation and periodic table.
- Describe these chemical reactions: exothermic and endothermic.
- Analyze the scientific principles of heat, heat transfer, and temperature.
- Perform simple temperature conversions between °F, °C, K and °R.
- Examine the principles of fluid flow in process equipment.
- Compare the hazards associated with pump and compressor system operations.
- Describe the safety hazards associated with a heat exchanger and cooling tower system.
- List the hazards associated with steam generation and furnace operation.
- Explain the relationship of science and chemistry to hazard recognition.
- Compare the hazards associated with the principles of reaction and distillation.

Key Terms

- **Absorbed heat**—effects include increase in volume and temperature, change of state, electrical transfer, and chemical change.
- **Acid**—a bitter-tasting chemical compound that has a pH value below 7.0, changes blue litmus to red, yields hydrogen ions in water, and has a high concentration of hydrogen ions.
- **Atom**—the smallest particle of a chemical element that still retains the properties of an element. An atom is composed of protons and neutrons in a central nucleus surrounded by electrons. Nearly all of an atom's mass is located in the nucleus.
- **Atomic mass unit (AMU)**—the sum of the masses in the nucleus of an atom.
- **Atomic number**—the total number of protons in an atom; determines the position of the element on the periodic table.
- **Balanced equation**—the sum of the reactants (atoms) equals the sum of the products (atoms).
- **Base**—a bitter-tasting chemical compound that has a soapy feel and a pH value above 7.0. It turns red litmus paper blue and yields hydroxyl ions.
- **Boiling point**—the temperature at which a liquid changes to a vapor.
- **Cavitation**—the formation and collapse of gas pockets around the impellers during pump operation; results from insufficient suction head (or height) at the inlet to the pump.
- **Centrifugal pump**—a dynamic pump that accelerates fluid in a circular motion. Commonly used in automatic control with fluid flow and level control.
- **Chemical bond (covalent)**—occurs when elements react with each other by sharing electrons. This forms an electrically neutral molecule.
- **Chemical bond (ionic)**—occurs when positively charged elements react with negatively charged elements to form ionic bonds through the transfer of valence electrons. Ionic bonds have higher melting points and are held together by electrostatic attraction.
- **Chemical equation**—numbers and symbols that represent a description of a chemical reaction.
- **Chemical reaction**—a term used to describe the breaking, forming, or breaking and forming of chemical bonds. Types include exothermic, endothermic, replacement, and neutralization.
- **Chemistry**—the science and laws that deal with the characteristics or structure of elements and the changes that take place when they combine to form other substances.
- **Compound**—a substance formed by the chemical combination of two or more substances in definite proportions by weight.
- **Electron**—a negatively charged particle that orbits the nucleus of an atom.
- **Element**—composed of identical atoms.

- **Fractional distillation**—a process that separates the components in a mixture by their individual boiling points.
- **Heat**—a form of energy caused by increased molecular activity.
- **Heat transfer**—heat is transmitted through conduction (heat energy is transferred through a solid object; for example, a heat exchanger), convection (requires fluid currents to transfer heat from a heat source; for example, the convection section of the furnace or economizer section of boiler), and radiation (the transfer of energy through space by the means of electromagnetic waves; for example, the sun).
- **Hydrocarbons**—a class of chemical compounds that contains hydrogen and carbon.
- **Liquid pressure**—the pressure exerted by a confined fluid. Liquid pressure is exerted equally and perpendicularly to all surfaces confining the fluid.
- **Material balancing**—a method for calculating reactant amounts versus product target rates.
- **Matter**—anything that occupies space and has mass.
- **Mixture**—composed of two or more substances that are only physically mixed. Mixtures can be separated through physical means such as boiling or magnetic attraction.
- **Molecule**—the smallest particle that retains the properties of the compound.
- **Neutron**—a neutral particle in the nucleus of an atom.
- **Oxidizer**—a chemical that yields oxygen.
- **Oxygen deficiency**—atmospheres with less than 20% oxygen.
- **Periodic table**—provides information about all known elements (for example, atomic mass, symbol, atomic number, boiling point).
- **pH**—a measurement system used to determine the acidity or alkalinity of a solution.
- **Pressure**—force or weight per unit area (Force ÷ Area = Pressure). Pressure is measured in pounds per square inch.
- **Proton**—a positively charged particle in the nucleus of an atom.
- **Pyrophoric**—a chemical that ignites spontaneously in air below 130°F.
- **Reactants and products**—raw materials or reactants are combined in specific proportions to form finished products.
- **Reaction (combustion)**—an exothermic reaction that requires fuel, oxygen, and heat to occur. In this type of reaction, oxygen reacts with another material so rapidly that fire is created.

- **Reaction (endothermic)**—a reaction that requires heat or energy.
- **Reaction (exothermic)**—a reaction that produces heat or energy.
- **Temperature**—the hotness or coldness of a substance.
- **Valve line-up**—a term used to describe opening and closing a series of valves to provide fluid flow to a specific point or tank before starting a pump.

Operating Hazards

Inside the chemical processing industry, a variety of operating hazards exist, and these should be carefully identified before working unsupervised. Each operating system has a unique set of hazards that can be identified and controlled. Operating hazards can be classified as equipment and system related, weather related, and chemistry and chemicals related. The types of equipment and systems that can cause an operating hazard include valves, piping, pumps, compressors, turbines, heat exchangers, cooling towers, boilers, furnaces, reactors, and distillation columns. This list could also include electrical items, instruments, rotating equipment, plastic plant equipment, and many other devices. Weather-related hazards such as lightning, tornadoes, hurricanes, hail, snow, rain, heat, and other phenomena may cause serious damage inside a chemical complex. The operating hazards associated with chemicals include the basic chemistry of how various components mix to form new products under a wide range of temperatures, pressures, and other variables.

Equipment- and System-Related Hazards

Pumps

On the outside, a simple pump system appears to have few, if any, operating hazards; however, looks may be deceiving. For example, most pumps are equipped with seals and bearings. Seals need to be lubricated in order to function efficiently. This is typically done using pumped liquid from the discharge port or from an independent source of oil. If a mechanical seal fails, an excessive amount of fluid will flow out around the rotating shaft. Checking pump seals, general operations, and vibration is part of a technician's routine checklist. Figure 15-1 shows a leaking pump seal. The most common problem associated with pump operation includes improper line-up and lack of understanding about how a centrifugal pump operates. **Centrifugal pumps** are used frequently in automatic process control. This means that a throttling type valve is installed in the system and works in conjunction with the pump. Centrifugal pumps require a certain amount of pressure on the suction side to push the liquid into the suction eye. The natural drawing action

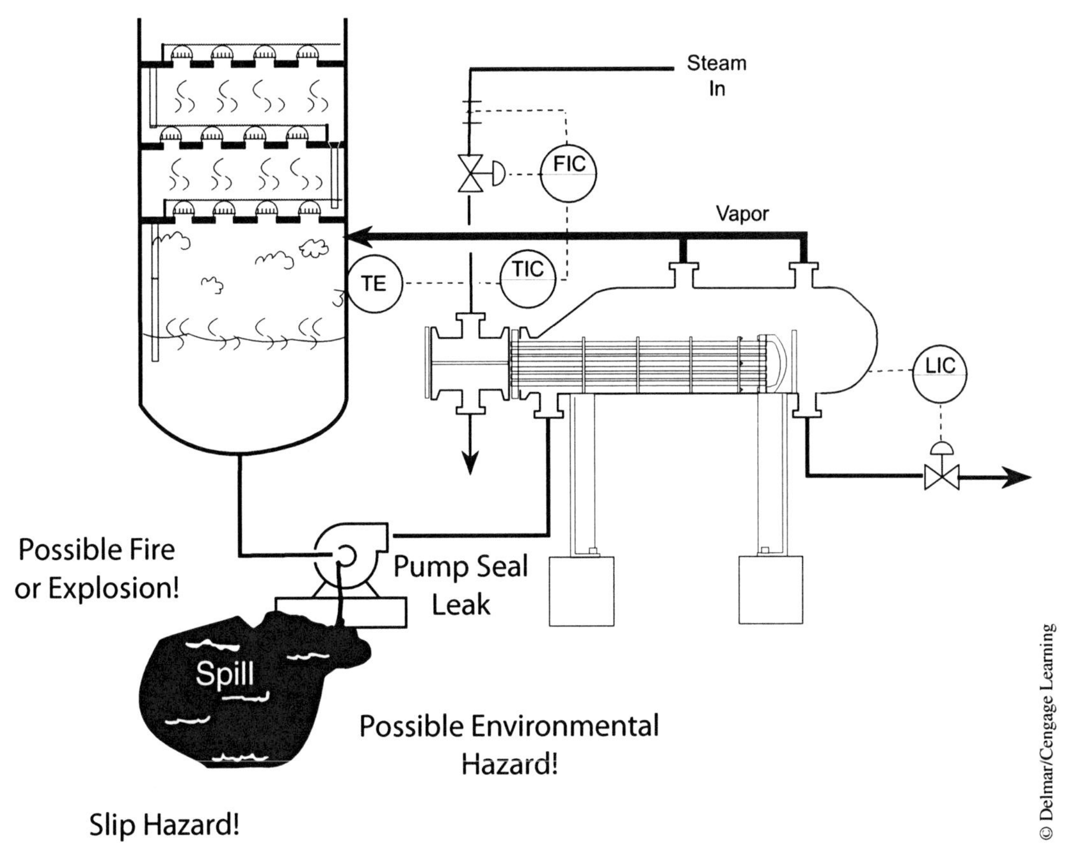

Figure 15-1 *Pump Seal Leak*

of the impeller creates a low-pressure phenomenon inside the impeller. Problems associated with centrifugal pumps include the following:

- Cavitation
- Vapor lock
- Improper line-up
- High discharge pressure variations or NPDH
- Variations in suction pressure or NPSH
- Feed composition changes
- Gear box problems
- Seals and bearings problem
- Broken suction and discharge gauges
- Breaker trips on motor
- Motor problems
- Gasket leaks
- Seal flush tubing plugs ups

Cavitation is defined as the formation and collapse of air pockets inside the pumping chamber. It can also be described as boiling, a process which can be very violent, with rapid pressure increases and decreases. Cavitation can damage the impeller, shaft, casing, or wear rings. This phenomenon can break the pump loose from the piping or foundation and sounds like marbles being agitated in a large blender. Cavitation can be prevented by simply increasing the NPSH or pinching down on the discharge valve. It appears to be caused when the pump out runs the liquid entering the suction eye, forming a serious vacuum, reducing the boiling point of the liquid to a point where it violently expands and then collapses as the pressure builds.

The boiling point of a substance is the temperature at which the vapor pressure exceeds atmospheric pressure, bubbles become visible in the liquid, and vaporization begins.

Molecular motion in water vapor produces pressure and increases as heat is added to the liquid. The vapor pressure of a substance can be linked directly to the strength of the molecular bonds of a substance.

Pressure directly affects the boiling point of a substance. As the pressure increases the following occur:

- Boiling point increases
- Escape of molecules from the surface of the liquid is reduced proportionally
- Gas or vapor molecules are forced closer together
- Vapor phase above a liquid could be forced back into solution

This is an important fact for a process technician to understand. A change in pressure shifts the boiling points of raw materials and products. Pressure problems are common in industrial manufacturing environments and must be controlled. Atmospheric pressure is 14.7 psi, so any pressure below this is referred to as a vacuum. Vacuum affects the boiling point of a substance in the opposite way that positive pressure does.

Pressure is defined as force or weight per unit area (Force ÷ Area = Pressure). The term *pressure* is typically applied to gases or liquids. Pressure is measured in pounds per square inch (psi). Atmospheric pressure is produced by the weight of the atmosphere as it presses down on an object resting on the surface of the earth. Pressure is directly proportional to height: The higher the atmosphere, gas, or liquid, the greater the pressure. At sea level, atmospheric pressure equals 14.7 psi. The principles of **liquid pressure** are the following:

- Liquid pressure is directly proportional to its density.
- Liquid pressure is proportional to the height of the liquid.
- Liquid pressure is exerted in a perpendicular direction on the walls of a vessel.
- Liquid pressure is exerted equally in all directions.
- Liquid pressure at the base of a tank is not affected by the size or shape of the tank.
- Liquid pressure transmits applied force equally, without loss, inside a closed container. Here is a common equation for solving pressure problems:

Height of liquid × 0.433 × specific gravity = pressure

Other hazards associated with a simple pump system include the buildup of static electricity in moving fluids. Accidental ignition of flammable gases may occur if equipment and piping are not correctly grounded or bonded. In pump systems that include heat exchangers, product contamination is possible if a tube ruptures or breaks. In situations like this, **hydrocarbons** show up in the cooling tower basin or product streams.

In general, any rotating equipment poses the risk of a serious injury if a process technician does not exercise respect and caution. Loose clothing, long hair, and exposed human tissue can be seriously injured in rotating equipment. Safeguards are typically placed around these potential hazards to prevent injury. These devices should never be removed during operational conditions. Examples of rotating equipment are pumps, compressors, fans, blowers, turbines, agitators, blenders or mixers, extruders, drills, feeders, and conveyors.

Figure 15-2 illustrates the relationship between a tank, piping, valves, control instruments, and a heat exchanger. When these pieces of equipment are arranged in a system, a hazard analysis reveals all of the possible safety situations. Some of these hazards include the following:

- Improper line-up turns the heat exchanger into a bomb.
- Tube leakage and inadvertent mixing of chemicals.
- Runaway temperature on the hot oil system.
- Motor catches on fire.
- Tank explodes.
- Pipe breaks.
- Tank ruptures.
- Manway bolts or gasket fails.
- Instrument leaks or fails.

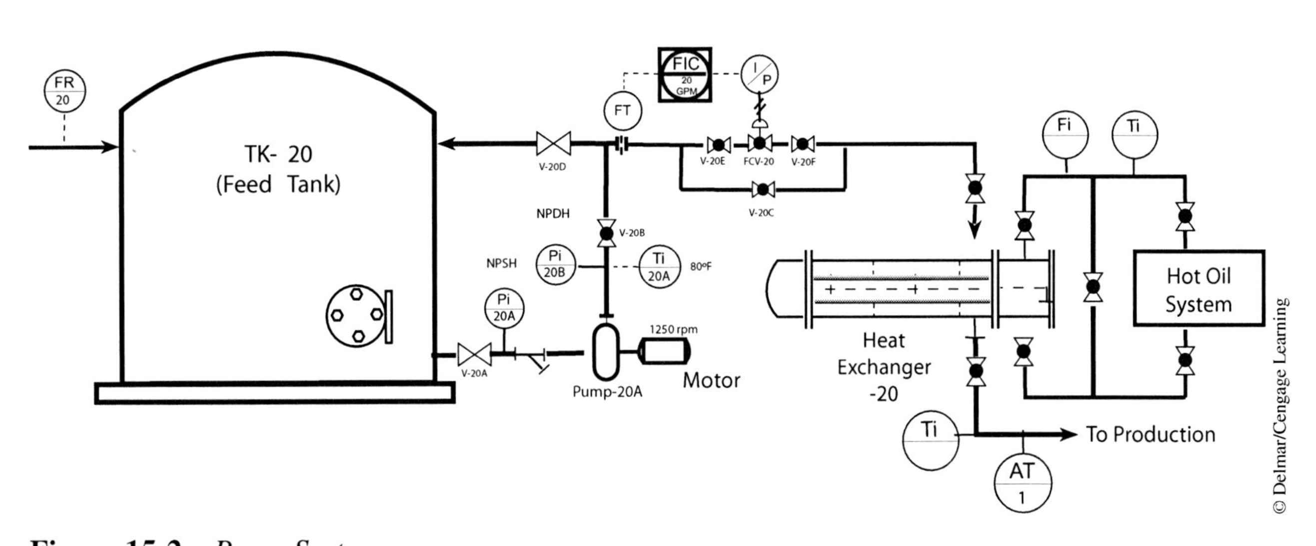

Figure 15-2 *Pump System*

- Grounding comes loose and fluid flow produces a static electric spark.
- Cathodic protection fails and bottom of tank corrodes.
- Pessure surges in the tank.

Compressors

Another type of operating hazard includes the use of tanks, piping line-ups, reciprocating compressors, and a procedure called purging. Each plant has its own set of procedures associated with purging out equipment. Watching the flammability limits is important when equipment containing air is purged using gas. If the gas enters too slowly, a flammable mixture is formed, and if it enters too quickly, it might produce a spark that will ignite the system. Natural gas is generally used to purge air because it is lighter and will tend to rise. In this type of system, the pipe line-up is located on the top of the tank and air is discharged out the bottom. The opposite procedure is used if propane is used for purging, because it is heavier than air. Figure 15-3 illustrates how natural gas is used to purge a tank.

Reciprocating gas compressors combine all the elements needed to create an explosion if air is present in the system. Air enters the compressor after extended shutdowns or equipment maintenance. Most procedures purge the air from the discharge and suction piping prior to startup. If this procedure is not carefully followed, air, gas, and heat from friction are present in the system. Figure 15-4 shows the basic layout for a reciprocating

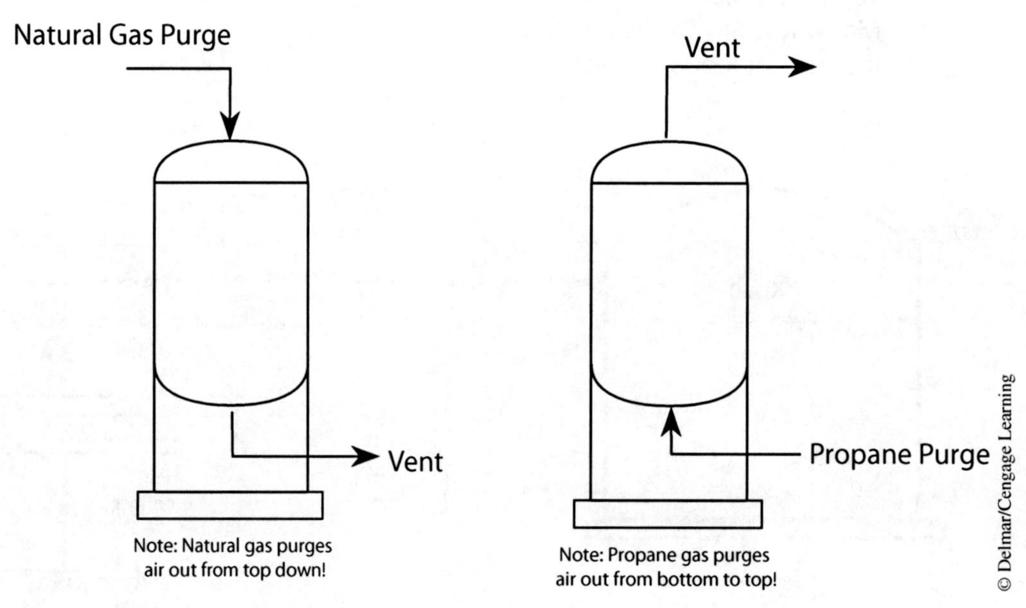

Figure 15-3 *Purging Procedure*

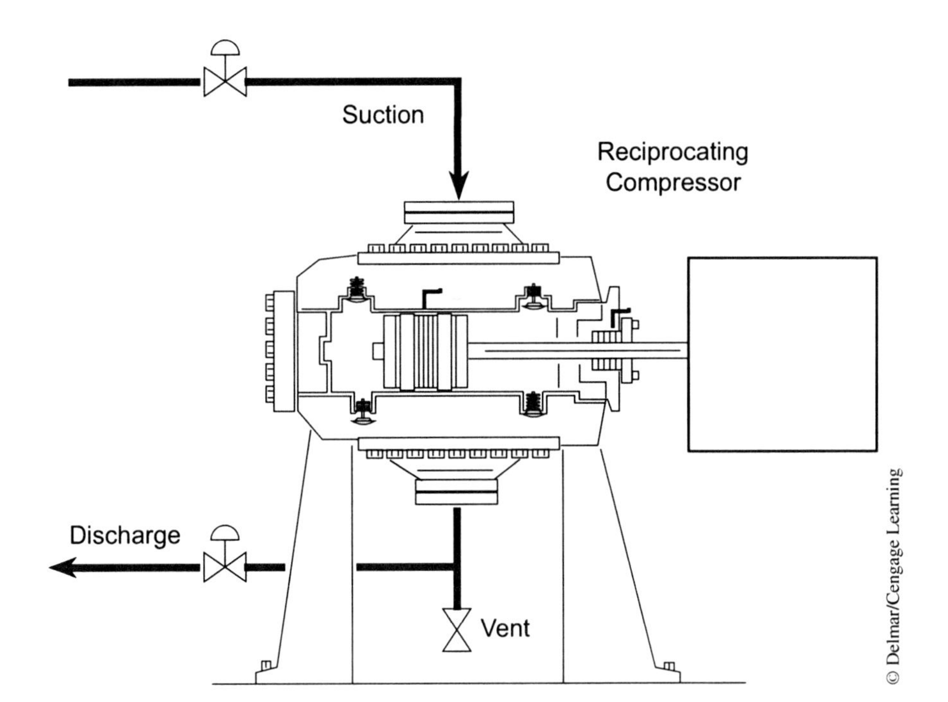

Figure 15-4
Reciprocating Compressor

compressor. Another hazard in gas compressors is the possibility of a fire if the packing leaks and finds an ignition source.

The compressor system is a vital part of modern process control. Compressors come in two basic designs: dynamic and positive displacement. Minor problems are occasionally experienced with compressor systems. Figure 15-5 shows a centrifugal multi-stage compressor on the left of the receiver and a positive displacement compressor on the right. Each system has different hazards. These problems are usually the result of dirt, adjustment problems, liquid in the receiver, improper line-up, or inexperience in operating the system. A number of safety issues should be addressed prior to operating a compressor system. Some of these safety issues include the following:

- Noise hazards
- High-pressure hoses blowing loose
- Hazards associated with compressed gas systems
- Hazards associated with rotating equipment.
- Mixing air and hydrocarbons into flammable or explosive concentrations
- Avoiding high pressure releases; eyes, ears, nose, skin
- Fires or explosions
- Incorrect line-up
- Loss of cooling water or lubrication
- Lifted safety

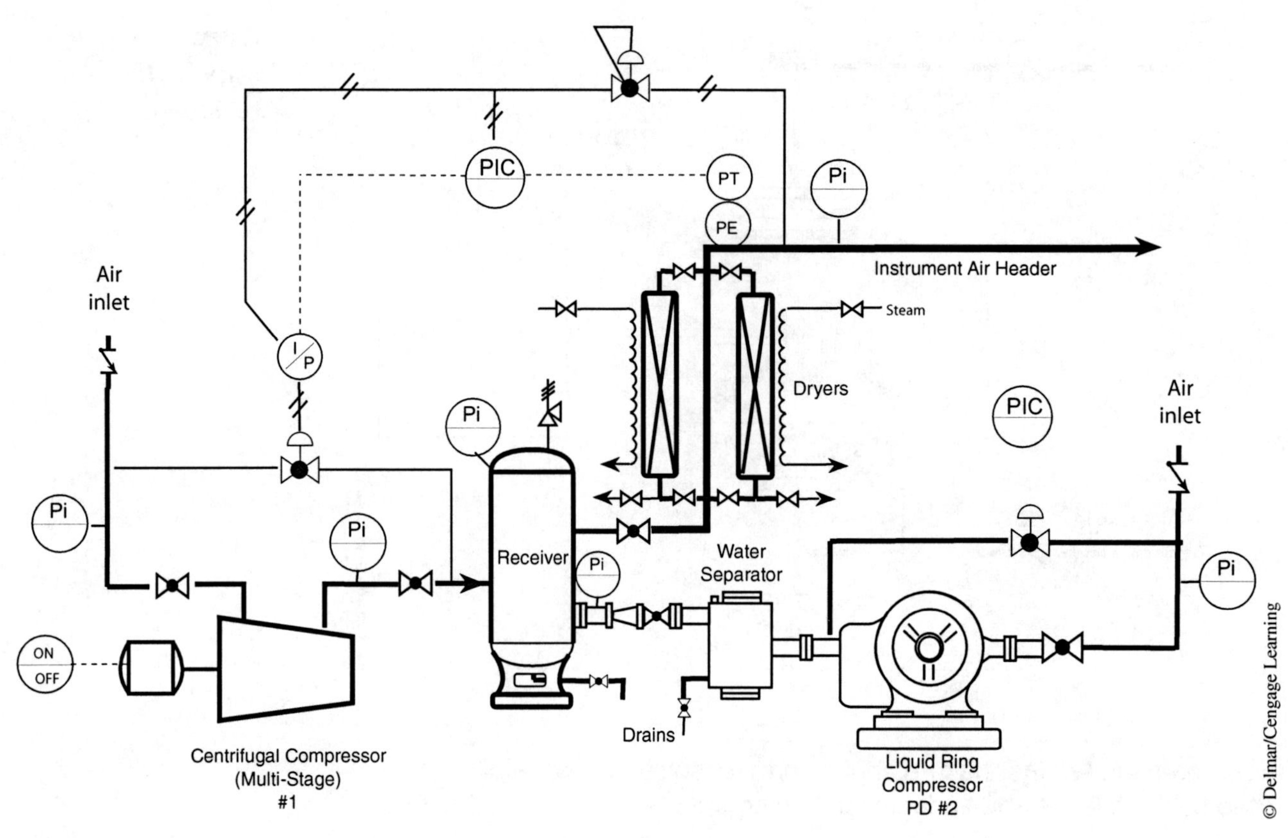

Figure 15-5 *Compressed Air System*

Experienced technicians can quickly fix compressor problems by making proper adjustments, cleaning the equipment, and ensuring lubrication or auxiliary systems are maintained at correct guidelines, temperatures and pressures.

Operating temperatures and pressures have been engineered for the safe operation of the system. At one time or another, most technicians are faced with situations that cause them to wonder how much heat and pressure the system can take. There is not a simple answer to this question; however, there are significant safety factors built into most process systems. This factor is typically four times the required number for metal thickness and pressures. In Figure 15-6 an operating pressure of 55 bars (800 psi) and a design pressure of 60 bars (880 psi) are needed on the drum. The required plate thickness of the vessel to hold this pressure is 5.8 mm (0.23 inch). A safety factor of four (4) is used to calculate the required plate thickness. Because the closest standard plate to this factor is 23.9 mm (0.94 inch), it is selected as the correct building material. When the vessel is completed, it will be tested at 1.5 times the design pressure or 1.5×60 bars $= 90$ bars. This type of vessel is also equipped with a safety valve or rupture disc which is designed to relieve pressure. These devices are generally set at 10% above the designed specifications.

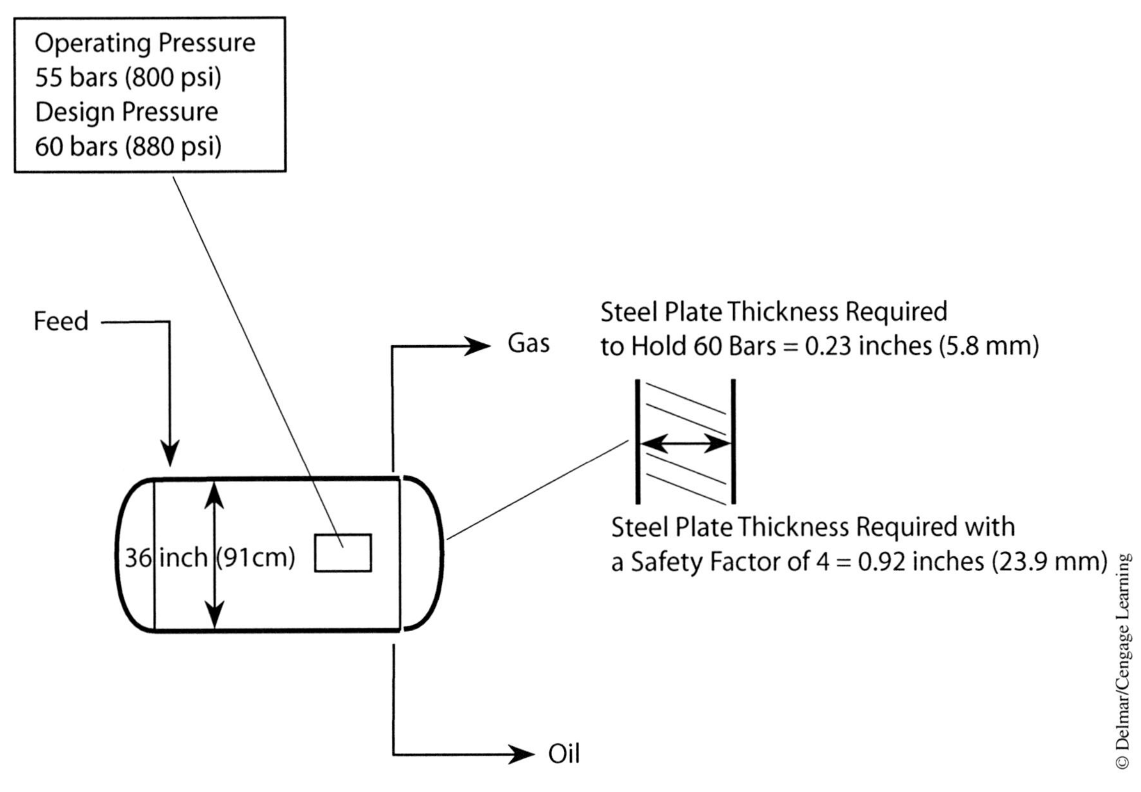

Figure 15-6 *Factor of 4 Safety Design*

Heat Exchangers

Heat exchangers are commonly used to heat or cool process flows in the chemical processing industry. A shell and tube heat exchanger has a cylindrical shell that surrounds a tube bundle. Fluid flow through the exchanger is referred to as *tubeside flow* or *shellside flow*. A series of baffles support the tubes, direct flow, decrease tube vibration, increase velocity, create pressure drops, and protect the tubes.

The effects of **absorbed heat** are the following:

- Increase in molecular activity
- Change of state—solid, liquid, gas
- Chemical change—matches
- Energy movement from hot to cold
- Radiant heat transfer
- Conductive and convective heat transfer
- Electrical transfer—thermocouple
- Increase in volume
- Increase in temperature

Heat and **heat transfer** can be described in the following terms:

- Conductive heat transfer—when heat energy is transferred through a solid object, for example, tubes in a heat exchanger

- Convective heat transfer—requires fluid currents to transfer heat from a heat source, for example, a convection section of furnace
- Radiant heat transfer—the transfer of energy through space by the means of electromagnetic waves, for example, the sun
- Evaporation—a form of convective heat transfer, for example, a cooling tower
- Sensible heat—heat that can be sensed or measured; increase or decrease in temperature, for example, a thermometer
- Latent heat—hidden heat, does not cause a temperature change
- Latent heat of fusion—heat required to melt a substance; heat removed to freeze a substance
- Latent heat of vaporization—heat required to change a liquid to gas
- Latent heat of condensation—heat removed to condense a gas
- Specific heat—the BTUs required to raise one pound of a specific substance 1° F.

The primary purpose of the heat exchanger system in Figure 15-7 is to transfer heat to the feed. This is accomplished in a two-step process where the liquid is heated up in the first exchanger and increased in the second heat exchanger. Heat exchangers can be used to heat or cool a substance. The chemical processing industry utilizes a variety of unique approaches in the process. Before being asked to operate a heat exchanger system, a technician spends significant time memorizing and learning the various types of heat exchangers and how they can be arranged in different systems. By measuring the hotness or coldness of a substance, we determine temperature. Process technicians use a variety of temperature systems. The four most common are K- °C- °F- °R, as described in Table 15-1.

K- °C- °F- °R

Key points about temperature and heat include the following:

- Heat is a form of energy caused by increased molecular activity that cannot be created or destroyed, only transferred from one substance to another.
- The hotness or coldness of a substance determines the **temperature**.
- Heat is measured in BTUs, and temperature is measured in K, °C, °F and °R.
- Temperature and heat are not the same.

The safety aspects associated with the operation of a heat exchanger system include the following:

- Chemical hazards associated with spills and leaks (see chemical list and material safety data sheet (MSDS)
- Hazards associated with burns
- Hazards associated with fires
- Hazards associated with explosions and boiling liquid expanding vapor explosion *(*Bleve)

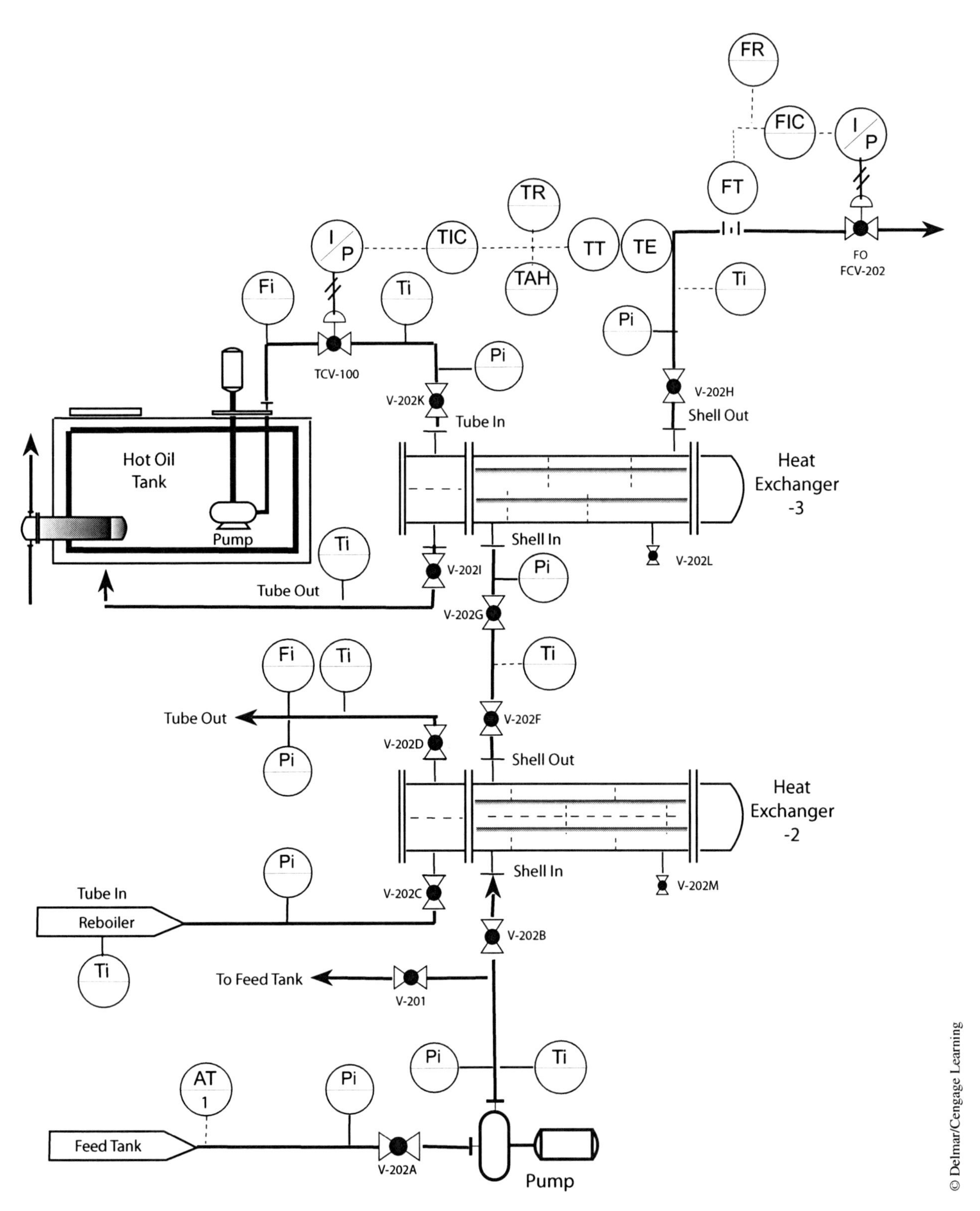

Figure 15-7 *Heat Exchanger System*

Table 15-1
Common Temperature Systems

Scale	Water boils	Water freezes	Conversion Formula
Kelvin (K)-	373 K	273 K	K = °C + 273
Celsius (°C)-	100 °C	0 °C	°C = (°F − 32) ÷ 1.8
Fahrenheit (°F)-	212 °F	32 °F	°F = 1.8 × °C + 32
Rankin (°R)-	672 °R	492 °R	°R = °F + 460

- Confined space entry; larger exchangers with tube bundles removed
- Equipment failure; tube leak, gasket leak, shell puncture or leak
- Error with valve line-up resulting in explosion or fire
- Pump failure resulting in overheating in heat exchanger
- Gauge failures
- Sampling, purging, or venting the shell
- Exceeding pressure and temperature ratings on heat exchanger code stamp for tubes and shell
- Utilizing incompatible materials with chemicals
- Working with hot materials under pressure

In the preceding list, a large number of items can be categorized as *operator error*. Unfortunately a high number of safety incidents can be attributed to mistakes made by process technicians or engineering. A process hazard analysis should always be performed prior to allowing technicians to operate a heat exchanger system. Proper training is also critical for new technicians assigned to heat exchanger systems. Some serious industrial accidents have been linked to a lack of training for new employees.

Cooling Tower

A cooling tower is a device used by the chemical processing industry to cool water and to provide a steady supply of cool water for industrial uses. Cooling towers are classified by how they produce airflow and how they produce airflow in relation to the downward flow of water. A cooling tower can produce airflow mechanically or naturally. Airflow can enter the cooling tower and cross the downward flow of water or run counter to the downward flow of water. Figure 15-8 shows the basic components of a cooling tower system. Before operating the cooling tower system, a technician needs to be familiar with the scientific principles associated with heat transfer, evaporation, fluid flow, equipment relationships with heat exchangers, instrument systems, safety, and the basic components of the cooling tower system.

The safety aspects of the cooling tower system include the following:

- Chemical additives: liquid, solid and gas (see chemical list and material safety data sheet (MSDS)
- Rotating equipment.
- Hazards of hot water
- Equipment failures (tube leak on condenser)

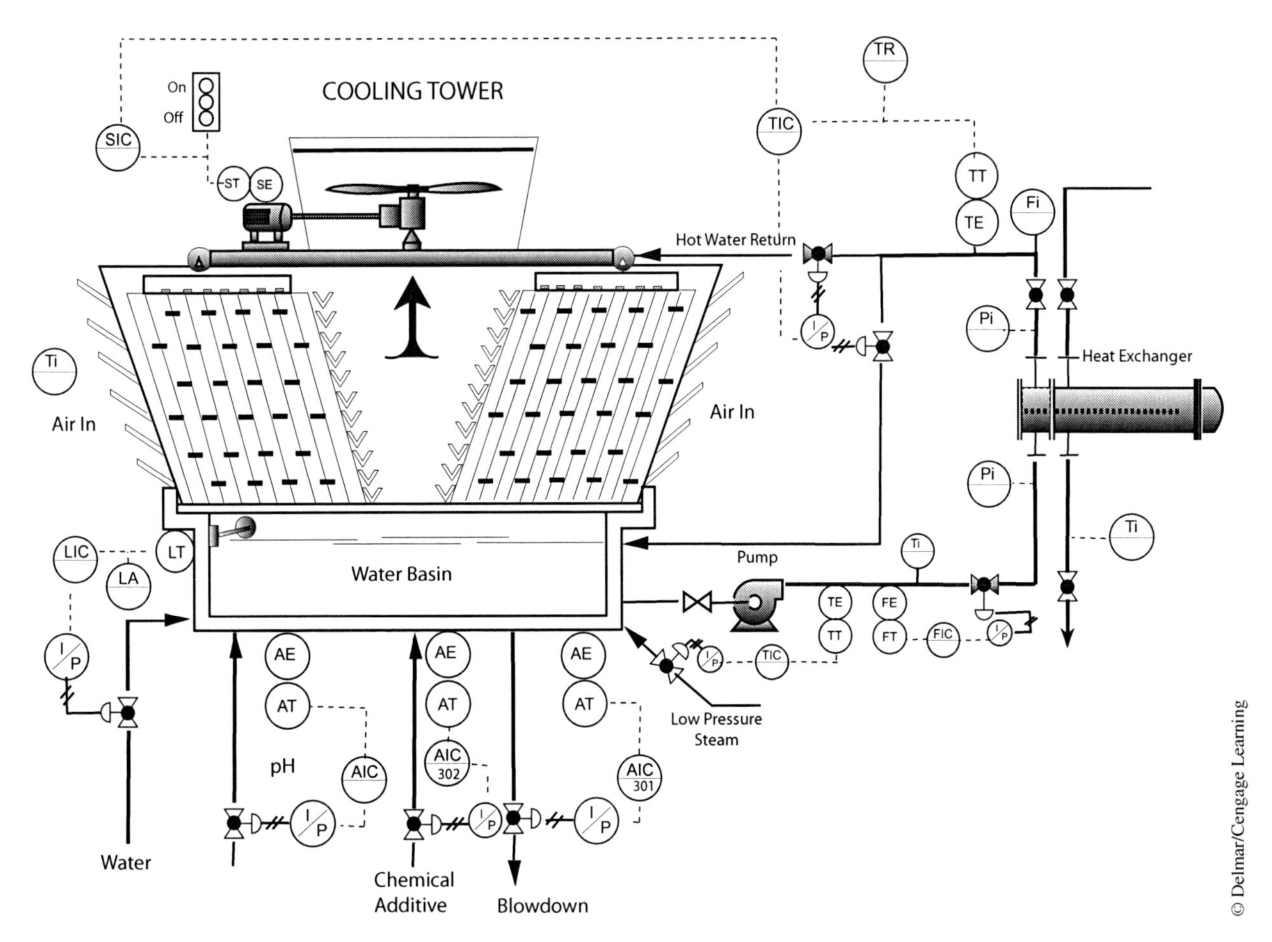

Figure 15-8 *Cooling Tower System*

- Working at heights
- Hazards of working with **acid** (see MSDS)
- Confined space entry (water basin empty)
- Hazardous energy
- Rotten wood
- Poor visibility due to vapor and foam

Furnace or Fired Heater

Another potential operating hazard can be found in a fired furnace system. This type of system is easy to operate, but maintenance of the shutdown and control system is often forgotten. Due to the possibility of a flame-out, a furnace can be a hazardous device. If this occurs, the concentration of hazardous gases builds up in the furnace until it explodes. Another common problem found in furnaces is a tube or coil rupture. Most tube leaks start out as small pin holes that cannot be easily seen through the inspection doors, but black smoke can be seen coming from the stack. Tube failures are typically related to flame impingement, high heat loads, or erosion. Overfiring the furnace causes the metal to weaken and crack. Flame patterns can be

Figure 15-9 *Typical Fired Heater*

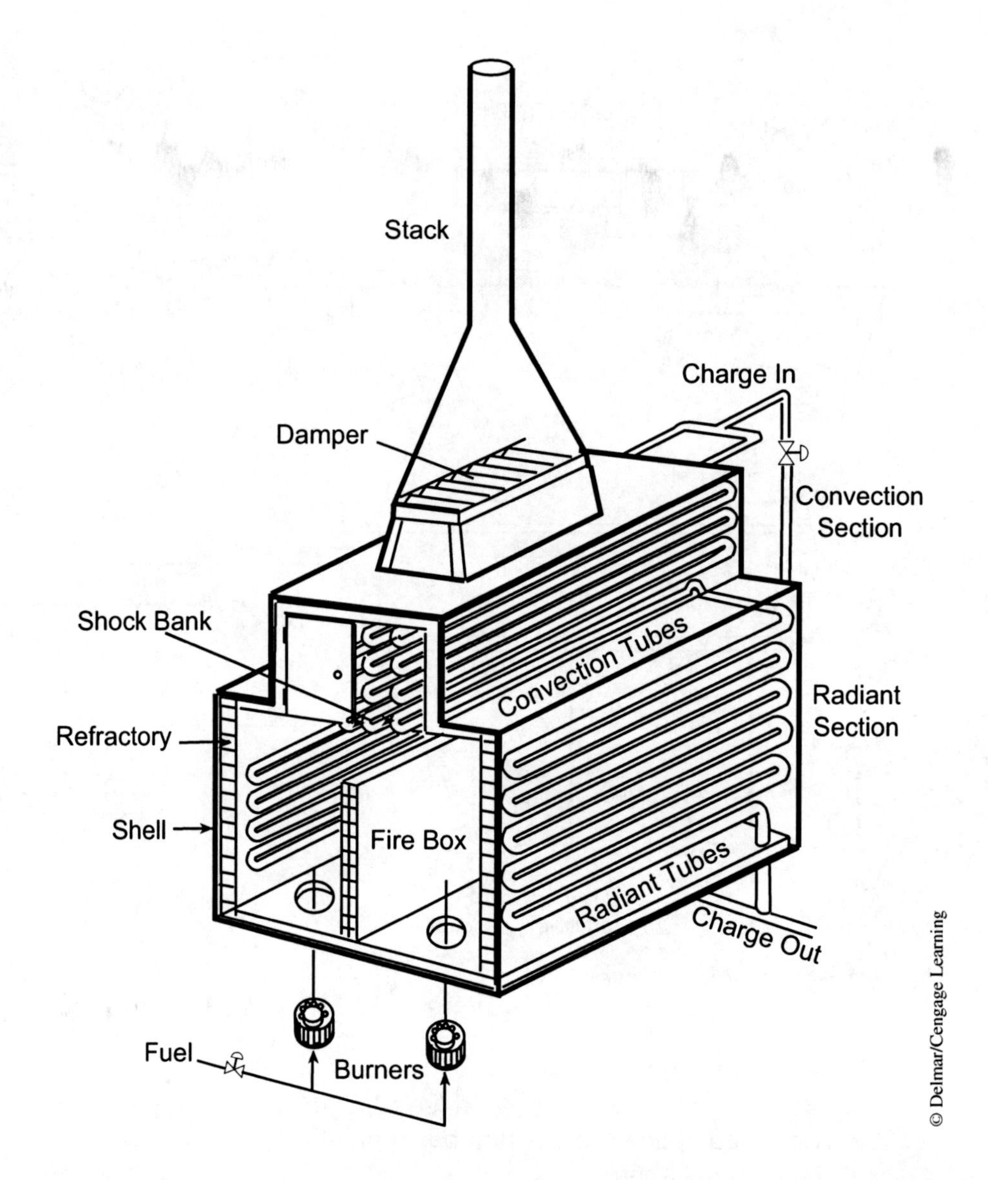

controlled by adjusting primary and secondary air registers. Internal tube erosion is the result of sustained, high flow rates or due to a phenomenon called *blasting*. This happens when a bubble of liquid boils. Figure 15-9 shows the layout of a typical fired heater.

Steam generation is another common operation found in the chemical processing industry. Although extremely useful, the production of steam is laced with hazards. Some of these hazards include accidental releases of high-, medium-, and low-pressure steam. High-velocity steam can be destructive when directed in a narrow stream, like those present in a pipe or vessel rupture. This presents a serious hazard to anyone or anything in the direct path of this rupture. High-velocity steam can cut through solid objects. It will also eat away the inner surfaces of industrial equipment used to contain it. As small particles of suspended solids are picked up in the steam, a shotgun effect

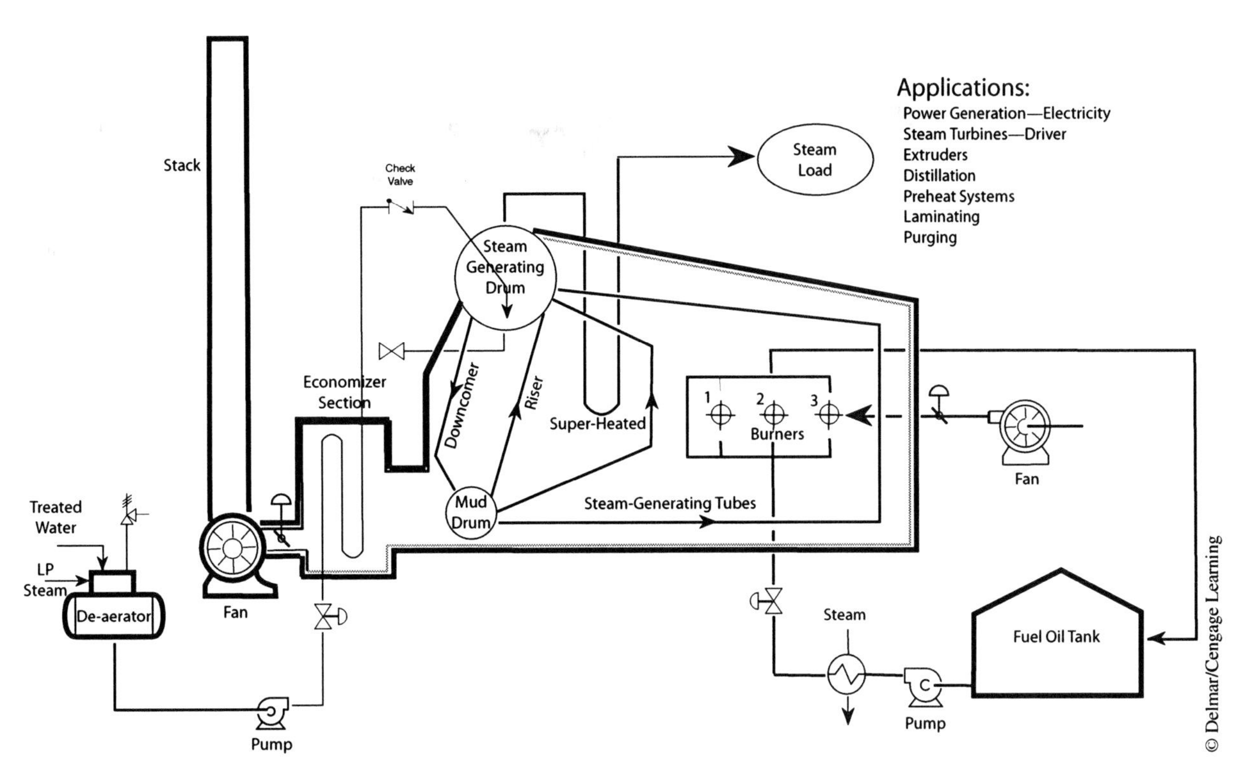

Figure 15-10 *Steam Generation—Uses and Applications*

occurs at every bend in the piping or at every valve. Over a period of time, this pitting process will weaken the metal. In steam systems, valving that is not designed for throttling service should not be used to regulate flow. This will destroy the trim, and a positive seal inside the valve will be impossible.

Another hazard associated with steam is the rapid increase in temperature as thermal energy is quickly transferred through the metal. This occurs under the principles of conduction and convection. Non-uniform heating of piping and equipment can cause expansion problems that will rupture equipment. Thermal warp, water hammer, and thermal expansion are all natural phenomena of which a process technician should be aware. Thick metals require more time to warm up than thinner metals. Devices such as steam turbines need to be slow-rolled and heated up gradually before being introduced to high pressure and high velocity steam. Water trapped in the system can expand to many times its original volume, causing severe problems. Figure 15-10 shows a typical steam generation system and the variety of ways it is used.

Steam Generation

Electricity is widely used in the day-to-day operation of a normal industrial facility. Power requirements are so high at some process units that special notification procedures are in place for the power company during startups.

Power generation typically starts outside the plant; however, some process facilities generate their own power depending on need. Electrical power generation facilities vary in design from nuclear- to steam-generated. This electricity is sold in high voltages to the chemical process industry. Inside the plant or facility, this high-voltage energy is stepped down into usable voltages at the central power station. Electrical power is distributed to various motor control centers (MCCs) in the plant. Voltage from the MCCs is used to operate most of the equipment inside a single operating unit. A typical refinery or chemical plant will have a large number of local MCCs. The hazards associated with electrical shock are severe and require specific training. Process technicians frequently start and stop equipment, lock out equipment, tag out equipment, and prepare electrically driven devices to be worked on. Local MCCs are inspected on each rotating shift. Pump and compressor motors are checked during routine rounds as technicians walk through their units. High temperatures and unusual sounds or vibrations are quickly found and reported to maintenance. Electrical equipment repair should be performed only by qualified plant electricians. Even problems that may appear routine or simple may have serious associated hazards. Figure 15-11 shows a typical power generation system.

Minor maintenance of equipment and instrumentation is required from time to time. One of the hazards associated with this is piping and equipment

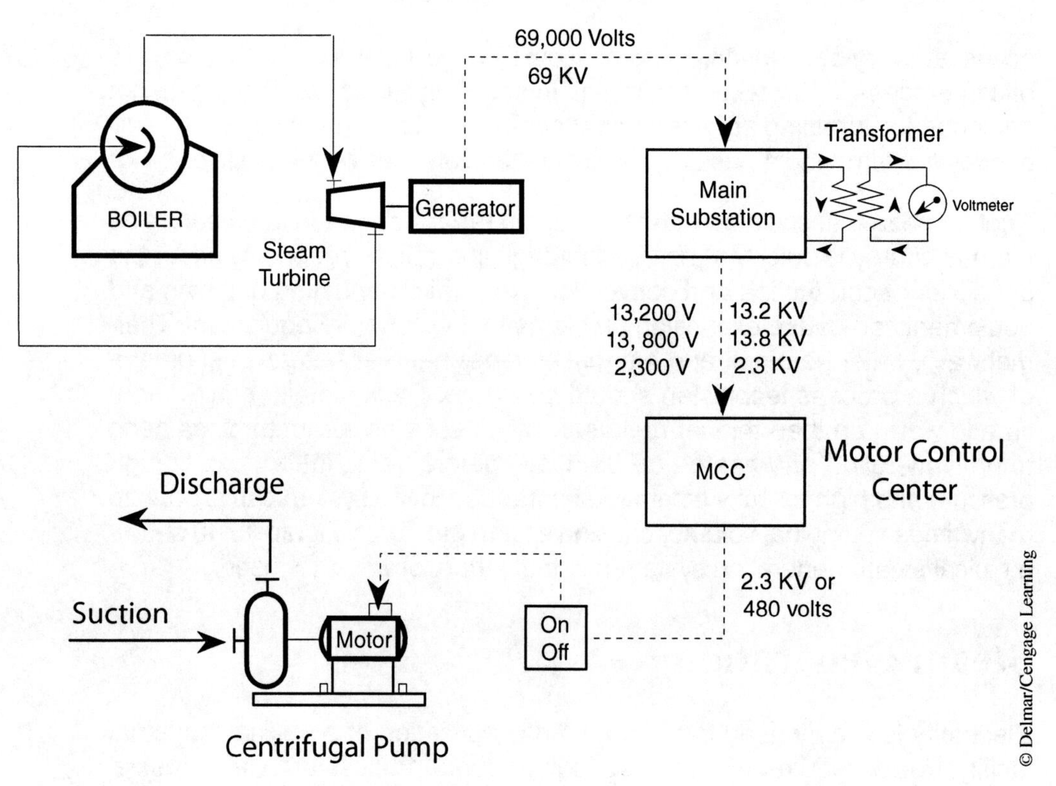

Figure 15-11 *Power Generation System*

under pressure. Pressure may be generated by gases or liquids and may be hazardous, flammable, or toxic. Special concern should be exercised when attempting to stop a leak in a screwed pipe or fitting. Older piping and fittings may suffer the effects of time and exposure to various chemicals and may break off when tightened. Systems under pressure should not be tightened until the pressure is removed. Only under rare and extreme situations should this rule be bypassed. It is easier to deal with a minor leak for a few days than to respond to a serious emergency situation.

Flare System

Flare systems in a plant pose a number of operating hazards that should be considered. A flare system is composed of a complex series of piping that weaves through all the primary connections in the plant. The plant header is a collection point for all of the emergency pressure releases that occur in the plant. This pipe is filled with flammable and toxic materials that flow toward a flare drum and flare. The flare is designed to burn gases at the highest point on the flare tip. Steam is provided at this point to disperse the hydrocarbons and mix in air, so the mixture can burn cleanly. When a flare is smoking badly, this is a strong indication that the plant is having serious problems. Somewhere in the plant a unit or series of units is in an emergency situation, and the last line of defense has been activated. Process technicians are elevated to their highest level of activity as the vast resources of the company are applied to solving the problem. A quick solution is needed to save valuable company resources. Black smoke from the flare clearly shows unburned hydrocarbons and the loss of company profits. Figure 15-12 is an example of a flare system. At the base of the flare, a large expanse is found, typically surrounded by a fence. This provides a safe barrier between personnel and equipment and the hazards of the flare under catastrophic conditions. The hazards associated with flare operation include the following:

- Flare pilot light goes out and heavy hydrocarbon vapors drop to ground and seek ignition source.
- Flare header flow overwhelms flare system and produces black smoke (unburned hydrocarbons).
- Fan motor catches on fire.
- Pump goes out on knock-out drum liquid seal and pressure increases on header.
- Steam ring fails.
- Support cable breaks during high winds and flare falls.

Weather-related Hazards

Major chemical complexes carefully monitor the weather so they can anticipate and prepare for any potential weather-related problems. Lightning is a common occurrence that rarely finds an unprotected or ungrounded system

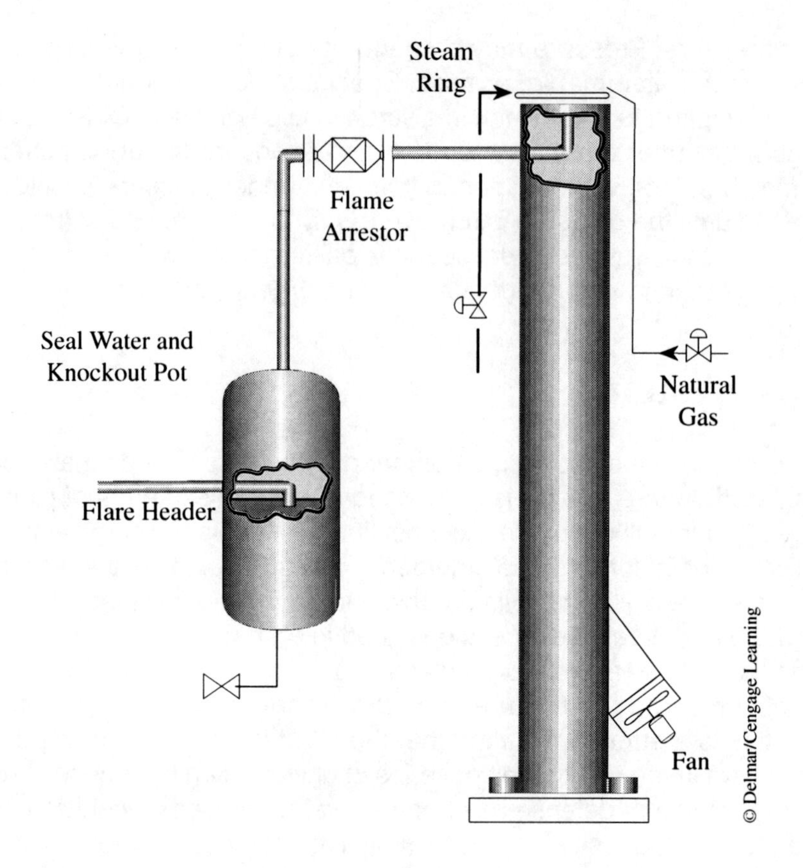

Figure 15-12 *Flare System*

in a plant. If it does, the consequences can be catastrophic, resulting in fires, explosions, and equipment failure. Process technicians working in this environment should be aware that there are brief periods of time when it is not safe to be on the tallest vessel or distillation column in the plant. A combination of high winds, rain, and lightning should alert a technician to the level of safety precautions that should be applied.

Heavy rainstorms can quickly drop the temperature in a chemical complex and cause wide variation swings in process variables. This can create a serious situation when process conditions begin to swing. Freezing weather can also create problems inside a plant. Falling ice, ice-covered walkways, slippery ladders, and frozen water and process lines can present serious operating hazards.

Hurricanes present a potentially dangerous situation laced with numerous operating hazards: high winds, flooding, power outage, and flying debris. As hurricanes form and begin to pick up speed and energy, plant personnel closely watch projected landfall. Chemical plants and refineries are designed to operate continuously. Shutdowns are expensive and are postponed until absolutely necessary. As a hurricane approaches a chemical facility, only key personnel are left to operate the plant in standby mode. Non-critical operations are shut down, and critical operations are secured.

The severity of the storm will determine the action plan the plant will use and how much of the facility will actually be shut down. Chemical processing units are designed to withstand high winds; however, a hurricane has a number of twists and variations that make it hazardous to operate.

Chemicals- and Chemistry-related Hazards

Chemistry-related hazards are associated with the way two or more chemicals will respond to each other under specific operational conditions. A process or chemical technician works with a wide assortment of chemicals and hazards. Each chemical has a set of characteristics that set it apart from any other. Ethylene is produced in an Olefins Unit and is used in the production of high density polyethylene (HDPE). It is a colorless, flammable gas with a wide explosive range that has the density of air. Propylene (polymer and chemical grade) is produced in an Olefins Unit. It is a colorless, explosive, flammable gas that is heavier than air. Butadiene is a colorless, explosive, flammable gas that is a liquid at temperatures below 68°F. Another gas produced at an Olefins Unit is hydrogen. This odorless, colorless, highly flammable gas is difficult to contain because it can easily escape due to its small molecular size. Hydrogen sulfide (H_2S) is a sulfur compound that smells like rotten eggs. It is toxic to humans at small doses measured in parts per million. It quickly overpowers the sense of smell, and death results from prolonged exposure.

Reactors

Reactors are used to make new products by making or breaking **chemical bonds**. Reactors come in a variety of designs and shapes and have numerous applications. **Exothermic** reactions produce heat, and **endothermic** reactions require heat in order to progress. Figure 15-13 shows a typical stirred reactor.

- Reactors utilize the basic principles of chemistry. **Chemistry** is described as the study of the characteristics or structure of elements and the changes that take place when they combine to form other substances. Reactor technicians play a major role in the production and manufacture of raw materials that are used to make finished products. Some of the materials found in fluidized gas bed reactors are pyrophoric. Pyrophoric is a term used to describe how a substance ignites spontaneously in air below 130°F.

Modern chemistry is an essential part of the process environment and for this reason a vital part in the initial training for reactor technicians.

An **atom** is the smallest particle of an element that still retains the characteristics of an element. Atoms are composed of positively charged particles

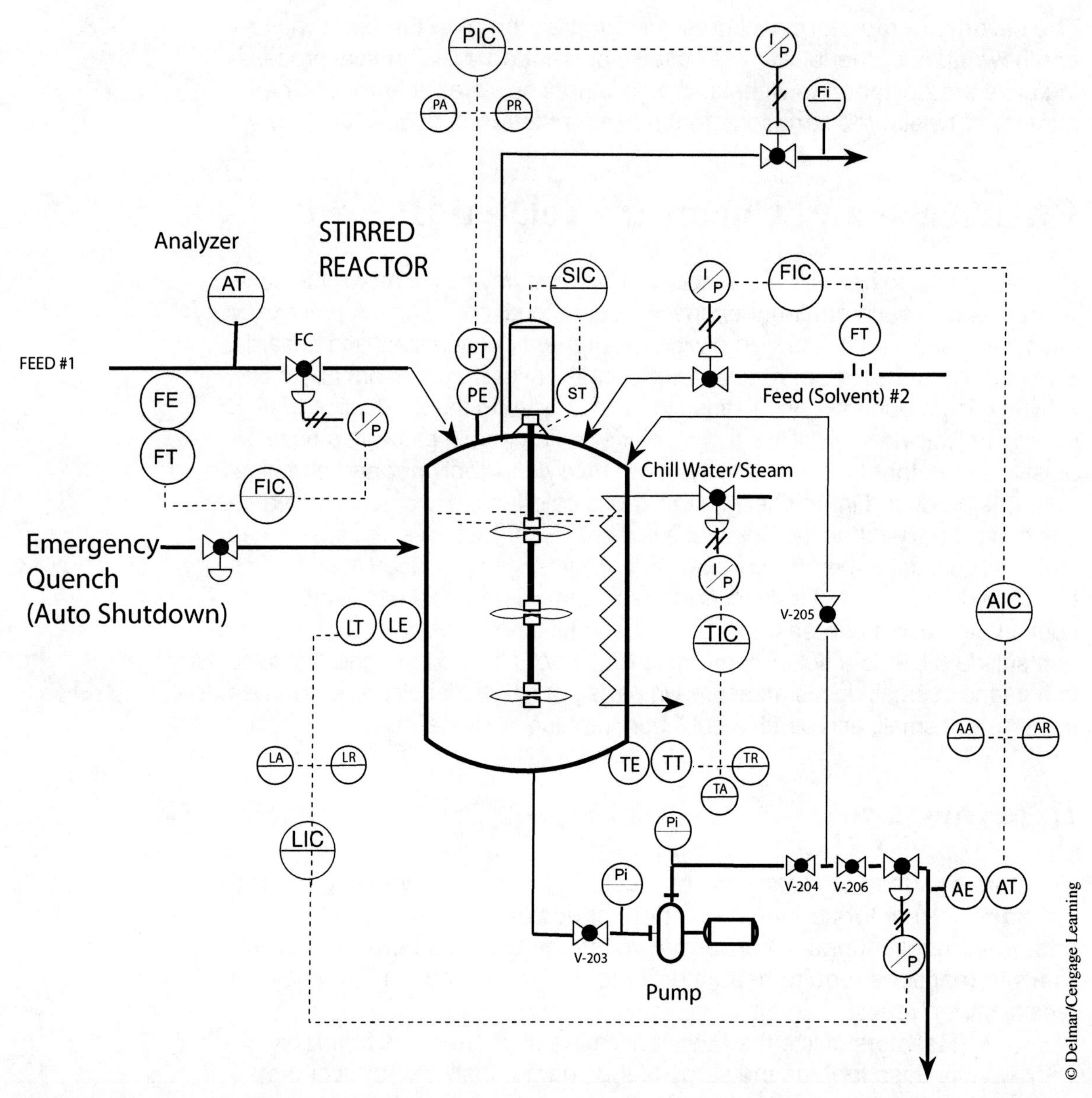

Figure 15-13 *Stirred Reactor*

called **protons**, an equal number of neutral particles called **neutrons**, and negatively charged particles called **electrons**. Protons and neutrons make up the majority of the mass in an atom and reside in an area referred to as the nucleus. The sum of the masses in the nucleus (protons and neutrons) is called the **atomic mass unit (AMU)**. The **atomic number** of an element is determined by the number of protons in its nucleus. The atomic number of the element determines its position on the periodic table. Electrons are negatively charged particles that orbit the nucleus. Protons and electrons are typically equally balanced in an atom. This is important because it ensures that the element is equally neutral. Valence electrons reside in the outermost shell of an atom. Valence electrons are important to chemistry because they provide the links in virtually every **chemical reaction**. Atoms share or transfer their valence electrons to form chemical bonds.

Chemical bonding is typically classified as **ionic** or **covalent**. Covalent bonds occur when atoms react with each other by sharing electrons. This forms an electrically neutral molecule because the protons and electrons electrically balance each other. Alkylation stirred reactors utilize an acid and a base to perform different functions. An acid is a chemical compound that has a pH value below 7.0 that changes blue litmus to red and yields hydrogen ions in water. A base is a bitter-tasting chemical compound that has a soapy feel and a pH value above 7.0. It turns red litmus paper blue and yields hydroxyl ions.

Matter is anything that occupies space and has mass. The four physical states of matter are solid, liquid, gas, and plasma. Plasma can be found in powerful magnetic fields. Molecules and compounds are products of chemical reactions. A **compound** is defined as a substance formed by the chemical combination of two or more substances in definite proportions by weight. A **molecule** is the smallest particle that retains the properties of the compound. Solutions are a type of homogenous mixture. The term homogenous refers to the evenly mixed composition of the solution. A common example of a homogenous solution is red dye and water. As the contents of the red dye come into contact with the water, it is evenly dispersed throughout the solution. Mixtures do not have a definite composition. A **mixture** is composed of two or more substances that are only mixed physically. Because a mixture is not chemically combined, it can be separated through physical means, such as boiling or magnetic attraction. Crude oil is a simple example of a mixture. It is composed of hundreds of different hydrocarbons. Reactor technicians separate the different components in the crude oil by heating it to the boiling point in a distillation column.

Chemical elements found in nature are listed on **the periodic table** (Figure 15-14). These **elements** can be described as the building blocks of all substances. Each element is composed of atoms from only one kind of element. Elements are represented by letters on the periodic table. The letter symbol for hydrogen is H. The letter symbol for carbon is C. A list of all known chemical symbols can be found on a periodic table. A good understanding of the periodic table is required for all reactor technicians. A chemical **reaction** can be described by associated numbers and symbols. The atomic number indicates the number of protons in an atom, and the atomic mass unit indicates the total number of electrons along with the protons and neutrons in an atom. In a **chemical equation**, the raw materials or **reactants** are placed on the left side. As the reactants are mixed together, they yield predictable products. A yield sign or arrow immediately follows the reactants. The products are placed on the right side of the equation. Because atoms cannot be created or destroyed, a common rule of thumb is, *what goes*

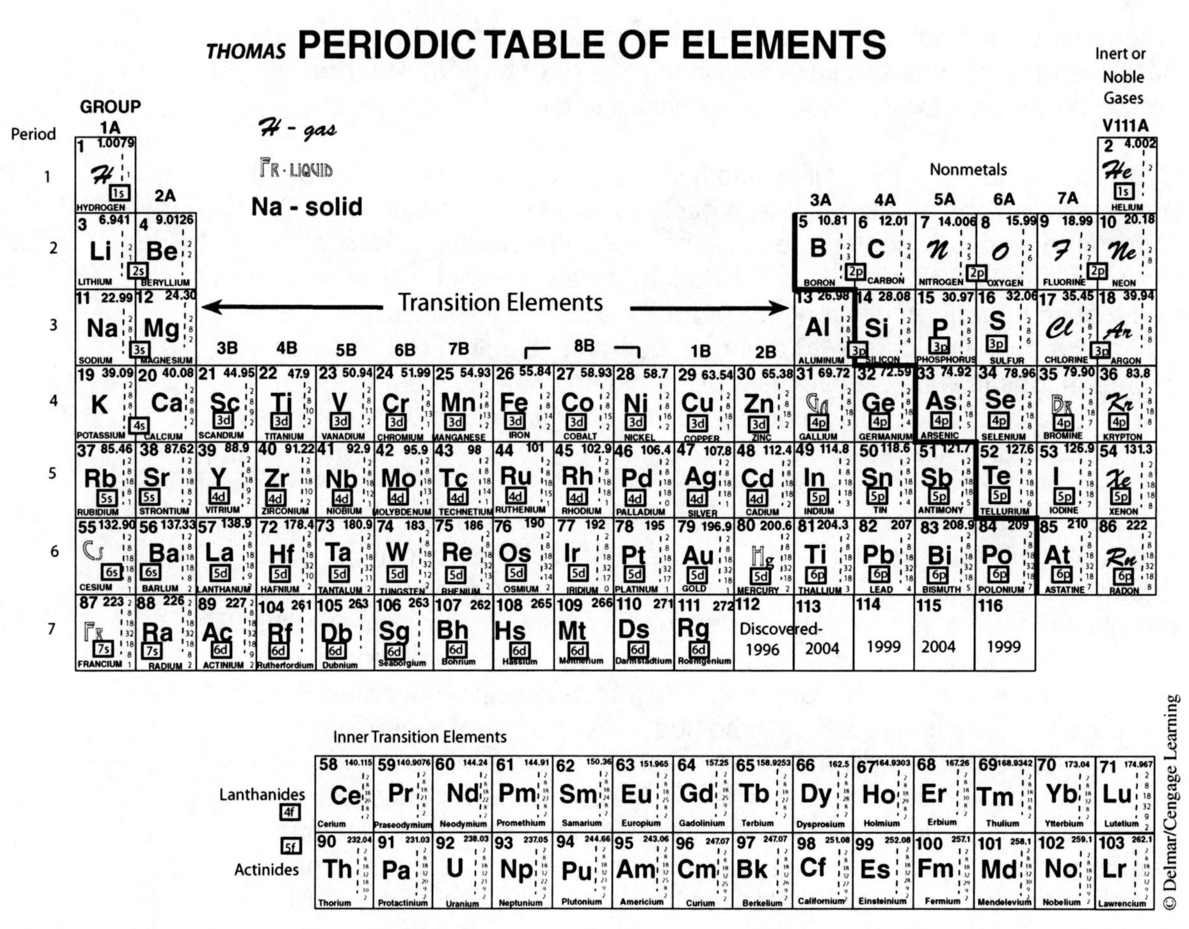

Figure 15-14 *Thomas Periodic Table*

into a chemical equation must come out. The sum of the reactants must equal the sum of the products. The term *balanced equation* describes how the sum of the reactants (atoms) equals the sum of the products (atoms).

Example:

(Butane) (Pentane)

C_4H_{10} + C_5H_{12} + Liquid Catalyst + Solvent → New Product

Butane has:		*Pentane has:*	
4 carbon atoms	= 48 AMU	5 carbon atoms	= 60 AMU
10 hydrogen atoms	= 10 AMU	12 hydrogen atoms	= 12 AMU
	58 AMU		72 AMU

In order for the reaction to occur in the stirred reactor, a specific amount of each chemical must be mixed at a specific RPM, temperature, and pressure. Under these conditions, new products will be chemically formed.

An important aspect of reactor operation is **material balance**, as shown in this example:

$Cu + H_2SO_4 \rightarrow CuSO_4 + H_2$

1 Copper	1 Copper
2 Hydrogen	2 Hydrogen
1 Sulfur	1 Sulfur
4 Oxygen	4 Oxygen

Another important aspect of reactor operation is *mass relationships*. When working out mass relationships, you need to have a good understanding of the periodic table. Certain elements combine to form chemicals that are easily recognized. Water (H_2O) or carbon dioxide (CO_2) is a good example of elements and proportions and atomic mass units that can be found on the periodic table. Consider the following chemical equation for working a mass relationship:

$H_3PO_4 + 3NaOH \rightarrow Na_3PO_4 + 3H_2O$.

In this problem phosphoric acid and sodium hydroxide react to form sodium phosphate and water. By looking at the reactants and products relationship and total weights, you can see more easily how a chemical reaction works.

$H_3PO_4 + 3NaOH \rightarrow Na_3PO_4 + 3H_2O$

Phosphoric acid H_3PO_4 (**reactant #1**)

3 hydrogen atoms	$= 3 \times 1.008$	$= 3.024$ AMU
1 phosphorus atom	$= 1 \times 30.98$	$= 30.98$ AMU
4 oxygen atoms	$= 4 \times 16$	$= 64$ AMU
		98 grams, pounds, or tons

Sodium hydroxide 3NaOH (**reactant #2**)

3 sodium atoms	$= 3 \times 23$	$= 69$ AMU
3 oxygen atoms	$= 3 \times 16$	$= 48$ AMU
3 hydrogen atoms	$= 3 \times 1.008$	$= 3.024$ AMU
		120.02 grams, pounds, or tons

$98 + 120 = 218$ AMU. *Reactants' total molecular weight.*

Sodium phosphate Na_3PO_4 (**product #1**)

3 sodium atoms	$= 3 \times 23$	$= 69$ AMU
1 phosphorus atoms	$= 1 \times 30.98$	$= 30.98$ AMU
4 oxygen atoms	$= 4 \times 16$	$= 64$ AMU
		163.98 grams, pounds, or tons

Water $3H_2O$ (**product #2**)

6 hydrogen atoms	= 6 × 1.008 =	6.048 AMU
3 oxygen atoms	= 3 × 16 =	48 AMU
		54.05 grams, pounds, or tons

163.98 + 54.05 = 218.03 AMU. Products' total molecular weight.

AMU can be substituted for other weights like grams, pounds, or tons.

Reactor technicians are required to look carefully at mass relationship problems. For most of these situations a chemist has already calculated the correct reactant amounts and specified the correct operating conditions. Technicians see mass relationship problems in actual weights, grams, pounds, or tons. An example of this is found in the following reaction between nitrogen and hydrogen, yielding ammonia.

N_2	+	$3H_2$	→	$2NH_3$
120 lbs		39 lbs		159 lbs

This is the normal operation of the unit; however, at the end of the month production needs to be increased. When this occurs the nitrogen is increased to 420 lbs. This will require an adjustment on the hydrogen in order for the product to remain within operational guidelines. When this occurs the reactor technician is required to make a small calculation.

N_2	+	$3H_2$	→	$2NH_3$
120 lbs		39 lbs		159 lbs
420 lbs		? lbs		? lbs.

The relative weight is 420 lbs. The original actual weight is 120 lbs. To solve this problem simply divide the relative weight by the actual weight. This will give you a factor of 3.5. By multiplying the hydrogen feed rate of 39 lbs by 3.5, you can calculate the new flow rate. 420 ÷ 120 = 3.5.

$$3.5 \times 39 = 136.5 \text{ lbs.}$$
$$3.5 \times 159 = 556.5 \text{ lbs.}$$

This process will take production rates from 159 lbs to 556.5 lbs per hour. This simple process can be used to calculate most reactor problems that deal with relationships.

Safety hazards associated with the operation of a stirred reactor include the following:

- Runaway exothermic or endothermic reaction.
- Feed composition changes cause rapid pressure changes.
- Feed concentration increase—reaction becomes unstable.

- Agitation problems will reduce reaction.
- Loss of cooling water—temperature and pressure increase.
- Bearings fail resulting in fires and loss of containment.
- Seals or containment fails resulting in acid or caustic burns.
- Loss of pressure control—increase or decrease.
- Reaction time in reactor—reaction incomplete.
- Explosions.
- Temperature increase—doubles reaction rate for every 10°C increase.
- Loss of catalyst—reaction will stop.
- Vapor cloud or release.

Distillation System

Distillation is a process that separates a substance from a mixture by its boiling point. During the distillation process, a mixture is heated until it vaporizes, and then is recondensed on the trays or at various stages of the column where it is drawn off and collected in a variety of overhead, sidestream, and bottom receivers. The condensed liquid or overhead product is referred to as the *distillate*, whereas the liquid that does not vaporize in a column is called the *residue* or *bottom product*. The composition of the feedstock is important. Variations in the feedstock will cause significant changes to the operation of the distillation system. A variety of chemicals are typically introduced to a distillation column and separated. Two-component mixtures are referred to as *binary*, whereas three-component mixtures are called *ternary*.

Distillation columns come in two designs: packed or plate. During tower operation, raw materials are pumped to a feed tank and mixed thoroughly. Mixing is usually accomplished with a pump-around loop or a mixer. This mixture is pumped to a feed preheater or furnace where the temperature of the fluid mixture is brought up to operating conditions. Preheaters are usually shell and tube heat exchangers or fired furnaces. The fluid enters the feed tray or feed section in the distillation column. Part of the mixture vaporizes as it enters the column while the rest flows downward, flashing and condensing over and over again until, based upon temperature and pressure, the liquid stops on a tray where the molecular structure of the components are similar.

A distillation tower is a series of stills placed one on top of the other. As vaporization occurs, the lighter components of the mixture move up the tower and are distributed on the various trays. The lightest component goes out the top of the tower in a vapor state and is passed over the cooling coils of a shell and tube condenser. As the hot vapor comes in contact with the coils, it condenses and is collected in the overhead accumulator. Part of this product is sent to storage while the other is returned to the tower as reflux. Most distillation systems use a procedure called *total reflux*. Reflux is used to control product purity and temperature. Distillation systems are designed to separate the components in a mixture by **boiling point**. The lightest component typically boils first and is found leaving the column as a vapor. This vapor is typically cooled and condensed and sent out for further processing. In a total reflux situation, all of the reflux streams are sent back to the column. In some situations this process can be hazardous. In a C3 splitter, debutanizer and depropanizer column, methyl acetylene (MAC), and propadiene (PD) can build up to dangerous concentrations (60% mole weight) and detonate or explode. Figure 15-15 shows a distillation column on total reflux. In order for this to occur the system would need to be on total reflux for several days and some product removed from the bottom and overhead. Operational hazards associated with distillation include the following:

- Fires and explosions
- Leaks—gasket, valve, instruments, piping, and pumps
- Pressure surges—composition changes
- Loss of cooling water—results in flaring
- Steam valve sticks in open position—overheats tower
- Feed valve fails in open position—floods trays or column
- Lightning strikes column
- High winds, tornado, or hurricane
- Tube leak mixes hydrocarbon with water
- Exothermic reaction occurs in column
- Object strikes column and damages structural integrity
- Water in feed causes rapid expansion problems
- Computer failure
- Power failure
- Instrument failure
- Feed tank ruptures

Other operating hazards include the harmful effect of acids and caustics on human tissue and the effects of copper and acetylene to form explosive acetylides. Mercury and ammonia react to produce explosive compounds, and **pyrophoric** materials will burst into flame when exposed to air.

Human Factors

Human error has been identified as the most common cause of industrial accidents. A partial list of factors that contribute to human error includes fatigue, failure to follow procedures, poor communication, poorly designed equipment, lack of training, poor understanding by management of ergonomic issues, and the psychological state of employees.

Shift work and the hazards associated with fatigue have been studied for a number of years. *A technician's greatest nemesis is fatigue* and mental

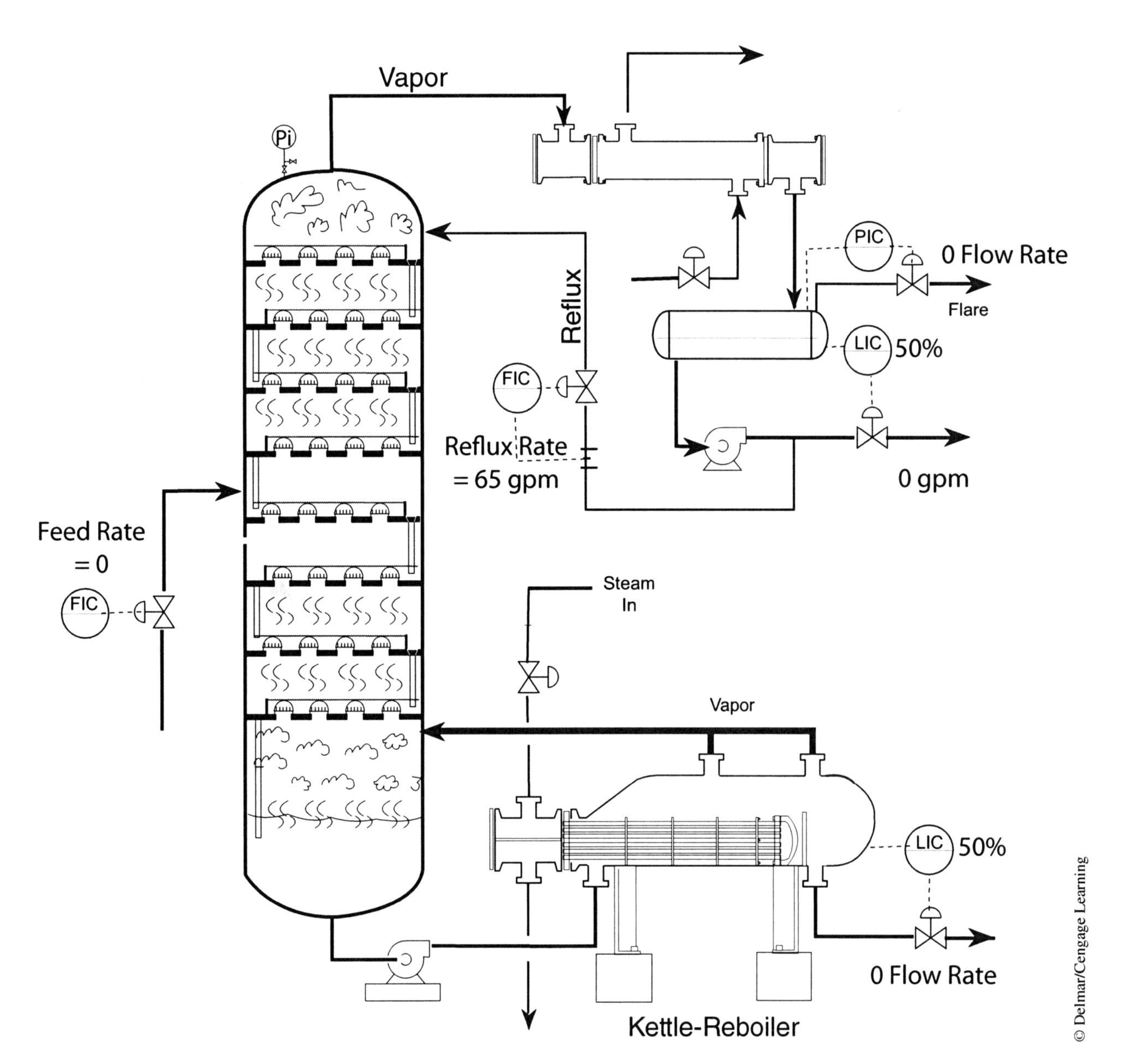

Figure 15-15 *Distillation Column on Total Reflux*

stress associated with various of reasons. Most of the biggest industrial disasters have taken place between the hours of 1:00 A.M. and 4:00 A.M. Most large chemical complexes have between 15% and 25% overtime.

As more studies are conducted on the human body's circadian rhythms, new rotating shift schedules have been developed. The new rules for safe shift design recommend clockwise or forward shift rotation, days to night to evening for 8-hour rotating shifts. A large number of plants use the popular 12-hour rotating shift model. In the United States, over 15 million people work rotating shifts. Shift workers suffer from more stress-related illnesses than those in other occupations. Some research indicates that shift worker performance problems have cost the global industries over $700 billion per year.

Some companies provide classes for shift workers' families to help them understand the job pressures and health hazards. Some of these organizations are providing 24-hour day care and transportation services for children and elderly parents. Other options include cross-training, opening an ownership of safety program, team building, opening communication to daytime management, staggered work shifts, opening facilities to family members, and company-sanctioned napping.

Many people in the chemical processing industry are focusing on the elements of behavioral safety. In this approach, process technicians take ownership of the program, define the major obstacles to safety, and create a framework for reporting safety issues and work with management to develop short- and long-term preventive solutions.

Operator training has long been a critical factor in developing a competent operating team. Senior technicians possess a wealth of information and experience that is difficult to package into technical manuals. As these baby boomers retire, a severe shortage in technical competence will occur. These problems can be addressed through systematic hiring practices that focus on graduates of local community colleges with process technology programs. Unfortunately, most companies use human resource hiring practices from the 1980s. These human resources development programs do not reflect the emergence of the industry-driven process technician training program. The Associate of Applied Science degree has become the standard entry-level credential sought after by individuals seeking employment in the CPI. Human resources development professionals need to focus on transcripts and the interview process in connection with systematic hiring practices. The bank of preemployment tests from the 1950s and 1960s should be forcibly retired, and a more professional approach should be applied to hiring this new workforce.

Failure to follow established guidelines and poor communication contribute to a number of human errors in a plant. Typically this is a training issue, but it can be linked to a number of other things. Some of these items include attitude, stress, time management, and concentration. Attitude and motivation are important factors in job performance. The psychological and emotional state of process technicians will determine how effective they will be in completing typical job assignments. Teamwork includes the development of the complex personal relationships required to complete company goals. Arguments between team leaders or members of a team can lead to human errors. Divorce or death in the family can also create a stressful situation.

Summary

Operating hazards can be classified as equipment- and system-related, weather-related, and chemistry- and chemicals-related. The types of equipment and systems that can cause an operating hazard include valves, piping, pumps, compressors, turbines, heat exchangers, cooling towers, boilers, furnaces, reactors, and distillation columns. This list could also include electrical items, instruments, rotating equipment, plastic plant equipment, and many other devices. Weather-related hazards may cause serious damage inside a chemical complex. These include lightning, tornadoes, hurricanes, hail, snow, rain, heat, and other phenomena related to the weather. The operating hazards associated with chemicals include the basic chemistry of how various components mix to form new products under a wide range of temperatures, pressures, and other variables.

Human error has been identified as the most common cause of industrial accidents. A partial list of factors that contribute to human error includes fatigue, failure to follow procedures, poor communication, poorly designed equipment, lack of training, poor understanding by management of ergonomic issues, and the psychological state of employees.

Shift work and the hazards associated with fatigue have been studied for a number of years. *A technician's greatest nemesis is fatigue* and mental stress associated with a various reasons.

Review Questions

1. Describe the various operating hazards found in the chemical processing industry.
2. Explain how shift work affects a process technician.
3. Describe the hazards associated with a pump system.
4. Explain the safety hazards associated with the operation of a compressor system.
5. Explain how the principle of pressure applies to compressors and pumps.
6. Describe the hazards associated with cooling towers.
7. Describe the safety hazards associated with heat exchangers.
8. List the operational hazards associated with boilers and furnaces.
9. Explain the safety features associated with reactor operation.
10. Describe the safety hazards associated with distillation operation.
11. Describe the most common cause for industrial accidents.
12. Explain the relationship between chemicals and the periodic table.
13. List the methods and scientific principles associated with heat transfer.
14. List the four states of matter.
15. Compare the following terms: *mixture*, *solution*, and *compound*.
16. Define the term *chemistry*.
17. Describe *cavitation* and the hazards associated with it.
18. Explain the principles of distillation.
19. Explain how a cooling tower operates and transfers heat energy.
20. Explain the hazards associated with acids and bases.

Weapons of Mass Destruction, Hurricanes and Natural Disasters

OBJECTIVES

After studying this chapter, the student will be able to:

- Describe the hazards associated with the use of weapons of mass destruction on the chemical processing industry.
- Identify the potential risks and hazards facing the chemical processing industry.
- Review the use of nuclear, conventional, and dirty bombs on civilian and industrial facilities.
- Describe the use of chemical weapons by terrorists.
- Describe the hazards associated with chemical asphyxiates.
- Explain the hazards associated with the use of biological weapons.
- Compare the hazards associated with biological toxins and viruses.
- Describe the destructive forces and hazards associated with tornados.
- Review the Fujita Scale for tornado wind speed, F-scale, intensity phrase, and type of damage.
- Describe the hazards associated with hurricanes and the Saffir-Simpson Hurricane Wind Scale.
- Describe the hazards associated with the use of military weapons on the CPI by terrorists.

Key Terms

- **Biological agents and weapons**—includes hazardous bacteria like anthrax, cholera, pneumonic and bubonic plague, tularemia, and Q-fever. Includes the use of viruses like smallpox, Venezuelan equine encephalitis, and viral hemorrhagic fever. Includes the use of biological toxins like botulism, ricin, staphylococcal enterotoxin B, and tricholthecene mycotoxins.
- **Chemical asphyxiants**—are formulated to remove or displace oxygen and interfere by not allowing the victim to utilize oxygen after it enters the respiratory system.
- **Emergency preparedness**—the prior preparation of essential items and plans in the event of an emergency situation. Examples include 72-hour emergency essential kits, evacuation plans, and food and water storage.
- **Fujita scale**—used to classify a tornado using a F-scale number, intensity phrase, wind speed, and type of damage possible.
- **Hurricane**—is a powerful, swirling storm with tentacles reaching out from a singular eye. Hurricanes have sustained wind speeds between 74–155 mph, heavy rainfall and flooding, and tremendous storm surges from 4 to 18 feet, spin off tornados. They cause downed trees and broken limbs, damaged homes, loss of electricity, utilities, and basic commodities. Hurricanes that form north of the equator rotate counterclockwise, whereas the opposite rotation occurs for storms that develop south of the equator. The spinning circulation is due to the Coriolis effect, which responds to the earth's rotation.
- **Saffir-Simpson hurricane wind scale**—a system designed to categorize a hurricane using sustained winds and type of damage possible.
- **Terrorism**—an attempt to change a belief or point of view through the use of a violent act. Terrorist acts are designed to generate fear and focus world-wide media attention to their cause.

Weapons of Mass Destruction

Benjamin Franklin once stated, "Failure to prepare is preparing to fail." The use of weapons of mass destruction can have a catastrophic effect on the chemical processing industry and the communities that surround them. Weapons of mass destruction can be classified as **biological agents**, chemical agents, and bombings. In light of the events that occurred on September 11, the world has changed and the chemical processing industry has revised its approach to safety preparedness. Other incidents like the Bhopal, India disaster, the Phillips explosion, Exxon Valdez oil release, and the BP Gulf oil release of 2010 have illustrated how vulnerable the industry is to unexpected events. In light of the information available to the industry, significant work remains incomplete. A series of options are available to the chemical processing industry including the use of *what-if* analyses. Using a modified brainstorming session can identify a number of potential risks. Some of these include the use of weapons of mass destruction.

Potential Risks include the following:

- Terrorist airplane flight into a chemical plant or refinery
- Terrorist entry into CPI workforce
- Use of conventional bomb
- Use of nuclear weapon
- Suicide bomber
- Use of chemical weapons
- Use of **biological weapons**
- Tornado hits
- Fast-forming hurricanes
- High winds
- Use of military weapons
- Flooding

Terrorist Flies an Airplane into Chemical Plant, Refinery, or Oil Rig

On September 11 it became clear to the United States that terrorism was not something that happens only in other countries and is reported by local news media. The methods used by the 9-11 terrorists differed from any previous attack aimed at the heart of the United States. For weeks after the incident, the world viewed the unfolding drama and recognized that a new age of terrorism was dawning. America was vulnerable. Four commercial airplanes were used as weapons of mass destruction. The consequences of the attack were psychological, emotional, spiritual, mental, and physical. Terrorism uses the insidious approach of doing horrible things in order to change people's beliefs and viewpoints. What are the effects of the events on September 11? Ten years have passed since the towers fell. Were the terrorists successful in changing the viewpoint and beliefs of the American people? These questions can be studied economically, politically, and through shifts in our government leaders, trading partners, and traditional allies. Figure 16-1 shows the damage inflicted by terrorists on the twin towers.

The Houston gulf coast area is home to over 600 companies associated with the chemical processing industry. Each of these companies produces different products and presents different hazards to the community. The key to attacking one or more of these companies is strategic knowledge about vital nerve centers. Many of these companies store raw materials in liquid state that at atmospheric pressure will turn into a vapor. Some of these chemicals will explode violently when exposed to an ignition source. Others are extremely hazardous to humans. Terrorists would need to know the following about a chemical processing facility:

- Location
- Quantity of materials

Figure 16-1 *World Trade Center*

- Physical and chemical hazards
- Direction of the wind
- Safety instrumentation
- Potential for catastrophic effects
- Domino effect with other industries
- Economic effect on community

In order to gather this type of information, a terrorist would need to work in the intended target company and have a sound understanding of engineering and process operations. Specific safety controls would need to be removed and GPS coordinates provided. In light of changing demographics in the Gulf-coast area and the need for better trained employees to replace the baby-boomers, this scenario is a strong possibility.

A good example of these rising tensions involves recent events in Texas when 17 Afghan soldiers went AWOL from Lackland Air Force base in 2010. On June 3, 2010, a man from Hempstead, Texas was charged with trying to help Al-Qaeda, and has been connected to the infamous Fort Hood shootings in 2009 by Major Nidal Hasan. A plane flown into a chemical plant or refinery or an oil rig in the Gulf of Mexico represents the potential for catastrophic effects.

Terrorist Enters CPI Workforce

Starting in 1994, local community colleges and universities began developing and implementing structured programs to train process technicians to replace the retiring baby-boomer generation. For the first time, prospective employees of the chemical processing industry were able to receive state-approved certificates and degrees, specialized training, and portable credentials to enter the workplace. During this time period, groups from Syria, Iraq, Iran, Kuwait, Jordan, Saudi Arabia, and other Arab countries began contacting these schools for specialized training. The Arab states have a large number of refineries, pipelines, and chemical plant operations. For many years, engineering programs have been producing chemical, mechanical, electrical, nuclear, petroleum, computer, and industrial engineers. At the same time degrees have been offered in chemistry at the bachelor, master's, and Ph.D. levels. Other areas of specialized training are offered in instrumentation, electrical, safety, and construction management. Each of these programs provides an educational avenue into the chemical processing industry. Typically, background checks are performed on all employees; however, this process may not uncover deep-seated beliefs and plans. Because these processes do not always work, the best line of defense is employee awareness and response. Most companies are proud of the rich diversity and cultural influences that make up their workforce. Many of the future terrorists will be home grown and will appear at first glance to be harmless. Companies will need to be proactive and inventive to head off any potential problems that could lead to terrorist acts. Figure 16-2 is a photo of a typical community college.

Use of Conventional Bomb

As mentioned earlier in this chapter, knowledge of critical nerve centers in the plant is essential to the long-term success of terrorist activities. This type of knowledge is not easily obtained. Conventional bombs cause large-scale destruction, death and injures, and pollution of the environment. Conventional weapons produce terrible injuries of human tissue. Conventional bombs release most of their energy in the initial blast. When a bomb threat is received by a chemical plant or refinery, a number of steps are followed:

- If the threat is considered real, the bomb threat alarm is sounded.
- Authorities are notified.
- Cell phones should not be used.

Figure 16-2
Community College

(a)

(b)

- A quick visual check is made of the area.
- The operating unit is placed in a safe condition.
- Return occurs after the all-clear is sounded.

Use of Nuclear Weapons

A one megaton nuclear bomb is capable of destroying the largest city in the world. When a nuclear bomb is detonated, it releases 35 percent of its energy as heat, 50 percent in the initial blast, and 15 percent as nuclear radiation. One kilogram of nuclear fission fuel is 20 million times more powerful than the same amount of TNT. If a nuclear bomb exploded, it would kill or seriously injure everyone within 3 to 30 miles. Typical temperatures in a nuclear explosion exceed 300,000°C. The shock wave from this type of explosion produces firestorms and hurricane-type winds that blow down everything in their path.

Suicide Bomber—Dirty Bomb

A terrorist may choose to use a dirty bomb, which is typically made from 100 pounds of conventional explosives, bundled with cobalt-60. The blast zone is over 300 feet in all directions, scattering or pulverizing the ground-up radioactive material. The wind will help in spreading the lethal fallout. When small particles are inhaled, the long-term effects can be serious.

Use of Chemical Weapons

The use of chemical weapons on innocent civilians is considered to be cruel and barbaric. The purpose of these weapons is to kill or incapacitate. Unfortunately, many terrorist organizations are sponsored by military forces that have expertise in the use of these chemical weapons. In order to respond to this type of threat, it is important to understand basic self-protection methods. Figure 16-3 is a photo of a typical chemical plant.

Chemical weapons are identified by how they affect the body and whether they are nerve toxins, **chemical asphyxiants**, respiratory irritants, skin irritants, or burning agents. In areas with large concentrations of chemical plants and refineries, terrorists may choose other chemicals to accomplish their goals. Examples of this could include railroad cars, tractor-trailer trucks, or bulk storage of industrial chemicals. Specific chemicals that may be targeted include chlorine, anhydrous ammonia, LPG, hydrocyanic acid, hydrogen fluoride, and sulfur dioxide. Hundreds of other chemical combinations could be included in this text.

Figure 16-3 *Chemical Plant*

Nerve Toxins

In 1995, terrorists from the extremist group Aum Shinriko decided to use the nerve agent sarin to kill as many people as possible in a Tokyo subway. Bags of a binary chemical were placed in six different locations and mixed together using wooden sticks. The mixed components formed the deadly nerve toxin, sarin, which filled the subway with toxic fumes. Because the sarin mixture was only 20 percent pure, it did not have the desired effect on the 5,522 people exposed. Only 12 died. Had the sarin been of a higher purity, over 5,000 people would have died.

The most widely known nerve agents or organophosphate compounds are sarin, tabun, soman, and VX, which have properties similar to industrial grade pesticides. Most of these agents are hazardous to the respiratory system. Tabun and sarin are easy to produce and appear to be the terrorist weapon of choice. Soman, sarin, and tabun evaporate at a rate slightly faster than pure water. The nerve agent VX must be heated in order to increase the evaporation rate. This makes VX dangerous because it lingers in the environment for days causing death and injury as it slowly releases its deadly respiratory poison.

The physical properties of nerve agents are as follows:

- Clear and colorless.
- Slight odor or odorless (pure product).
- Purity of the agent determines odor (fruity odor).
- Less pure VX has a sulfuric odor.
- Found in liquid state with different evaporation rates.
- Poisonous chemical vapors are typically introduced through lungs or inhalation.
- May also enter through skin absorption or ingestion.

Symptoms associated with nerve agents include excessive reactions:

- Salivation
- Diarrhea
- Mucous production
- Urination
- Seizures and muscle twitching

Decontamination procedures and treatment of nerve agents include the use of salt or fresh water, alkaline soap, and, in some cases, the use of Clorox bleach. The antidotes for this type of exposure include two drugs: atropine and 2PAM (Pralidoxime chloride, protopam chloride.) Soldiers typically have these in their antidote kits. In a large scale terrorist attack, the hospitals would not have enough antidotes in stock.

Chemical Asphyxiants

Any substance that displaces oxygen has the ability to suffocate a person; however, chemical asphyxiants are formulated to remove or displace oxygen and interfere by not allowing the victim to utilize oxygen after it enters the respiratory system. A wide assortment of these chemicals can be found in the chemical processing industry. Common asphyxiants include hydrogen sulfide (H_2S), carbon monoxide (CO), and nitrogen compounds. Where these chemicals are involved, the victim attempts to breathe in air but the cells are unable to get the oxygen they need to survive. The primary chemical asphyxiants used by terrorists and fanatics are hydrogen cyanide and cyanogen chloride. The 900 followers of Jim Jones in Jonestown, British Guiana used cyanide to commit suicide. Cyanide is a liquid at temperatures below 79°F; however, it vaporizes quickly when this temperature is exceeded. This chemical most often does its damage through inhalation; however, it can be ingested, injected, or absorbed. Early indicators of cyanide poisoning are agitation, confusion, headache, restlessness, dizziness, seizures, coma, and eventually death. Cyanide smells like bitter almonds but can be detected by only 70 percent of the general population. Women tend to be able to detect the smell of cyanide better than most men.

Treatment of cyanide poisoning is time sensitive and typically requires three drugs: amyl nitrite, sodium nitrite, and sodium thiosulfate. Most hospitals and EMS services keep a small supply on hand. If a large population were effected, local services would be inadequate and would need outside help. By the time the medicine arrives, most victims would be critically ill or dead.

Carbon Monoxide

Another deadly chemical is carbon monoxide, which bonds to hemoglobin and prevents oxygen absorption. This makes the blood appear to be bright red, and even the skin takes on a reddish hue in the late stages of carbon monoxide poisoning. Unfortunately, there is no known antidote to this type of poisoning. The common treatment is the use of 100 percent oxygen utilizing a respirator.

Nitrates and Nitrites

In nitrate or nitrite poisoning, the composition of the blood is chemically altered through the hemoglobin. This process forms a new compound called methemoglobin that inhibits the natural processes associated with the transportation of oxygen. In this type of poisoning the skin and lips appear to take on a bluish color. The drug *Methylene blue* can be administered at the hospital to turn the methemoglobin back to hemoglobin, and the victim is placed on 100 percent oxygen.

The chemical processing industry utilizes a number of chemicals that are classified as respiratory irritants. Examples of these chemicals include anhydrous ammonia, chlorine, and phosgene. Chlorine is typically used to chlorinate drinking water and as a fungicide agent and antimold product. Chlorine is used in cleaning swimming pools, showers, and toilets. Phosgene is another common chemical used in organic synthesis, insecticide and dye manufacture, and in polyurethane production. Phosgene's deadly properties were used in concentration camps by the Nazis in the second World War. This chemical smells like fresh-cut hay. Another common chemical used in the chemical processing industry is anhydrous ammonia. It is transported by railcar, trucks, and pipelines and stored in large quantities on site. Anhydrous ammonia causes severe respiratory irritation and could easily kill large populations if it was inadvertently released.

Other chemicals that could be used by terrorists include hydrochloric acid and alkalis. When these chemicals come into contact with human tissue, the results are extremely hazardous. Many industrial processes utilize these chemicals. At the present time, thousands of different chemicals pose a threat in the heavily industrialized sections of the Gulf coast. A variety of hydrocarbons exist in these facilities that have the ability to explode or catch on fire causing severe damage. Most gates or perimeter fencing are not equipped to stop a well organized terrorist group. Specifically, suicide bombers could inflict serious damage if they hit a vital nerve.

Use of Biological Weapons

Like a nuclear bomb attack, the use of biological agents sends fear into the innocent civilian population. Many people are totally unaware of the hazards associated with the biological arsenal hidden in the terrorists' Pandora's box. Biological agents are easy to produce by terrorists with limited knowledge and understanding of chemistry and microbiology. Biological agents are made from biological toxins and micro-organisms. Biological toxins can be classified as chemical compounds produced by poisonous microbes, animals, or plants. Microorganisms are living bacteria and viruses that have the ability to establish hazardous infections in humans that are easily spread from one population group to another. Biological terrorists have the ability to cultivate, harvest, and introduce these microorganisms into populated areas.

Bacteria are best described as single cell microorganisms with plant-like structures. Examples of hazardous bacteria include anthrax, cholera, tularemia and Q fever. Anthrax spores have the ability to lie dormant for 40 years before being inhaled or ingested, thus releasing a rapidly progressing infection that forms coal black skin lesions. Approximately 17 countries have developed an anthrax weapons program. Forty years ago the World Health Organization released a study that indicated the release of 110 pounds of aerosolized anthrax on a city of 5 million would kill 100,000 and injure 250,000. Public outcry and government and military concerns have put many of these programs in cold storage.

Cholera or vibrio cholerae typically enters the body through the consumption of food or water. Historically, cholera spreads when sewage systems are improperly set up. The bacterium attaches to the soft tissue of the small intestine that sets off a chain reaction of excess fluid production, inability of large intestine to absorb the fluid, followed by diarrhea, dehydration, and low fluid shock. Cholera infection takes 12 to 72 hours of exposure, followed by intestinal cramping, vomiting, diarrhea, headache, and the loss of 5 to 10 liters of fluid per day. The treatment for cholera includes the aggressive use of IV fluids and antibiotics.

Pneumonic and bubonic plague have left a permanent scare on the world because of the massive death toll it had in 541 A.D. (the first great plague) and 1346 A.D. (the second plague pandemic). It is estimated that the first plague killed 55 percent of the population of North Africa, Europe, and Central and Southern Asia. The second killed 33 percent of the population of European nations and 13 million in China. The phrase *God bless you* (from priests' inability to give last rites) and the nursery rhyme *Ring around the rosie, pocket full of posies, ashes, ashes, we all fall down* (from the red ring around infected lymph node, smell of death, and burning bodies) come from these time periods. Bubonic plague is transmitted through flea-borne bacteria associated with infected rodents. Pneumonic plague is spread person to person through sneezing, coughing, breathing, or talking. Without immediate attention or within 24 hours, the death rate is 100 percent; however, only 45 percent live who are treated soon after the onset of symptoms. The World Health Organization released a study that indicated the release of 110 pounds of aerosolized pneumonic plague on a city of 5 million would kill 36,000 and injure 150,000. Man-made plague (bacteria *Yersinia pestis*) presents in three ways: septicemic, bubonic, and pneumonic.

Viruses

In general terms, viruses are smaller than bacteria, live inside cells, utilize the host for reproduction and metabolism, are costly to produce, and are extremely difficult to cure. The military from a wide array of countries have conducted research on the following:

- Smallpox (*Variola virus*)
- Venezuelan equine encephalitis (VEE)
- Viral hemorrhagic fever (VHF)

Smallpox is the most serious virus because it has no known cure, is extremely contagious, and the vaccine is available in small quantities.

Biological Toxins

Another area of concern is with biological substances that have their origin in plants, animals, or microbes. Comparatively speaking, these types of toxins are more toxic than toxins produced by industry. Biological toxins are typically used to contaminate fresh water supplies, food sources, or specific populations or individuals. Examples of biological toxins include the following:

- Botulism (botulinum toxins)
- Ricin
- SEB (staphylococcal enterotoxin B)
- T2 (tricholthecene mycotoxins)

Tornado Hits

Tornados are classified as extremely violent, turbulent, rotating columns of air that maintain contact with the cumulonimbus cloud and the surface of the earth. Tornados produce wind speeds between 40 and 110 mph and tend to gather or collect debris, rocks, and dust (flying projectiles). These destructive forces of nature are around 250 feet across and travel for several miles on the ground before dissipating. Some tornados have

Figure 16-4 *Tornado*

achieved wind speeds of over 300 mph and have been measured at more than a mile wide while wrecking destruction and death over several states before jumping back into the clouds only to re-emerge somewhere else. The most extreme tornado in U.S. history was classified as an F5 and tore through Missouri, Illinois, and Indiana. It stayed on the ground for over 219 miles and killed 695 people on March 18, 1925. The United States has over 1,000 recorded tornados a year. The most deadly tornado on record occurred April 26, 1989, and was called the Daultipur-Salturia Tornado in Bangladesh; it killed 1,300 people. Tornados or twisters rotate counterclockwise in the northern hemisphere and the opposite in the southern. These destructive acts of nature have a conical shape and rotate cyclonically. Figure 16-4 shows the destructive nature of a tornado.

The chemical processing industry has a number of safety systems built into the construction of industrial facilities. Unfortunately, none of these precautions could handle a direct hit by an F3 to F6 twister. Every plant or refinery has a weak or vulnerable spot. In the event of a direct hit, only automatic shutdowns and quick thinking will prevent a catastrophic situation. Towers could be thrown down, lines broken, vessels ruptured, and hazardous product released. The CPI utilizes the **Fujita scale** to classify Tornados.

The Fujita Scale

F-Scale Number	Intensity Phrase	Wind Speed	Type of Damage Done
F0	Gale Tornado	40–72 mph	Some damage to chimneys; branches off trees; pushes over shallow-rooted trees; damages sign boards.
F1	Moderate Tornado	73–112 mph	The lower limit is the beginning of hurricane wind speed; peels surface off roofs; mobile homes pushed off foundations or overturned; moving autos pushed off the roads; attached garages may be destroyed.
F2	Significant Tornado	113–157 mph	Considerable damage; roofs torn off frame houses; mobile homes demolished; boxcars pushed over; large trees snapped or uprooted; light object missiles generated.
F3	Severe Tornado	158–206 mph	Roof and some walls torn off well constructed houses; trains overturned; most trees in forest uprooted.
F4	Devastating Tornado	207–260 mph	Well constructed houses leveled; structures with weak foundations blown off some distance; cars thrown and large missiles generated.
F5	Incredible Tornado	261–318 mph	Strong frame houses lifted off foundations and carried considerable distances to disintegrate; automobile-sized missiles fly through the air in excess of 100 meters; trees debarked; steel reinforced concrete structures badly damaged.
F6	Inconceivable Tornado	319–379 mph	These winds are unlikely. The small area of damage they might produce would probably not be recognizable along with the mess produced by F4 and F5 winds that would surround the F6 winds. Missiles, such as cars and refrigerators, would do serious secondary damage that could not be directly identified as F6 damage. If this level is ever achieved, evidence for it might only be found in some manner of ground swirl pattern, for it may never be identifiable through engineering studies.

Fast-Forming Hurricanes

One scenario has a mild tropical storm due to arrive in the early morning hours as your family prepares to go to sleep for the night. The news report indicates the arrival of much needed rain in the Houston, Gulf coast area.

Within four hours of the lights going out, the tropical storm rapidly begins to increase in strength. Shortly before daybreak, a Category-3 Hurricane makes land fall. In this example the chemical processing industry would have little warning as wind gusts increase in intensity. Heavy rains saturate the ground, and a series of small tornados are spun off, wreaking havoc on everything they touch. Within minutes of landfall, electrical power goes out, and the general population is unaware that a deadly hurricane is in their backyards. Shift workers are trapped as they struggle to make it to the plant. Process technicians on shift struggle to secure the unit in hurricane-force winds. Most plants can operate with backup power when the electric company service goes down; however, one of the small tornados hits the utility section of one large company. Backup systems fail, utility air goes down, steam generation is interrupted, different parts of the plant go down as dangerous hydrocarbons are directed toward the flares. High winds from the storm damage a series of critical cooling towers and topple several distillation columns. A large ammonia tank is ruptured by falling debris. A dangerous vapor cloud forms and flows into the local community.

Although this example or what-if scenario has not occurred, the industry has contingency plans in place to handle the unexpected. Fast-forming hurricanes present a danger to companies located near the Gulf of Mexico.

Hurricanes are powerful, swirling storms with tentacles reaching out from a singular eye. Hurricanes have sustained wind speeds between 74–155 mph. Hurricanes that form north of the equator rotate counter-clockwise, whereas the opposite rotation occurs for storms that develop south of the equator. The spinning circulation is due to the Coriolis effect, which responds to the Earth's rotation. Hurricanes form about 10 to 15 degrees north or south of the equator and move in a curved path toward the north or south pole. For a hurricane to develop, a number of conditions must exist. The first is the temperature of the surface water must be above 80°F (27°C). As the seawater evaporates, it adds moisture to the hurricane. This process intensifies as the storm continues the process of pulling in moisture, while returning it back to the Gulf in the form of rain. A hurricane can pick up as much as 2 billion tons of water a day as it literally pulls energy from the warmer water molecules suspended above the surface of the sea. The energy produced by an average hurricane over a two day period could supply the energy needs of the United States for a year, if it was possible to convert it to electricity. When a hurricane makes landfall, it is cut off from its primary source of energy and becomes a deep tropical depression. Figure 16-5 shows a high level photo of a hurricane.

These storms are characterized by pulsing wind gusts from 74 to over 155 mph, heavy rainfall and flooding, tremendous storm surges from 4 to 18 feet, spin-off tornados, and causing downed trees and broken limbs, damaged homes, loss of electricity, utilities, and basic commodities. In most cases, hurricanes are tracked carefully as they enter the Gulf of

Figure 16-5 *Hurricane*

Mexico. Most of these tropical depressions start out as storms that have wind speeds between 39 and 73 mph. As these storms enter the Gulf of Mexico, they gain strength from the warmer waters of the Gulf Stream. Hurricanes are graded by three factors: wind speed, eye pressure, and height. The chemical processing industry uses the **Saffir-Simpson hurricane wind scale**. This scale uses wind speed as its primary factor.

Chemical plants and refineries will closely monitor the development and direction of a hurricane and will attempt to operate until the last minute. Prior to the storm making landfall, contractors and technicians walk the unit and tie down any object that may become airborne. Established emergency procedures are put in place as tank levels are adjusted to required positions. All nonessential operations are shutdown and secured for high winds. Prior to hurricane Katrina, most operating facilities would keep a skeleton crew on shift and would attempt to keep the different processes operating at minimal rates. After hurricane Katrina, a large percentage of the chemical processing industry decided to keep a skeleton on shift and shutdown the operating units until it was safe to restart. Typically, Category 3-5 hurricanes will initiate a unit shutdown.

Skeleton crews are composed of technicians and supervisors who have advanced skills in working through the type of problems a hurricane would present. Most often these crews are made up of volunteers. These technicians and supervisors will stay on the unit 24 hours a day, seven days a week, until the crisis has passed. Employees are paid for each hour he or she is on the unit, even during scheduled sleep periods.

In order to keep the chemical plants and refineries operational, the utilities section must stay on line during the storm. Because electrical power brought in from outside companies tend to fail during a hurricane, the plant must operate on its own power generation system. Other utilities that need to be operational include steam, water, air, nitrogen, hot oil, computers, natural gas, and waste water treatment and storage. Every critical system has a series of safety features. It is possible for a technician to shut down an entire area by pushing a single button. Unfortunately, this may require extensive down time in order to repair the unit after the hurricane has passed. Chemical and refinery facilities work hard to bring the unit back up after the storm passes.

The Saffir-Simpson Hurricane Wind Scale

Category	Sustained Winds	Type of Damage Done
Tropical Depression	0–38 mph	Storm surge 0 feet. Rain and wind.
Tropical Storm	39–73 mph	Storm surge 0–3 ft. Wind gusts may be dangerous.
Cat-1	74–95 mph	***Dangerous winds will produce some damage***. People, pets, and livestock at risk. Mobile homes, frame homes, and homes built before 1994 may be damaged if not anchored correctly. Unprotected windows may break, masonry, vinyl siding, may fall, and tree branches may take out power lines. Storm surge: 4–5 feet above normal. Barometric pressure: 28.94 in.Hg. 980mb. 97.7 kPa.
Cat-2	96–110 mph	***Extremely dangerous winds will cause extensive damage***. Substantial risk of injury or death to people, pets, and livestock. Mobile homes, frame homes, and homes built before 1994 will be damaged or destroyed. Newer mobile homes may also be destroyed. Well constructed homes could sustain serious roof or siding damage. Shallow rooted trees will be uprooted or snapped. Tree limbs large and small will be broken off. Electric power will be totally lost, and potable water will be difficult to obtain. Storm surge: 6–8 feet above normal. Barometric pressure: 28.5-28.93 in.Hg. 965.1-979.7 mb. 96.2-97.7 kPa.

(continued)

(continues)

Cat-3	111–130 mph	***Devastating damage will occur***. High risk of injury or death to people, pets, and livestock. Everything from Cat-2 plus. Higher percentage of homes and property damaged. Complete failure of older metal buildings. High-rise windows blown out, fences, canopies, and commercial signs damaged. Loss of electricity and clean water for days to weeks. Storm surge: 9–12 feet above normal. Barometric pressure: 27.91-28.49 in.Hg. 945.1-964.8 mb. 96.2-97.7 kPa.
Cat-4	131–155 mph	***Catastrophic damage will occur.*** High risk of death or injury to people, pets, and livestock from flying and falling debris. Everything from Cat-3 plus. Most trees and power poles snapped or uprooted, power outages can last for weeks to months. Human suffering from long-term food and water shortages will be severe. Most of the area will be uninhabitable for weeks or months. Storm surge: 13–18 feet above normal. Barometric pressure: 27.17-27.9 in.Hg. 920.1-944.8 mb. 91.7-94.2 kPa.
Cat-5	155+ mph	***Catastrophic damage will occur.*** People, pets, and livestock are at high risk of injury or death from flying debris, even if indoors in framed home, mobile home, or barn. Large amounts of windborne debris. A high percentage of industrial buildings and apartments destroyed. Most unprotected windows will be damaged or blown out of homes and high-rise buildings. Human suffering will be at a high level as heat and humidity returns with mosquitoes, and no modern conveniences are available. Storm surge: 19 plus feet above normal. Barometric pressure: < 27.17 in.Hg. <920.1 mb. <91.7 kPa.

Use of Military Weapons

The chemical processing industry is aware that terrorist organizations have connections with well funded foreign military. These groups often have access to military weapons, cash, and training. In reviewing the number of scenarios available for hazard preparation, it has been determined that it would be difficult to prevent the use of a military weapon or weapons on facilities operated by the CPI. Weapons can be grouped into four areas: Air Force, Navy, Marines, and Army weapons. The Air Force utilizes the following weapons: AC-130 gunship, F-22 Raptor, F-15 Eagle, unmanned aerial vehicles (UAV), Magic fighter helmets, B-1 Lancer, U-2, air force munitions and much more (31 total).

The Navy utilizes naval destroyers, aircraft carriers, F-14 Tomcat, high speed vessels, boats that fly, submarines, F/A-18 Hornet, and much more (22). The Marines use the Harrier, M-16A4 rifle, the Osprey, M-40A3 rifle, Riverine craft, and much more (9). The Army uses M203 grenade launcher, M9 pistol, armored personnel carrier, GPS, hand grenades, night vision devices, rocket-propelled grenades, AH-64D Apache longbow helicopter, smoke grenades, Abrams tank, M4 carbine rifle, M110 sniper rifle, Striker armored vehicle, Bradley fighting vehicle, body ventilation system, the Crusher, Future battlefield communications, Future Army warrior technology, M107 rifle, Raven UAV, M109A6 155mm Howitzer, Future infantry 2025 uniform, Gator jeep, helicopters, army aircraft carrier, army catamaran, lightweight MLRS, liquid body armor, XM29 assault rifle, Humvee, mortars, auto assault 12 shotgun, stinger missile targeting aircraft, flares, Rhino runner bus, gun trucks, landmines, flamethrowers, M242 bushmaster, and others (72).

Some of these weapons could fall into the wrong hands and be used by terrorists against the chemical processing industry. Because of the large number of military weapons to choose from and the sophistication and training of many terrorist organizations, the future is difficult to predict.

Flooding

Some of the most common hazards in the world are storm surges, flash floods, mud slides, and localized flooding. Awareness of flood hazards is helpful if you live in low-lying areas, near rivers or creeks, culverts, gullies, dams, or dry streambeds. Floods may develop slowly, over days or weeks. Flash floods develop quickly and tend to sweep clear everything in their paths creating dangerous walls of silt, debris, and rocks. Flooding is a common hazard in the United States, resulting in mudslides and dangerous walls of debris filled with raging water. Figure 16-6 shows the hazards associated with flooding.

When floods occur in locations near the chemical processing industry, a number of hazardous situations exist. Most chemical plants and refineries have environmental control units designed to clean up and recycle hazardous waste. These systems include lagoons, retention ponds, clarifiers, filters, aeration basins, and a number of other systems. Over a long period of time, hazardous materials accumulate in the sewer systems that are connected to the environmental control or water treatment unit. When flood waters encroach on the sewer systems and waste treatment facilities, every hazardous chemical is picked up and carried away into the community. Local rivers, streams, and drainage facilities may carry this material into close contact with the public or out to sea.

Figure 16-6 *Flooding*

High waters make roads and highways difficult to negotiate and create a number of hazardous driving conditions. Flood waters over culverts form hazardous currents that can drown children, adults, pets, and livestock. Many people attempting to get out of flooding conditions accidentally drive into deep water and must be rescued from stranded vehicles or must swim to safety. Hurricanes often bring in large amounts of rain and variable storm surge levels that tend to add to the destruction caused by high winds.

Emergency Preparedness

Emergency preparedness requires that each family have a 72-hour emergency kit, evacuation plans, contact names, and meeting places for family members in case of separation. In most emergency situations, a number of essentials are required to sustain life. Prior preparation is important before access to food and water is cut off. Adequate home food storage in case of an emergency takes months or years to develop. Examples of emergency kits can be found on the Internet. Items like generators, ice, water, food, medical supplies, prescriptions, flash lights, chain saws, emergency radios, small air conditioners, extension cords, and communication devices were helpful during the aftermath of hurricanes Katrina, Rita, Ike and others. Many of the situations mentioned in this chapter that result from biological weapons, chemical weapons, nuclear or conventional bombs, tornados, hurricanes, and terrorist attacks can be anticipated and prepared for. Although many of these things will never occur, it is important to plan and prepare as if they will happen.

Summary

Terrorism uses the insidious approach of doing horrible things in order to change people's beliefs and viewpoints. Weapons of mass destruction can be classified as biological agents, chemical agents, and bombings. Benjamin Franklin once stated, "Failure to prepare is preparing to fail." The use of weapons of mass destruction can have a catastrophic effect on the chemical processing industry and the communities that surround them. The Houston Gulf Coast area is home to over 600 companies associated with the chemical processing industry. Each of these companies produces different products and presents different hazards to the community. The key to attacking one or more of these companies is strategic knowledge about vital nerve centers. Many of these companies store raw materials in liquid state that at atmospheric pressure will turn into vapor. Some of these chemicals will explode violently when exposed to an ignition source. Others are extremely hazardous to humans. Terrorist would need to know the:

- Location
- Quantity of materials
- Physical and chemical hazards
- Direction of the wind
- Safety instrumentation
- Potential for catastrophic effects
- Domino effect with other industries
- Economic effect on community

In order to gather this type of information, a terrorist would need to work in the intended target company and have a sound understanding of engineering and process operations. Specific safety controls would need to be removed and GPS coordinates provided. In light of changing demographics in the Gulf Coast area and the need for better trained employees to replace the baby-boomers, this scenario is a strong possibility.

Potential Risks include the following:

- Terrorist airplane flight into a chemical plant or refinery
- Terrorist entry into CPI workforce
- Use of conventional bomb
- Use of nuclear weapon
- Suicide bomber
- Use of chemical weapons
- Use of biological weapons
- Tornado hits
- Fast-forming hurricanes
- High winds
- Use of military weapons
- Flooding

A one megaton nuclear bomb is capable of destroying the largest city in the world. When a nuclear bomb is detonated, it releases 35 percent of its energy as heat, 50 percent in the initial blast, and 15 percent as nuclear radiation. One kilogram of nuclear fission fuel is 20 million times more powerful than the same amount of TNT. If a nuclear bomb exploded, it would kill or seriously injure everyone within 3 to 30 miles. Typical temperatures in a nuclear explosion exceed 300,000°C. The shock wave from this type of explosion produces firestorms and hurricane-type winds that blow down everything in their path.

A terrorist may choose to use a dirty bomb, which is typically made from 100 pounds of conventional explosives, bundled with cobalt-60. The blast zone is over 300 feet in all directions, scattering or pulverizing the ground-up radioactive material. The wind will help in spreading the lethal fallout. When small particles are inhaled, the long-term effects can be serious.

The use of chemical weapons on innocent civilians is considered to be cruel and barbaric. The purpose of these weapons is to kill or incapacitate. Unfortunately, many terrorist organizations are sponsored by military forces that have expertise in the use of these chemical weapons. In order to respond to this type of threat, it is important to understand basic self-protection methods.

Chemical weapons are identified by how they affect the body and whether they are nerve toxins, chemical asphyxiants, respiratory irritants, skin irritants, and burning agents. In areas with large concentrations of chemical plants and refineries, terrorists may choose other chemicals to accomplish their goals. Examples of this could include railroad cars, tractor-trailer trucks, or bulk storage of industrial chemicals. Specific chemicals that may be targeted include chlorine, anhydrous ammonia, LPG, hydrocyanic acid, hydrogen fluoride, and sulfur dioxide.

The most widely known nerve agents or organophosphate compounds are sarin, tabun, soman, and VX, which have properties similar to industrial grade pesticides. Most of these agents are hazardous to the respiratory system. Tabun and sarin are easy to produce and appear to be the terrorist weapon of choice. Soman, sarin, and tabun evaporate at a rate slightly faster than pure water. The nerve agent VX must be heated in order to increase the evaporation rate. This makes VX dangerous because it lingers in the environment for days, causing death and injury as it slowly releases its deadly respiratory poison.

The physical properties of nerve agents are as follows:

- Clear and colorless.
- Slight odor or odorless (pure product).
- Purity of the agent determines odor (fruity odor).

- Less-pure VX has a sulfuric odor.
- Found in liquid state with different evaporation rates.
- Poisonous chemical vapors are typically introduced through lungs or inhalation.
- May also enter through skin absorption or ingestion.

Symptoms associated with nerve agents include excessive reactions:
- Salivation
- Diarrhea
- Mucous production
- Urination
- Seizures and muscle twitching

Decontamination procedures and treatment of nerve agents include the use of salt or fresh water, alkaline soap, and, in some cases, the use of Clorox bleach. The antidotes for this type of exposure include two drugs: atropine and 2PAM (Pralidoxime chloride, protopam chloride.) Soldiers typically have these in their antidote kits. In a large scale terrorist attack, the hospitals would not have enough antidotes in stock.

Any substance that displaces oxygen has the ability to suffocate a person; however, chemical asphyxiants are formulated to remove or displace oxygen and interfere by not allowing the victim to utilize oxygen after it enters the respiratory system. A wide assortment of these chemicals can be found in the chemical processing industry. Common asphyxiants include hydrogen sulfide (H_2S), carbon monoxide (CO), and nitrogen compounds. Where these chemicals are involved, the victim attempts to breathe in air, but the cells are unable to get the oxygen they need to survive. The primary chemical asphyxiants used by terrorist and fanatics are hydrogen cyanide and cyanogen chloride.

The chemical processing industry utilizes a number of chemicals that are classified as respiratory irritants. Examples of these chemicals include anhydrous ammonia, chlorine, and phosgene.

Like a nuclear bomb attack, the use of biological agents sends fear into the innocent civilian population. Many people are totally unaware of the hazards associated with the biological arsenal hidden in the terrorist's Pandora's box. Biological agents are easy to produce by terrorists with limited knowledge and understanding of chemistry and microbiology. Biological agents are made from biological toxins and micro-organisms. Biological toxins can be classified as chemical compounds produced by poisonous microbes, animals or plants. Microorganisms are living bacteria and viruses that have the ability to establish hazardous infections in humans that are easily spread from one population group to another. Biological terrorists have the ability to cultivate, harvest, and introduce these microorganisms into populated areas.

Bacteria are best described as single cell microorganisms with plant-like structures. Examples of hazardous bacteria include anthrax, cholera, tularemia and Q fever. Anthrax spores have the ability to lie dormant for 40 years before being inhaled or ingested, thus releasing a rapidly progressing infection that forms coal black skin lesions.

Cholera or vibrio cholerae typically enters the body through the consumption of food or water. Historically, cholera spreads when sewage systems are improperly set up. The bacterium attaches to the soft tissue of the small intestine that sets off a chain reaction of excess fluid production, inability of large intestine to absorb the fluid, followed by diarrhea, dehydration and low fluid shock. Cholera infection takes 12 to 72 hours of exposure, followed by intestinal cramping, vomiting, diarrhea, headache, and the loss of 5 to 10 liters of fluid per day. The treatment for cholera includes the aggressive use of IV fluids and antibiotics.

Pneumonic and bubonic plague have left a permanent scare on the world because of the massive death toll it had in 541 A.D. (first great plague) and in 1346 A.D. (second plague pandemic). Bubonic plague is transmitted through flea-borne bacteria associated with infected rodents. Pneumonic plague is spread person to person through sneezing, coughing, breathing, or talking. The World Health Organization released a study that indicated that the release of 110 pounds of aerosolized pneumonic plague on a city of 5 million would kill 36,000 and injure 150,000. Man-made plague (bacteria Yersinia pestis) presents in three ways: septicemic, bubonic, and pneumonic.

Viruses are smaller than bacteria, live inside cells, utilize the host for reproduction and metabolism, are costly to produce, and are extremely difficult to cure. The military from a wide array of countries have conducted research on the following:

- Smallpox (*Variola virus*)
- Venezuelan equine encephalitis (VEE)
- Viral hemorrhagic fever (VHF)

Smallpox is the most serious virus because it has no known cure, is extremely contagious, and the vaccine is available in small quantities.

Biological toxins have their origin in plants, animals, or microbes. Comparatively speaking, these types of toxins are more toxic than toxins produced by industry. Biological toxins are typically used to contaminate fresh water supplies, food sources or specific populations or individuals. Examples of biological toxins include:

- Botulism (botulinum toxins)
- Ricin
- SEB (staphylococcal enterotoxin B)
- T2 (tricholthecene mycotoxins)

Tornados are classified as extremely violent, turbulent, rotating columns of air that maintain contact with the cumulonimbus cloud and the surface of the earth. Tornados produce wind speeds between 40 and 110 mph and tend to gather or collect debris, rocks, and dust (flying projectiles). These destructive forces of nature are around 250 feet across and travel for several miles on the ground before dissipating. Some tornados have achieved wind speeds of over 300 mph and have been measured at more than a mile wide while wreaking destruction and death over several states before jumping back into the clouds only to reemerge somewhere else.

Fast-forming hurricanes present a danger to companies located near the Gulf of Mexico. A hurricane is defined as a counterclockwise rotating storm with sustained winds in excess of 74 mph. Hurricanes are powerful, swirling storms with tentacles reaching out from a singular eye. These storms are characterized with having pulsing wind gusts from 74 to over 155 mph, heavy rainfall and flooding, tremendous storm surges from 4 to 18 feet, spin-off tornados, and causing downed trees and broken limbs, damaged homes, loss of electricity, utilities, and basic commodities. The chemical processing industry uses the Saffir-Simpson Hurricane Wind Scale to categorize hurricanes. This scale uses wind speed as its primary factor.

Review Questions

1. Describe the hazards associated with fast-forming hurricanes in the waters around the United States.
2. List the wind speeds, classifications, and hazards associated with tornadoes.
3. Explain the things a terrorist would need to know in order to inflict severe damage on the chemical processing industry.
4. List examples of terrorist type activities in the United States that could indicate possible future activities.
5. Describe the Saffir-Simpson Hurricane Wind Scale.
6. Describe the Fujita scale and explain how it works.
7. Describe how terrorists could infiltrate the chemical processing industry.
8. Explain what would happen if an airplane was randomly crashed inside a chemical plant or refinery.
9. List the steps that are taken when a bomb threat occurs inside an operating chemical facility.
10. Explain the events that would take place in your community if a suicide bomber set off a dirty bomb.
11. Describe the effects of a nuclear bomb going off inside a major city.
12. Explain how chemical weapons are classified.
13. Explain the following terms: nerve toxins, chemical asphyxiants, respiratory irritants, skin irritants, and burning agents.
14. List specific chemicals in your community that could be used by terrorist as chemical weapons.
15. Describe the physical properties associated with nerve agents.
16. Describe the risks and hazards associated with biological weapons.
17. Explain what the results would be if a Category-5 hurricane made land fall in the Houston, Texas area.
18. Explain what the consequences would be if a F5 tornado made a direct hit on a large chemical plant and refinery.
19. Describe the damage a terrorist using military weapons on a chemical plant or refinery could inflict and explain how he or she would be captured or stopped.
20. List examples of biological toxins.

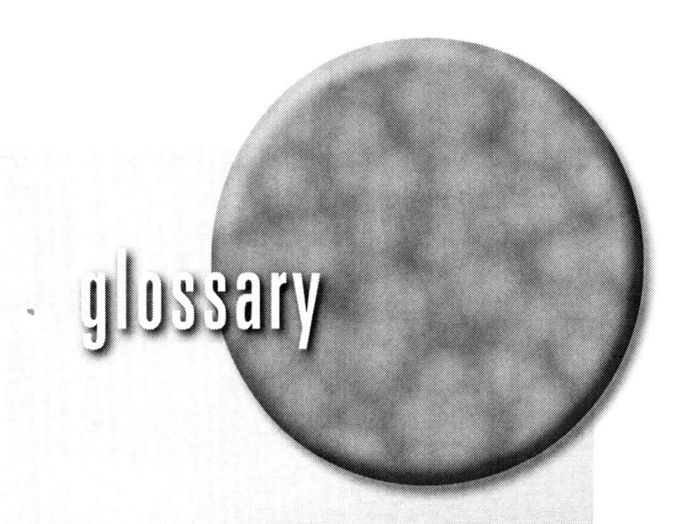

Absorbed heat, effects of—increase in volume, temperature, change of state, chemical change, or electrical transfer.

Accident—an unplanned disruption of normal activity resulting in an injury or equipment damage.

Accident prevention, basic principles of—include a safe working environment, safe work practices, and effective management.

Accident prevention, keys to—(1) determining the cause and (2) preventing its recurrence.

Acid—a bitter-tasting chemical compound that has a pH value below 7.0. Changes blue litmus to red, yields hydrogen ions in water, and has a high concentration of hydrogen ions.

ACM—asbestos-containing material.

Acute effect—a chemical that has an immediate adverse effect on biological tissue.

Acute hazards—produce symptoms that develop rapidly.

Acute (immediate) poisons—examples of these chemicals include chlorine, acids, and caustics. May be ingested, inhaled, injected, or absorbed.

Administrative controls—programs and activities used to control industrial hazards.

Administrative noise abatement—reduction of employee exposure time to industrial noise.

Air permits—permits that must be obtained for any project which has the possibility of producing air pollutants.

Air pollution—the presence of contaminants or pollutant substances in the air that interferes with human health or welfare or produces other harmful environmental effects.

Air-purifying respirator—mechanically filters or absorbs airborne contaminants.

Air-supplying respirator—provides the user with a contaminant-free air source.

Ambient—surrounding or encircling.

Anacusis—total hearing loss.

Anchor point—a tie-off connection device used to secure the free end of a full body harness lanyard.

Anesthetic gases—have a numbing effect and will cause the loss of sensation and unconsciousness. Affected workers will become dizzy, lose coordination, and fall asleep because the central nervous system has been chemically depressed. Death may occur from respiratory paralysis. All organic gases are considered anesthetics.

Aqueous—water-based solvent system. Examples include acids, alkalis, and detergents. Properties include low vapor pressure at ambient temperatures and low system toxicity.

Asbestos—a fire-retardant material used in brake linings, hair dryer components, insulation, and shingles. It was used by the ancient Greeks for lamp wicks and by the Egyptians as mummy wrap.

Atom—the smallest particle of a chemical element that retains the properties of an element. An atom is composed of protons and neutrons in a central nucleus surrounded by electrons. Nearly all of an atom's mass is located in the nucleus.

Atomic mass unit (AMU)—the sum of the masses in the nucleus of an atom.

Atomic number—identifies the position of the element on the periodic table and the total number of protons in the atom.

Audiometry—the science of testing, measuring, and recording hearing ability.

Auto-ignition—the temperature at which a liquid will spontaneously ignite without a spark or flame.

Balanced equation—the sum of the reactants (atoms) equals the sum of the products (atoms).

Base—a bitter-tasting chemical compound that has a soapy feel and a pH value above 7.0. It turns red litmus paper blue and yields hydroxyl ions

Basic principles of accident prevention—include a safe working environment, safe work practices, and effective management.

Biological agents and weapons—include hazardous bacteria like anthrax, cholera, pneumonic and bubonic plague, tularemia, and Q-fever. Includes the use of viruses like smallpox, Venezuelan equine encephalitis, and viral hemorrhagic fever. Includes the use of biological toxins like botulism, ricin, staphylococcal enterotoxin B, and tricholthecene mycotoxins.

Biological hazards—any living organism capable of causing disease in humans. This includes insects, bacteria, fungi, and molds.

Biological system—designed to remove hydrocarbons from wastewater. A biological system includes an aeration basin, clarifiers, lagoons, sewer systems, pumps, and so on.

Blind tracking—keeps a record of all blind installations. The unit blind book prevents unit startup upsets and improves efficiency.

Blinding—a term applied to the installation of slip blinds between pipe flanges. Blinding isolates a process stream and allows a mechanical craftsperson to work on a piece of equipment safely.

Boiling point—the temperature at which a liquid changes to a vapor.

Bonding—physically connecting two objects together with a copper wire.

Buckles, D-rings and snap hooks—auxiliary equipment found on the full body harness, lanyard, and anchor point.

Bulk containers—have a rated design for liquids of 119 or more gallons, 882 pounds for solids, and a water capacity greater than 1,000 pounds for gases.

Burnable materials—have flash points above 100°F and are referred to as combustibles. This includes kerosene, No. 6 fuel, brake fluid, and antifreeze.

Cavitation—the formation and collapse of gas pockets around the impellers during pump operation. Results from insufficient suction head (or height) at the inlet to the pump.

Ceiling level "C"—the maximum allowable human exposure limit for an airborne substance. This amount cannot be exceeded, even momentarily.

Central nervous system—the brain and spinal cord.

Centrifugal pump—a dynamic pump that accelerates fluid in a circular motion. Commonly used in automatic control with fluid flow and level control.

CFR—Code of Federal Regulations.

Chemical asphyxiants—substances such as carbon monoxide and hydrogen cyanide that prevent cells from using oxygen or prevent the blood from supplying oxygen.

Chemical bond (covalent)—occurs when elements react with each other by sharing electrons. This forms an electrically neutral molecule.

Chemical bond (ionic)—occurs when positively charged elements react with negatively charged elements to form ionic bonds through the transferring of valence electrons. Ionic bonds have higher melting points and are held together by electrostatic attraction.

Chemical equation—numbers and symbols that represent a description of a chemical reaction.

Chemical hazards—can be categorized as carcinogens, mutagens, teratogens, reproductive toxins, asphyxiation, anesthetic, neurotoxic, allergic response, irritants, sensitizers, corrosives, toxic, highly toxic, and those that target organ effects.

Chemical Manufacturing Association (CMA)—organized in 1988, Responsible Care® addresses the public's concern about the use and distribution of chemicals.

Chemical process evaluation—analyzes the hazardous properties of reactants, products, products that might be formed under certain conditions, and environmental effects.

Chemical processing industry (CPI)—broad term used to describe chemical plants and refineries, power plants, food processing plants, paper and pulp plants, and city utilities.

Chemical reaction—the breaking, forming, or breaking and forming of chemical bonds. Types include exothermic, endothermic, replacement, and neutralization.

Chemistry—the science and laws that deal with the characteristics or structure of elements and the changes that take place when they combine to form other substances.

Chemtrec—the Chemical Transportation Emergency Center provides information on chemicals around the clock. It is a service of the Chemical Manufacturers Association and can be reached by calling 1-800-424-9300.

Chronic effect—a slow-developing adverse effect (cancer) that may take 10 to 20 years to appear. Typically results after long-term exposure to a chemical.

Chronic (delayed) hazards—symptoms develop over a long time period. Chronic (delayed) hazards include asbestos fibers, coal dust, and toxic metals such as lead or manganese.

Clean Air Act (CAA) of 1970—designed to (1) enhance the quality of the nation's air, (2) accelerate a national research-and-development program to prevent air pollution, (3) provide technical and financial assistance to state and local government agencies, and (4) develop a regional air pollution control program.

Clean Water Act—passed in 1898, initially focused on toxic pollutants, In 1972, adopted the Best Available Technology (BAT) strategy for all cleanups. In 1987, provisions for funded sewage treatment plants were provided as well as citizen suit provisions. Regulates the release of pollutants into lakes, streams, and oceans.

Code of Federal Regulations (CFR)—contains all of the permanent rules and regulations of OSHA and is produced in paperback format once a year.

Community Awareness and Emergency Response (CAER) programs—designed to respond to the community's concerns about the manufacture and use of chemicals.

Community right-to-know—increases community awareness of the chemicals manufactured or used by local chemical plants and business. Involves community in emergency response plans, improves communication and understanding, improves local emergency response planning, and identifies potential hazards.

Compliance audits—conducted by OSHA auditors to ensure compliance with governmental rules and regulations.

Compound—a substance formed by the chemical combination of two or more substances in definite proportions by weight.

CPI—see chemical processing industry.

Cutaneous hazards—chemicals that affect the dermal layer of the body.

Degradation—the breakdown or loss of physical properties.

Dermatitis—skin irritation or inflammation.

Designated equipment owner—the process technician who operates a piece of equipment or process.

Don a respirator—put on a respirator.

Donning and doffing—terms used to describe putting on personal protective equipment and taking off personal protective equipment.

DOT—see U. S. Department of Transportation.

Earth Day, 1970—held to educate the U.S. public about environmental concerns.

Electrical lockout log—includes the name of the person locking out the equipment, the date, time, and location. This same information is recorded when the lock is removed.

Electrical shock—causes ventricular fibrillation and paralysis of the respiratory system.

Electron—a negatively charged particle that orbits the nucleus of an atom.

Element—composed of identical atoms.

Emergency preparedness—the prior preparation of essential items and plans in the event of an emergency situation. Examples include

72-hour emergency essential kits, evacuation plans, food, and water storage.

Emergency response—a procedure initiated by the loss of containment for a chemical or the potential for loss of containment that results in an emergency situation requiring an immediate response. *Emergency response* drills are carefully planned and include preparations for worst case scenarios. Examples include the following: vapor releases, chemical spills, explosions, fires, equipment failures, hurricanes, high winds, loss of power, and bomb threats.

Emergency response plan—a written plan that documents how specific individuals should respond during an emergency situation.

Enclosed space—any space that has restricted entry.

Engineering and environmental controls—hard technology improvements designed to make the work environment safer. Examples of this include sound-proofing installing guards to rotating equipment, automating product transfers, and so on.

Engineering noise abatement—reducing noise through new equipment design and innovation.

Environmental Protection Agency (EPA)—established to develop environmentally sound policies, establish national standards, support research and development, and enforce environmental regulations.

Equipment design evaluation—is designed so that the failure of one or more devices will not result in a disaster. Process equipment must comply with safety codes, government regulation, standards, and current industry practices.

Ergonomic hazards—activities that require chemical technicians to work in unusual or awkward positions for extended periods of time.

Ergonomics—the science of how people interact with their work environment.

Extinguishers, types of—Halon, CO_2, dry chemical, water, and foam.

Federal Highway Administration—regulates the transportation of hazardous materials and truck traffic.

Federal Railroad Administration—regulates railroad traffic including the transportation of hazardous materials.

Federal Register—a publication that (1) produces information on current OSHA standards, and (2) shows all adopted amendments, deletions, insertions and corrections to government standards.

Filter life—the estimated amount of service time an air purifying filter will function properly.

Fire, classes of—Class A (wood, paper), Class B (gas, LPG), Class C (electrical), and Class D (combustible metals).

Fire, elements of—fuel, oxygen, and heat source.

Fire hydrants—located throughout the plant to provide water for fire trucks.

Fire monitors—used to cool down exposed facilities and equipment and limit the spread of the fire. The fire monitor is equipped with an adjustable 500-gallon-per-minute nozzle that has three distinct spray patterns: fog, straight steam, and power cone.

Fire, reporting of—reporting requires supplying (1) name and phone number, (2) fire location and extent of fire, and (3) products involved in fire.

Fire, stages of—the incipient stage (no smoke or flame, little heat, combustion begins), smoldering stage (increased combustion, smoke, no visible flame), flame stage (flames become visible), and heat stage (excessive heat, flame, smoke, toxic gases).

Fire tetrahedron—the four sides of the tetrahedron are represented by fuel, oxygen, chemical reaction, and a source of ignition. The removal of any one of these components will extinguish the fire.

First aid—the immediate, temporary care given to an accident victim.

First responder—awareness level—individuals who are trained to respond to a hazardous substance release, initiate an emergency response, evacuate the area, and notify proper authorities.

First responder—operations level—an individual who has been trained to respond with an aggressive posture during a chemical release by going to the point of the release and attempting to contain or stop it.

Fit testing—a critical part of a respiratory protection program that provides respirator training, matches respirators to technicians' face structure, identifies specific chemical hazards in assigned areas, and provides information on how to locate and dispose of filters, cartridges, and respirators.

Flammable gas—any gaseous material that forms an ignitable mixture when combined with 13 percent air at 68°F and 14.7 psi.

Flammable liquid—has a flashpoint below 100°F (37°C).

Flammable substances—chemicals with flashpoints below 100°F (38°C). Examples of this include benzene, toluene, gasoline, ethyl alcohol, and ethyl ether.

Flashpoint—The lowest temperature at which a flammable liquid will produce a rich enough vapor concentration to ignite in the presence of an ignition source.

Flow diagram—a simplified process drawing that uses standard symbols and diagrams to identify equipment and flows.

Four classes of fires—see fire, classes of.

Four levels of PPE—see PPE, four levels of.

Fractional distillation—a process that separates the components in a mixture by their individual boiling points.

Fujita scale—used to classify a tornado using an F-scale number, intensity phrase, wind speed, and type of damage possible.

Gas—a formless fluid that assumes the shape of the vessel it is in, exerting pressure equally in all directions at normal pressures, temperature, and conditions. Examples include ammonia, argon, helium, hydrogen, methane, nitrogen, and oxygen.

Grounding—a procedure designed to connect an object to the earth with a copper wire and a grounding rod.

Hazard Communication Standard (HAZCOM) of 1983—increased plant worker awareness of chemical hazards and gave instructions on appropriate safety measures for handling, storing, and working with these chemicals.

Hazard, subsidiary—material other than primary hazard.

Hazardous chemical—a chemical that has been determined to be a physical hazard to humans.

Hazardous materials evaluation—includes a detailed analysis of all the properties of the materials handled, stored, and processed in the plant. This process looks at: (1) quantities, (2) physical properties, (3) toxicity, (4) stability hazards, (5) corrosiveness, and (6) impurities.

Hazardous Materials Identification System (HMIS)—a labeling system frequently used by industrial manufacturers to identify the hazards associated with chemicals.

Hazards and Operability Study (HAZOP)—a system designed to identify hazards to technicians, equipment, operations, and the environment.

Hazards that initiate an emergency response—have been determined by the chemical processing industry to be: (1) an explosion, (2) fire, (3) vapor release, (4) toxic chemical release, (5) large product or chemical spill, and (6) loss of containment for radioactive material.

HAZCOM standard—Hazard Communication Standard (HCS), known as the Workers' Right-to-Know. Hazard Communication 29 CFR 1910.1200 training is required upon initial assignment.

HAZMAT—hazardous materials.

HAZMAT response team—the hazardous materials response team falls under the operations level of emergency response. This team receives specialized training to control and handle a hazardous chemical spill or release.

HAZWOPER—Hazardous Waste Operations and Emergency Response.

HCP—hearing conservation program.

Health hazard—a chemical that has been statistically proven by one or more scientific studies to have acute or chronic health risks for humans.

Heat—a form of energy caused by increased molecular activity. Heat has a variety of forms: sensible, latent heat of vaporization, latent heat of fusion, latent heat of condensation, and specific heat. Measured in BTUs or calories.

Heat exhaustion—dizziness, weak pulse, cool, moist skin, and painful heat cramps.

Heat transfer—the three modes of heat transfer are radiant, conduction, and convection. Radiant heat transfers energy through space by means of electromagnetic waves. Conduction occurs as the molecules that make up a solid begin to vibrate and transfer energy across their matrices. Convection occurs as the molecules that make up a liquid or a gas speed up as heat energy is added and move naturally from areas of hot to cold.

Heatstroke—a dangerous condition in which the body is no longer able to cool itself. When this happens, the internal body temperature rises sharply and constitutes a medical emergency.

Hematopoietic system—a term applied to the blood-forming system in the human body.

Hepatotoxin—a chemical suspected of causing liver damage.

HMIS—see Hazardous Matierals Identification System.

Hose reel—designed to take the hose to the fire. Most hose reels have an adjustable nozzle and approximately 100 feet of 1 1/2 inch hose.

Hot work—welding, cutting, or using a spark-producing device.

Hurricane—powerful, swirling storm with tentacles reaching out from a singular eye. Hurricanes have sustained wind speeds between 74 to 155 mph, heavy rainfall and flooding, and tremendous storm surges from 4 to 18 feet, and spin off tornados. They cause downed trees and broken limbs, damaged homes, loss of electricity, utilities, and basic commodities.

Hydrocarbons—a class of chemical compounds that contains hydrogen and carbon.

IDLH—immediately dangerous to life and health.

Ignition temperature—the lowest temperature at which a substance will automatically ignite.

Improper filter—a term applied to the selectivity characteristics of a respirator. Process technicians must select the correct respirator to remove a specific contaminant.

Incident command system—a military-type system designed to respond to an emergency.

Incident commanders—persons responsible for organizing and coordinating response activities. They are surrounded by formal organizations with defined lines of authority and responsibility that provide information and carry out orders.

Incompatible—a term applied to chemicals that react violently when they come into physical contact.

Industrial hygienists—collect samples from the work environment to determine hazardous conditions.

Interlock—a device that will prevent an operational action unless a specific condition has been satisfied.

Ionizing radiation—cannot be detected by any of the five human senses and are classified as alpha particles, beta particles, gamma rays, x-rays, and neutron particles.

Keys to accident prevention—see accident prevention, keys to.

Legionnaires' disease—this biological hazard is a form of pneumonia caused by inhaling Legionellae bacteria. This strain of bacteria has been found in cooling towers and heat exchangers.

Lethal concentration 50 (LC_{50})—when the concentration of an airborne chemical reaches the level where it kills 50 percent of the test animals.

Levels of response—the chemical processing industry has two levels of emergency response: (1) first responder—awareness level, and (2) first responder—operations level.

Lightning—has an electrical discharge equivalent to 1,000,000,000,000,000 hp.

Liquefied petroleum gas (LPG)—a liquid-gas mixture that has a gauge pressure of 40 psig at 70°F (21.1°C).

Liquid pressure—the pressure exerted by a confined fluid. Liquid pressure is exerted equally and perpendicularly to all surfaces confining the fluid.

Loss of containment—the chemical processing industry has defined the following situations as the primary causes for loss of containment: (1) pipe or flange failure, (2) pump seal failure, (3) explosions, (4) fires, (5) overfilled tanks, (6) overpressured tanks, and (7) overturned drums or containers.

Lower explosive limit (LEL)—the lowest concentration at which a vapor or gas will produce a rich enough vapor concentration in air to ignite in the presence of an ignition source.

Material balancing—a method for calculating reactant amounts versus product target rates.

Material safety data sheet (MSDS)—has ten sections: Product Identification and Emergency Information, Hazardous Ingredients, Health Information and Protection or Hazards Identification, Fire and Explosion Hazard, Data and Chemical Properties, Spill Control Procedure, Regulatory Information, Reactivity Data, Storage and Handling, and Personal Protective Equipment.

Matter—anything that occupies space and has mass.

Mechanical integrity—a term that applies to the soundness of a plant process.

Mechanical permit acceptor—the person who accepts and returns a work permit.

Mechanical person—the person who performs the work.

Milligrams per liter (mg/l)—milligrams of material per liter of air.

Mixture—composed of two or more substances that are only physically mixed. Mixtures can be separated through physical means such as boiling or magnetic attraction.

Molecule—the smallest particle that retains the properties of the compound.

MSDS—see material safety data sheet.

Mutual aid agreement—provides a formal accord between industry and outside emergency response organizations in the event of a catastrophic release or situation.

N.O.S.—not otherwise specified.

National Council of Industrial Safety—was formed in 1913 to promote safety in the workplace.

National Fire Protection Association—has a standardized system used in chemical hazard identification.

National Institute for Occupational Safety and Health (NIOSH)—one of the three primary agencies created under the Occupational Safety and Health Act. Responsibilities include testing and certifying protective devices and research in safety and health issues in the workplace.

Negative fit test—to perform a negative fit test, you must pull the straps on the respirator snug to ensure a tight seal. By placing the open palm over the respirator's inlet(s) a vacuum should be created when you inhale. Hold the vacuum for 10 seconds to ensure a good seal.

Nephrotoxin—a chemical suspected of causing liver damage.

Neurotoxin—a chemical suspected of causing nerve damage. Some linkages exist between neurotoxins and behavioral and emotional abnormalities.

Neutron—a neutral particle in the nucleus of an atom.

NIOSH—see National Institute for Occupational Safety and Health.

Noise—valueless, unwanted sound.

Non-bulk containers—have rated capacities for liquids (119 gallons or less), solids (882 pounds or less or total capacity of 119 gallons) and gases (a water capacity of 1000 pounds or less).

Non-flammable compressed gas—a material that does not conform to the definitions of a flammable or poisonous gas and exerts an absolute pressure of 41 psia at 68°F.

Nuclear Regulatory Commission (NRC) of 1974— established to regulate the nuclear devices used in the chemical processing industry. This includes x-ray and measuring devices used to inspect vessels and equipment.

Occupational safety and health—deals with items like personal protective equipment, HAZCOM, permit systems, confined space entry, hot work, isolation of hazardous energy, and so on.

Occupational Safety and Health Act of 1970—the purpose of OSHA is to (1) remove known hazards from the workplace that could lead to serious injury or death and (2) ensure safe and healthful working conditions for American workers. The Occupational Safety and Health Act applies to four broad categories: agriculture, construction, general industry, and maritime. There are three primary agencies responsible for the administration of the Occupational Safety and Health Act: National Institute for Occupational Safety and Health (NIOSH), Occupational Safety and Health Administration (OSHA), and Occupational Safety and Health Review Commission (OSHRC).

Occupational Safety and Health Administration (OSHA) of 1970—one of the three primary agencies created under the Occupational Safety and Health Act. Independent inspectors are allowed to enter and inspect the workplace, cite violations, and set deadlines.

Occupational Safety and Health Review Commission (OSHRC)—one of the three primary agencies created under the Occupational Safety and Health Act.

Olfactory—a term used or associated with the sense of smell.

Operator practices and training evaluation— includes operational procedures, training for operating technicians and supervisors, startup and shutdown procedures, permit system, housekeeping and inspection, chemical hazard recognition, emergency response, use of PPE, and auditing.

OSHA regulation CFR 29 1910.95—requires all process technicians who are exposed to over 85 decibels (dB) over an eight-hour TWA period be placed in a hearing conservation program.

Overpowering—a term applied to atmospheric concentrations that exceed the limitations of the respirator.

Oxidizer—a chemical that yields oxygen.

Oxygen deficiency—atmospheres with less than 20 percent oxygen.

Parts per million (ppm)—parts of material per million parts of air.

Periodic table—provides information about all known elements (for example, atomic mass, symbol, atomic number, and boiling point).

Permissible exposure limit (PEL)—regulatory limits set by OSHA on the amount or concentration of substance in the air, based on an eight-hour time weighted average (TWA) exposure.

Permit system—a regulated system that uses a variety of permits for various applications. The more common applications are cold work, hot work, confined space entry, opening blinding, permit to enter, lock-out and tag-out.

Personal protective equipment (PPE)—is used to protect a technician from hazards found in a plant. OSHA and EPA have identified four levels (A-B-C-D) of PPE that could be required during an emergency situation. Level A provides the most protection; level D requires the least.

pH—a measurement system used to determine the acidity or alkalinity of a solution.

Physical hazard—a danger related to electricity, noise, radiation, or temperature.

Physical hazards associated with chemicals—categorized as combustible liquid, compressed gas, explosive, flammable gas, flammable liquid, organic peroxide, oxidizer, pyrophoric, unstable, and water reactive.

Physical operations evaluation—includes chemical processes that change state. Examples of this include distillation, absorption, agitating, centrifuging, crushing and grinding, crystallization, evaporation, extraction, filtering, granulation, leaching, spraying, mixing, and milling.

Pipelines—regulated by the U.S. Department of Transportation, these are lines of pipe that convey liquids, gases, or finely divided solids.

Pittsburgh Survey 1907—provided the first statistical data on how many people were being killed or injured in the United States.

Plant location and layout evaluation—key elements include drainage and runoff control, climatic conditions, affects of uncontrolled releases, community capability and emergency response, plant accessibility, available utilities, gate security, hazardous unit placement, and spacing of equipment. Additional elements to this evaluation include NEC-regulated electrical installations, clearly marked exits, building ventilation, firewalls, fire spread considerations, foundation and subsoil loadings, and administrative building location.

Polyester full body harness—a safety device designed to evenly distribute the forces of an accidental fall.

Polyester lanyard—a tie-off rope that is attached to a full body harness.

Positive fit test—performed by exhaling and covering the exhaust ports of a respirator. If the respirator has a good, seal a positive pressure will fill the face piece.

Potable water—drinkable water.

PPE—see personal protective equipment.

Pressure—force or weight per unit area (force ÷ area = pressure). Pressure is measured in pounds per square inch.

Primary hazard—the hazard classification of the material with the greatest risk percentage component being shipped. See 172.101 table for CFR. The Code of Federal Regulations contains all of the permanent rules and regulations of OSHA and is produced in paperback format once a year.

Process hazards analysis (PHA)—designed to identify the causes and consequences of fires, vapor releases, and explosions. Also, a structured brainstorming system used to identify hazards. HAZOP is an example of a PHA.

Process representative—the first line supervisor or designated representative who owns the equipment.

Process safety—the application of engineering, science, and human factors to the design and operation of chemical processes and systems.

Process safety management (PSM) standard— designed to prevent the catastrophic release of toxic, hazardous, or flammable materials that could lead to a fire, explosion, or asphyxiation.

Process technology—the study and application of the scientific principles associated with the operation and maintenance of the chemical processing industry.

Proton—a positively charged particle in the nucleus of an atom.

Pyrophoric—ignites spontaneously in air below 130°F.

Qualified first aid provider—an employee qualified by a certified medical group to administer first aid.

Radiation sickness—the breakdown of cells in the body due to radiation exposure.

Radioactive substances—include metallic uranium, x-ray, and strontium 90. These hazardous materials break down the cells of exposed tissue.

Reactants and products—raw materials or reactants are combined in specific proportions to form specific products.

Reaction (combustion)—an exothermic reaction that requires fuel, oxygen, and heat to occur. In this type of reaction, oxygen reacts with another material so rapidly that fire is created.

Reaction (endothermic)—a reaction that requires heat or energy.

Reaction (exothermic)—a reaction that produces heat or energy.

Reducing noise at its source—a technique that utilizes engineering controls and limits employee access to high noise areas.

Redundancy—a process that uses two or more devices to shut down a system.

Resource Conservation and Recovery Act (RCRA)—enacted as Public Law in 1976. The purpose of RCRA is to protect human health and the environment. A secondary goal is to conserve our natural resources. RCRA completes this goal by regulating all aspects of hazardous waste management: generation, storage, treatment, and disposal. This concept is referred to as cradle-to-grave.

Respiratory protection—a standard designed to protect employees from airborne contaminants.

Risk evaluation—a process that is used to consider all the potential risk factors found in a chemical process. Primary areas of concentration include hazardous materials, chemical process, physical operations, equipment design, layout and location, and training.

Routine hot-work area—mechanical or fabrication shops.

Safe haven—a designated area that is safe from vapor releases.

Safety—an attitude that includes careful planning, following safety rules, safe work practices, and the use of personal protective equipment.

Saffir-Simpson hurricane wind scale—a system designed to categorize a hurricane using sustained winds, and type of damage possible.

Selectivity—the specific compound or contaminant that a respirator is designed to remove.

Simple asphyxiants—gases such as nitrogen, helium, hydrogen, carbon dioxide, and methane that will displace the oxygen content in air.

Solid waste—non-liquid, non-soluble material ranging from municipal garbage to industrial waste that contains complex and sometimes hazardous substances.

Solvent—used to dissolve another material. Includes aqueous and non-aqueous.

Spark—occurs when electricity jumps a gap. Sparks and arcs occur during the normal operation of electrical equipment.

Stages of a fire—see fire, stages of.

Standard Threshold Shift (STS)—an audiometric test that is compared to prior tests to determine a measurable hearing loss.

Standby—a technician who is certified and trained to support and warn technicians who have entered a confined space.

STS—see Standard Threshold Shift.

Subsidiary hazard—see hazard, subsidiary.

Systemic poisons—formed when toxic gases enter the bloodstream through the lungs and migrate toward specific body organs and tissues.

Target organ toxin—a chemical that selectively targets a specific organ in the body

TECP—see totally encapsulating chemical protective suit.

Temperature—the hotness or coldness of a substance. Measured in degrees Fahrenheit or Celsius.

Terrorism—an attempt to change a belief or point of view through the use of a violent act. Terrorist acts are designed to generate fear and focus world-wide media attention to their cause.

Totally Encapsulating Chemical Protective (TECP) suit—A suit that provides the maximum amount of protection when engineering controls are not sufficient. Has a self-contained air system and can be used for a limited time.

Toxic hazards—fuels, metal fumes, solvents, products, and byproducts.

Toxic substance—a chemical or mixture that may present an unreasonable risk of injury to health or the environment.

Toxic Substances Control Act (TSCA)—a Federal Law enacted in 1976. TSCA was intended to protect human health and the environment. TSCA was also designed to regulate commerce by (1) requiring testing and (2) imposing necessary restrictions on certain chemical substances. TSCA imposes requirements on all manufacturers, exporters, importers, processors, distributors, and disposers of chemical substances in the United States.

Toxicology—the science of noxious or harmful effects of chemicals on living substances.

TWA—time weighted average.

Unsafe act—any act that increases a person's chance of having an accident.

Unsafe condition—a condition in the working environment that increases a technician's chance of having an accident.

Upper explosive limit (UEL)—the highest concentration at which a vapor or gas will produce a rich enough vapor concentration in air to ignite in the presence of an ignition source.

U.S. Department of Transportation (DOT)—shipments of hazardous materials are regulated by the U.S. Department of Transportation. These regulations contain specific information on how hazardous materials are identified, placarded, documented, labeled, marked, and packaged.

Valve line-up—a term used to describe opening and closing a series of valves to provide fluid flow to a specific point or tank before starting a pump.

Vapor—a gas that is formed when a chemical vaporizes.

Vapor pressure—the pressure exerted in a confined space by the vapor above its own liquid.

Water pollution—the man-made or human-induced alteration of physical, biological, chemical, or radiological integrity of water.

index

A

absorbed heat, effects, 108, 259
absorption of hazardous chemicals, 38
accidents
  classification systems, 28
  defined, 24
  hazard recognition and, 27–28
  prevention and investigation, 24, 27–28, 29, 31
acids
  and caustics (bases), 84–85
ACM (asbestos-containing material), 50
acute effect, 72
acute hazards, 24
acute poisons, 24
Adamson Act, 7
administrative controls
  Code of Federal Regulations (CFR), 150, 151, 153, 158
  community awareness, 209
  defined, 2, 10, 220
  emergency response, 209
  equipment for monitoring, 212–214
  first aid, 208, 214–215
  Hazards and Operability Study (HAZOP), 210
  housekeeping, 211
  job safety analysis, 209–210
  safety inspections, 211–212
  training, 210–211
administrative noise abatement, 136, 142, 201
age
  Age Discrimination in Employment Act, 7
air
  Clean Air Act (CAA), 36
  compressed air systems, equipment hazards, 83
  hazard of, 122
  pollution. *See* air pollution
Air Control Board (ACB), 41
air in equipment
  hazards, 83
air pollution, 40
  air permitting, 36, 41–42
  Clean Air Act (CAA), 36
  defined, 36
  See also dust and gases
alarms, 193–194
American Chemical Society, 12–13, 14
anacusis, 136, 141
anchor points, 224, 236, 237
anesthetic gases, 50, 51
anesthetic gases hazard, 51
aqueous, 108
Arco chemical plant explosion, 115–118
Arrhenius, Svante, 84
asbestos, 50, 54
  ACM (asbestos-containing material)
  asbestosis, 50
  defined, 50
asphyxiants, 53, 75, 289–290
audiometry, 136, 141
auto-ignition, 241

B

bases and acids, 84–85
BAT (best available technology), 43
Bhopal, India, 55–60
biological agents
as weapons of mass destruction, 19, 291–292
defined, 2
biological hazards
defined, 19, 136
liquids, 74–75
weapons, 291–292
biological system, 36
blind tracking, 234, 244
boiling point, 250, 254–255
bonding
defined, 136, 137
diagram, 139
brazing, 245
dust. *See* dust and gases
respirators. *See* respirators
See also air pollution
buckles, 234, 237
bulk containers, 72
bunker gear, 245
Bureau of Apprenticeship, 9
burnable materials, 108

C

CAA (Clean Air Act of 1970), 36
CAER (Community Awareness and Emergency Response) programs, 208, 209, 216
cancer-producing agents. *See* carcinogens
carbon dioxide fire extinguishers, 124
carbon monoxide, 290
carcinogens, 75
asbestos. *See* asbestos
caustics and acids, 84–85
ceiling level "C," 50
central nervous system, 50, 51
CFR. *See* Code of Federal Regulations (CFR)
Chemical Manufacturing Association (CMA), 208
chemical processing industry (CPI), 13, 285
Chemical Transportation Emergency Center (Chemtrec), 72
chemical process evaluation, 168, 191
chemical processing industry (CPI), 13, 285
Chemical Transportation Emergency Center (Chemtrec), 72
fire, chemistry of, 121–122
physical hazards, 28
See also chemical hazards
chemical agents
as weapons of mass destruction, 19
chemical hazards, 24, 25
asphyxiant hazard, 2, 50, 51
chemtrec (Chemical Transportation Emergency Center), 72
defined, 2
explosions, 109
flashpoints, 120
operating hazards, 252
physical, 25, 28
physical and health hazards, 52
warning labels, 98–99
See also toxicology
chemtrec (Chemical Transportation Emergency Center), 72
chronic hazards and effects, 24, 72
Civil Rights Act, 7
Clayton Act, 7
Clean Air Act (CAA), 36
Clean Water Act (CWA), 42
closed systems/closed loop sampling, 197–198
combustibles
combustible liquids, 74
explosions, 109
See also fire; flammability
common industrial hazards, 25
community awareness
Community Awareness and Emergency Response (CAER) programs, 208, 209, 216

community right-to-know
  defined, 36, 45
  purposes, 45
compliance audits, 223, 228
  regulatory overview, 238
compressed air systems, equipment hazards, 83
condensed phase of explosions, 109
control of hazardous energy, operating hazards, 252
CPI (chemical processing industry), 2, 285
cutaneous hazards, 50
cutting hazards, 245

D

D-rings, 234, 235, 237
decontamination procedures, 243
degradation, 150, 154
delayed foamover, 80
deluge systems, 203
Department of Transportation. *See* U.S. Department of Transportation (DOT)
dermatitis, 50
designated equipment owner, 150, 234
dirty bombs, 287
displacement, water hazards, 82
donning and doffing, 234, 236, 243
dose response relationship, 38
DOT. *See* U.S. Department of Transportation (DOT)
Draeger pumps, 244
dry chemical fire extinguishers, 124–125
dust and gases, 61–64
  asbestos. *See* asbestos
  explosions, 109
  flammable gases, 108
  respiratory protection programs, 178–186

E

ear protection, 174
Earth Day, 1970, 36
elderly. *See* age
electricity and electrical equipment
  electrical lock-out log, 150, 158–160
  equipment hazards, 26, 136–137, 146
  grounding. *See* grounding
  operating hazards, 146
  shock, 136, 137
emergency planning and response
  operating hazards, 252
  personal protective equipment (PPE), 169–186
employees and employment relationship
  HAZCOM standard, 2
enclosed spaces, 150, 154
endothermic reaction, 252, 269, 274
engineering controls
  alarms and indicators, 193–194
  chemical process evaluation, 190, 191
  closed systems/closed loop sampling, 197–198
  defined, 197
  deluge systems, 203
  equipment design evaluation, 190, 192
  explosion suppression systems, 203
  flare systems, 267
  floating roof tanks, 198–199
  hazardous materials evaluation, 190, 191
  interlocks, 190, 195–196
  noise abatement, 136, 201
  operator practices and training evaluation, 190, 191–192
  physical operations evaluation, 190, 191
  plant location and layout evaluation, 190, 191–192
  pressure relief devices, 202–203
  process containment, 196–197
  risk evaluation, 190, 191–193
  ventilation systems, 198–199
  waste treatment, 199–201
environmental issues, 81–82
  environmental effects of industry, 39
  EPA. *See* Environmental Protection Agency (EPA)
  pollution. *See* air pollution; water pollution
  water hazards, 83–84
Equal Pay Act, 7

equipment hazards, 83, 252, 264–265
ergonomics
  defined, 136, 137
  hazards, 24, 146–148
exhaustion, heat, 136, 139–140
exothermic reactions, 252, 269
explosions, 110–111, 122, 203, 287
  S.S. Grandchamp, 110–111
  Texas City, 110–111
extinguishers, 124–128
  carbon dioxide, 124–125
  dry chemical, 124–125
  foam, 125
  halon, 127
  use of, 128
  water, 127
Exxon Valdez, 39

F

Fair Labor Standards Act, 7
fall protection, 235–236
federal agencies
  EPA. *See* Environmental Protection Agency (EPA)
  Federal Highway Administration, 36
  Federal Railroad Administration, 36
  Nuclear Regulatory Commission (NRC), 36
  OSHA. *See* Occupational Safety and Health Administration (OSHA)
  See also headings beginning "U.S."
Federal Clean Water Act, 42
Federal Register, 220, 222
federal regulations. *See* Code of Federal Regulations (CFR)
fire hydrants, 108
fire tetrahedron, 108
fired heaters, 263–265
fires
  alarms and detection systems, 193–194
  burnable materials, 108
  carbon dioxide fire extinguishers, 108
  chemistry of, 120–121
  Class D, 124
  classifications, 18
  control, 120–121
  definitions, 108
  elements of, 124
  fighting, 129–131
  fire extinguishers, 18, 108, 124–128
  fire hazards, 118–120
  fire monitors, 108, 129–131
  fire tetrahedron, 108, 119
  fire triangle diagram, 121
  flooding, 299–300
  foam fire extinguishers, 125
  halon fire extinguishers, 127
  Handbook of Fire Protection, 120–121
  hydrants, 108
  monitors, 108
  polymers, 118
  prevention, 120–121
  protection, 120–121
  stages of, 123
  triangle diagram, 121
  types of extinguishers, 108, 124–128
  water fire extinguishers, 127
  See also combustibles; flammability
first responders, 235, 238–241
flammability
  flammable substances, defined, 108
  gases, 50, 108
  hazards, 119–120
  liquids, 72, 74–75
  See also combustibles; fires
flare systems, 267
flashpoints
  chemicals, hazardous, 74
  defined, 72
  role in creating fire, 123
floating roof tanks, 198–199
flow diagram, 220, 224, 225
foam fire extinguishers, 125–126
foamover, 80

four levels of personal protective equipment (PPE), 168, 175–178
frostbite, 83
Fujita scale, 294
full body harnesses, 235, 236, 237

G

gamma rays, 26–27
gases
compressed gas cylinders, 64–65
flammable, 64, 108
health hazards
and vapors, particulates, toxic metals, 51–54
grounding
defined, 136, 137
diagram, 139
to prevent static charge, 139

H

halon fire extinguishers, 127
hand protection, 171–172
Handbook of Fire Protection (National Fire Protection Association), 121–122
hazardous gases, 193–194
hazardous materials evaluation, 190, 191
hazardous materials (HAZMAT), 234, 242
Hazardous Materials Identification System (HMIS), 92, 93, 103
hazardous waste operations, 241–242
hazards
air, 122
air in equipment, 83
biological, 25
chemical. *See* chemical hazards
electrical, 26, 136–137, 146
explosive, 119–120
fire, 108
flammable, 64
flammable substances, 108, 119–120
gases, 53, 108
health. *See* health hazards
heat exhaustion, 136, 139–140
industrial noise, 26
ingestion of, 38
initiating emergency responses, 234, 238–244
injection of, 38
light-ends, 82–83
liquids, 72–73
noise, 146
personal protective equipment (PPE), 169–186
physical, 25
physical, associated with chemicals, 25
radiation, 26–27
steam, 79
toxic, 53
water, 79–80
HAZCOM (Hazard Communication), 8–9, 92, 95, 103, 168–186
HAZMAT (hazardous materials), 234, 242
HAZMAT response team, 234
HAZWOPER (Hazardous Waste Operations and Emergency Response), 234, 237–239
HCP. *See* hearing conservation program (HCP)
head protection, 174
health hazards
associated with liquids, 74–75
carcinogens, 75
defined, 50
gases, 108
liquids, 74–75
hearing conservation program (HCP)
defined, 136
diagram, 140
hearing loss, 136, 140–143
heat
absorbed heat, effects, 108, 259
defined, 108
role in creating fire, 108
heat exhaustion
defined, 136
symptoms, 139–140

heat transfer
defined, 108, 250, 259–260
role in creating fire, 108
hematopoietic system, 72, 86
hepatotoxin, 72
HMIS (Hazardous Materials Identification System), 92, 93, 103
hoisting equipment, 252
hose reel, 108
hose reels
defined, 108
hot work
area, 150
defined, 220
permits, 151–153, 223, 227
human error and operating hazards, 73–74
hurricanes, 2, 20, 282, 294–298
hydrants, 108, 127

I

ice and light-ends, 82–83
water hazards, 79–80, 83–84
IDLH (immediately dangerous to life and health), 168, 181
ignition
sources of, 241
ignition temperature, 109, 123
ignition temperatures
defined, 109
described, 123
diagram, 123
implementation
Hazard Communication (HAZCOM) Standard, 92
HAZCOM (Hazard Communication Standard), 92
implementation of HAZCOM standard
OSHA, 220, 221
improper filter, 168, 181
incident commander, 235, 238–242
indicators
engineering controls, 168, 169, 174–175
industrial hygienists, 208, 212, 216
interlocks, 190, 195–196
investigations and accident prevention, 31
ionizing radiation, 24, 26–27
fires
spontaneous combustion, 78–79

J

job safety analysis, 17–18

L

labor movement, history, 4–10
LC (lethal concentration), 92
Legionnaires' disease, 136, 145–146
LEL (lower explosive limit), 50
lethal concentration (LC), 92
levels of response, 235, 240
light-ends, 82–83
lightning, 136, 137, 146
liquefied petroleum gas (LPG), 72
liquids
biological hazards, 136, 145
chemical hazards, 119–120
handling and storage, 76–78
hazards, 72–73
health hazards, 74–75
physical hazards, 74–75
unstable, 74
Ludlow Massacre, 8

M

manufacturer responsibility, HAZCOM standard, 92–95
material classification
DOT (U.S. Department of Transportation), 92, 93, 99–100
material safety data sheet (MSDS), 92, 95
mechanical integrity, 220, 223, 226, 227
mechanical persons, 150
metallic compounds, 67
military weapons, 298–299
coal. *See* coal mines

MSDSs (material safety data sheets)
defined, 92
as part of HAZCOM standard, 95
sample, 96
mutagen, 53, 75
mutual aid agreements, 208, 210

N

National Apprenticeship Act, 7
National Council of Industrial Safety, 2, 6
National Fire Protection Association (NFPA), 7–8, 24
as part of HAZCOM standard, 103
National Institute for Occupational Safety and Health (NIOSH), 168, 179, 220, 221, 222
National Safety Congress, 6
National War Labor Board, 7
National Water Quality Standards, 43
negative fit tests, 168, 181
nephrotoxin, 72
nerve toxins, 288–289
nitrates, 290–291
nitrites, 290–291
noise
abatement, 136
engineering noise abatement, 136, 201
hearing loss causes, 141–143
industrial noise hazards, 143
permissible levels, 142
physiological effects, 141
protection from, 141–143
See also sound
non-bulk containers, 72
non-flammable compressed gas, 50
NRC (Nuclear Regulatory Commission), 36
nuclear explosions, 287
Nuclear Regulatory Commission (NRC), 36

O

occupational safety and health, definition, 2
Occupational Safety and Health Administration (OSHA), 2, 8, 117–118, 220, 221
Occupational Safety and Health Review Commission (OSHRC), 8
regulatory overview, 220, 221, 222
older persons. *See* age
Olfactory system, 50, 51
operating hazards, 252
brazing, 245
bunker gear, 245
reactors, 269–270
compressed air systems, 83
cutting, 245
emergency response, 220
equipment hazards, 252–253
compressed air systems, 83
fired heaters, 263–265
steam generation, 79, 264–265
fall protection, 235–236
hazardous waste operations, 241–242
hoisting equipment, 237
permits. *See* permits
personal protective equipment (PPE), 169–186
pump systems, 255
reactors, 269–270
steam generation, 264–265
welding, 245
operations and emergency response (HAZWOPER), 237–244
operator practices and training evaluation, 190, 191–192
organic peroxide, 74
OSHA. *See* Occupational Safety and Health Administration (OSHA)
overpowering, 168, 181
oxidizers, 79
oxygen
concentrating, 122

P

paints, 85–86
particulates, 60–61
PEL (permissible exposure limit), 98

penalties
solid waste, 199–201
permissible exposure limit (PEL), 50, 98
permissible levels
permit systems, 19, 150, 162–163
permits
blinding, 234, 244
confined space entry, 19
hazardous chemical, 19, 44, 92
hot work, 19
opening or blinding, 234, 244
operating hazards. *See* permits
routine maintenance, 19
personal protective equipment (PPE)
body protection, 169
defined, 10–11
ear protection, 174
emergency response, 175–178
eye protection, 172–173
foot protection, 173
four levels of, 168, 175–178
hand protection, 171–172
hazards, 175
head protection, 174
operating hazards, *See* personal
outerwear, 170–171
regulatory overview, 178–186
respirators, 168, 171
skin protection, 174–175
Phillips Chemical Company explosion, 115–118
physical hazards
chemicals, 24
definitions, 50
liquids, 74–75
toxicology, 92, 97–98
physical operations evaluation, 190, 191
Pittsburgh Survey 1906
Pittsburgh Survey of 1907, 2
plant location and layout evaluation, 190, 191–192
plant safety rules, general, 16–17
polyester full body harness, 236, 237
polymers, 118
positive fit tests, 168, 181
potable water, 16
potable water, definition, 2
pressure
characteristics associated with light-ends, 82–83
and pressurized equipment, 75–76
pressure relief devices, 202–203
prevention
fire, 108, 120–121
prevention and investigation
accidents, 29
prevention of accidents, 29
primary
hazards, 72
primary hazard, 72
primary purpose
process safety, 2
process containment, 196–197
process hazard analysis
regulatory overview, 208, 210, 220
process hazards analysis (PHA), 208, 210, 220
process operators. *See* process technicians
process safety, 2, 3–4
process safety management (PSM)
defined, 9
HAZCOM, 3
process safety management (PSM) standard, 220, 223–228
Process Safety Management Standard Hot Work, 220, 223–228
process systems, 76
process technicians
hazard recognition responsibilities, 11–12
roles, 11
process technology, 76
process technology, definition, 2, 13
protection
fire, 194–195
protection from
noise, 141–143

protection from when burning
polymers, 118
protective equipment (PPE), 168–186
as part of HAZCOM standard, 92, 93
PSM. *See* process safety standard (PSM)
Pure Food and Drug Act, 7
pyrophoric, 74, 251, 269

Q

qualified first aid provider, 208, 215
quality standards, 45

R

Radiation hazards, 119–120
radiation sickness
cause of, 119–120
defined, 109
radioactive substances, 24
RCRA (Resource Conservation and Recovery Act)
solid waste control, 199–201
reactors
operating hazards, 252
recognition of hazards
reducing noise at its source
defined, 136
redundancy, 190, 195
Refuse Act
regulatory overview, 221–222
change management, 227
compliance audits, 223, 228
employees, 224, 226
Occupational Safety and Health Act of 1970, 8, 220, 221
operation procedures, 226
personal protective equipment (PPE), 169–186
process hazard analysis, 225
process safety information, 224–225
process safety management (PSM) standard, 223–228
trade secrets, 228
Reid vapor pressure (RVP), 150, 154
reproductive toxin, 75
Resource Conservation and Recovery Act (RCRA), 8, 36
purpose, 43
solid waste control, 199–201
respirators
air-purifying, 168, 178
air-supplying, 168, 178
caring for and using, 179–183
hose line, 178, 184–186
personal protective equipment (PPE), 169–186
respiratory protection programs, 178–186
selecting, 179
self-contained breathing apparatus (SCBA), 183–184
See also self-contained breathing apparatus (SCBA)
respiratory hazards. *See* breathing hazards
respiratory protection programs
dust and gases, 182–186
respirators, 178–186
restructure of chemical processing industry (CPI), 174
risk evaluation, 190, 191–193
routes of entry of hazardous chemicals, 37–38
routine hot work area, 150
routine maintenance permits, 162–164

S

Safe Drinking Water Act, 8
safe havens, 235, 241
safety
defined, 2
industrial safety programs, early, 3
Saffir-Simpson hurricane wind scale, 297–298
SCBA (self-contained breathing apparatus), 183–186
scene safety and control, 242
selectivity, 168, 181
self-contained breathing apparatus (SCBA), 178, 183–186
September 11, 2001, 19, 283–285
shipping labels, 98–99

shock, electrical, 136, 137
shutdown devices, 195–196
simple asphyxiants, 50, 51
skin
  personal protective equipment (PPE), 169, 174–175
snap hooks, 234, 235, 237
Social Security Act, 7
solid waste control
  defined, 36, 43
  disposal, 44
  RCRA (Resource Conservation and Recovery Act), 8
  Resource Conservation and Recovery Act (RCRA), 8
  waste site cleanup, 199–201
solvents, 85, 109
sparks,
  defined, 136
  ignition sources, 137–139
spontaneous combustion, 78–79
S.S. Grandchamp, 110–111
Standard Threshold Shift (STS), 168, 174
standby, 150, 153–154
static charge
  generation, 81
storage of hazardous materials, 76–78
subsidiary hazards, 72, 79
suicide bombers, 287
systemic poisons, 51
  target organ toxin, 54, 75
systemic poisons hazard, 51

T

target organ toxin, 7, 72
temperature, 109
teratogen, 75
terrorism, defined, 2, 282
Texas City fire and explosion, 110–111
thermal expansion, 83
thermal explosions, 109–111
time weighted average (TWA), 136, 140
tornadoes, 20, 292–294
Totally Encapsulating Chemical Protective (TECP) suit, 168, 171, 175
toxic substances, 8, 75
  alarms and detection systems, 193–194
  EPA (Environmental Protection Agency), 9, 220, 222, 226, 228–229
  Toxic Substances Control Act, 44
Toxic Substances Control Act (TSCA)
  defined, 37
  enactment, 44
  penalties, 45
toxicology
  defined, 92, 97–98
  physical and health hazards, 24, 74–75
training and administrative controls
transmission of HBV
Transportation Department. *See* U.S. Department of Transportation (DOT)
TSCA. *See* Toxic Substances Control Act (TSCA)

U

uncontrolled mixing, 80
Union Carbide, 55–60
unsafe act, 24, 28
unsafe condition, 24, 28
unstable liquids, 74
U.S. Bureau of Mines, 7
U.S. Clean Water Act, 42
U.S. Department of Transportation (DOT)
  defined, 92, 93
  HAZCOM standard, 92, 93
  material classification, 99–100
  as part of HAZCOM standard, 92, 93
  shipping labels, 98–99
  shipping papers, 99–100

V

vacuum towers and water hazards, 81
vapor, 109
ventilation systems, 198–199

viruses, 292
viscosity of light-ends, 82–83

W

warning labels for hazardous chemicals, 98–99
  See also solid waste control
water
  acids, 84
  Clean Water Act, 8
  displacement, 82
  environmental issues, 81
  fire extinguishers, 108, 124–128
  flashing, 81
  foamover, 80
  hazards, 79–80
  ice and light-ends, 82–83
  static generation, 81
  steam, 79
  uncontrolled mixing, 80
  vacuum, 81
  water and acid, 81
  water flashing, 81
  water pollution
defined, 42
  legislation, 42
  water permitting, 42
  water reactive, 74
  weapons of mass destruction, 19–20, 282–283
weather hazards, 282–305 *See also* hurricanes, tornadoes, flooding
welding hazards, 245
World War I and labor movement, 7

CPSIA information can be obtained
at www.ICGtesting.com
Printed in the USA
FFOW01n0009040817
38417FF